"Guaranteed to amuse, entertain, and educate all lovers of the game."

Golf Weekly

"A book which will be read and reread with delight."

David B. Fay, Executive Director,
United States Golf Association

"Bob Sommers has made learning golf history a lot of fun as well."

Ben Crenshaw,
1995 Masters Champion

"A cornucopia of tales about the game and its players."

Publishers Weekly

A golf writer for more than forty years, **Robert Sommers** is former editor and publisher of *Golf Journal* and the author of *The U.S. Open, Golf's Ultimate Challenge,* and *Great Shots* (written with Cal Brown).

Golf Anecdotes

Golf Anecdotes

From the Links of Scotland to Tiger Woods

Robert Sommers

OXFORD UNIVERSITY PRESS

OXFORD
UNIVERSITY PRESS

Oxford New York
Auckland Bangkok Buenos Aires
Cape Town Dar es Salaam Delhi
Hong Kong Istanbul Karachi Kolkata
Kuala Lumpur Madrid Melbourne
Mexico City Mumbai Nairobi
São Paulo Shanghai Taipei Tokyo Toronto

First published by Oxford University Press, Inc., 1995
First issued as an Oxford University Press paperback, 1996
Reissued in cloth and paperback by Oxford University Press, Inc., 2004
198 Madison Avenue, New York, New York 10016

www.oup.com

The Library of Congress has catalogued the first edition as
Sommers, Robert (Robert T.)
Golf anecdotes / Robert T. Sommers
p. cm.
Includex index.
ISBN 0-19-506299-X ISBN 0-19-510654-7 (pbk.)

ISBN 0-19-517266-3 (reissued, cloth) ISBN 0-19-517265-5 (reissued, pbk.)
I. Golf — Anecdotes. I. Title.
GV967.S76 1995 796.352–dc20 94-36541

1 3 5 7 9 10 8 6 4 2
Printed in the United States of America

For Helen, alone

Contents

Foreword
By Arnold Palmer

It's a very simple title—*Golf Anecdotes*—for a book that runs the gamut of the game virtually from its inception. It does so in an informal fashion that is interesting and informative with its parade of short stories and incidents—some serious, some light-hearted and humorous—that reflect the vast knowledge and personality of its author. Throughout his long career as a newspaperman and literary executive with the United States Golf Association, Bob Sommers has been one of the most prominent students and chroniclers of the game, a man of charming and often acerbic wit with a vast knowledge of golf who has never tolerated fools and transgressors of the sanctity of golf. He has brought this all to bear in this entertaining book, and, as I read through its original version and the fresh material that Bob has put together for this updated manuscript, it stirred some thoughts and memories of my career beyond those in these pages and what has happened to the game in my lifetime.

I consider myself a traditionalist insofar as golf is concerned, but not of the hidebound sort unwilling to bend with the inevitable changes that have come about during my half century of professional involvement with the game. That is to say that I have supported and, I guess, played a small role in the progress the game has made on many fronts over the last 50 years or so. Who would have thought in the days right after the second World War that, by the turn of the century, the United States alone would have more than 25 million serious golfers, that there would be a period during the final decades of the 1900s that we would be figuratively opening a new golf course every day of the year? Or that this same surge of interest in the game would spread to virtually every corner of the globe? It seems to me that this all happened because this fascinating, intoxicating game was exposed to the general public through the explosion of television into a vast majority of the homes and public places throughout the world. Our PGA Tour and the other professional tours rode television from the backwaters to the mainstreams of sports universally, and many of the people watching golf who had never thought about playing it themselves began to try it. To try golf is to like it and it's one of the few sports that can be enjoyed at any age. Nothing pleases me more than to receive a letter from senior citizens who never played golf until they were in their sixties and now love the game or to hear from youngsters who, like I was, are hooked on golf at a very young age.

The success of the game has brought good fortune to many of us who play the game professionally and created a vast new industry that has contributed measurably to the world economy.

The basic game has changed little in my time and, in fact, since the Scots devised it centuries ago. But, golf has changed technologically and golfers have changed physiologically. Metal replaced wood in the driving clubs, and the increased strength and stature of today's generations of young people made length a cardinal and at the same time controversial focus. Great concern has been expressed in various quarters in recent years that many courses aren't long enough these days to test all sides of the players' games. I have made it clear how I view all of this. I have no problem with club technology because I can see that it makes the game more playable and enjoyable for all golfers, not just us professionals. Why shouldn't we employ modern technology to make it easier for the average player to hit the ball and hit it a little farther? My feeling, one that I have found is shared by many other leaders in the world of golf, is that the length problem can be solved by putting manufacturing restrictions on balls on the market so that, no matter what is done with the clubs, the balls cannot be hit as far as the best players on tour hit them today. I and others who agree with me have been lobbying the ruling bodies of golf to mandate this before things get completely out of hand and are no longer resolvable.

One other trend in golf that bothers me is a discernible slippage in respect for the traditions and the etiquette of the game. I'm not sure why this is happening. I don't want to think that, at the professional level, it is because, considering the huge amounts of money out there, it has become too much a hard, cold business and less of a pleasurable game that is to be played with courtesy, dignity, and good will. Whenever I have had the opportunity to speak to the generations of professional players who are on center stage today, I have told them that they have a duty to live by golf's rules and to preserve, protect, and carry on its wonderful traditions, that they have a special obligation to conduct themselves at all times in a manner that would do credit to themselves, their families, and their sport. In fact, it behooves everybody who loves the game to help maintain its principles, its honor, its decorum, its clean competitive spirit.

All of this is second nature to me. My father brought me up in the game this way. Perhaps the incident that had the greatest influence on me in this respect occurred when I was a teenager. I had reached the championship round on a junior tournament in Pittsburgh, but, as my opponent and I came down the stretch, I missed a short putt and, infuriated, I hurled my putter into a stand of trees beside the green. I regrouped, though, and went on to win the match. My parents were in the gallery and, when I climbed into the car for the triumphant ride back to Latrobe, I was greeted not by congratulations from my father but rather by his strong admonition that if I ever threw another club in anger like that, I would never play golf again while I was living in his house.

I think I have made my point. Enough of the soapbox. Turn the page and enjoy this anecdotal journey through the game of golf.

Preface

I was sitting in my office one day when Peter Andrews, who writes for *Golf Digest*, stuck his head through the door and asked how my research was coming along. I groaned. An accomplished author deep into research himself for a projected biography of William T. Sherman, the Civil War general who ravaged Georgia on his march to the sea, Peter told me he had come across an amusing tale that seemed just right for my mission. I told him to send it along with all appropriate speed.

It was indeed a funny story, telling in tedious detail of a man named Albert Haddock who had persistent and maddening problems playing the twelfth hole at the Mullion Golf Club, an undersized course of little more than 5600 yards on the rocky coast of Cornwall, in England.

The second shot on this hole demanded a carry across an inlet bordered by cliffs reaching sixty feet above the roiling Atlantic. In a long and frustrating career Haddock had never cleared the cliffs. One very good shot missed by only a yard or so, but he'd never conquered the chasm. His perpetual failures preyed on his mind.

The normal procedure called for the player to hit two shots. If neither carried the chasm, he would then wave the white flag of surrender and skip ahead to the next hole. Never one to take the easy way, Haddock occasionally hit six or seven balls. Some of them rolled gently to their doom, others soared out to sea. After years of failure he gave up on a second ball. When his first fell inevitably onto the sands below, he would scramble down the cliffs and there, among the rocks and stones, attack his ball until some lucky blow carried it to the crest.

Haddock had achieved a state of celebrity by then, and inevitably drew a gallery, which, in the perverse way of crowds, found satisfaction in his despair. They watched his futile labors and blushed at his flaming language. On one particular day, after Haddock nearly cleared the cliffs but failed—he claimed the hole played into the teeth of a raging gale—he clambered down among the rocks and carried out his usual exercise. But something seemed different this day.

For one thing, a larger than usual gallery collected on the rim of the hill, although Haddock claimed he knew nothing of them, as driven by an inner fury he struck blow after blow, before, after, and during which he cried a number of oaths of complex character that, he claimed, he had never used before and had indeed shocked even himself. Strangely, his caddie, evidently impressed with their originality, wrote them down.

As a consequence he was arrested and charged with violating the Profane Oaths Act, an obscure law dating to 1745 that prohibited swearing and cursing on Cornish golf courses.

Hauled before a magistrate, Haddock was fined five shillings, the highest that could be assessed under British law. The act provided as well that the fine should be assessed against each oath, and since the caddie had counted 400, Haddock owed the crown £100.

The British have never had a totally classless society; social status was reflected even in certain punishments. Had Haddock been a day-laborer, for example, he would have been fined no more than one shilling for each oath. Everyone else under the rank of gentleman would have been fined two shillings. Haddock was a gentleman, so he was given maximum sentence.

Naturally he objected. Standing squarely before the magistrate he challenged the amount of the fine on the novel basis that he could not he considered a gentleman while he played golf. He argued that whereas the prosecution used the term gentleman in the sense of rank, it should be used in the sense of character. Haddock maintained that many a day-laborer is of so high a character that under the act he should rightly be considered the same as gentlemen.

After pondering the merits of this innovative defense, the judge decided that by gentleman the law means a personal quality, not a social status. In his verdict he declared, "I find therefore that this case is not governed by the Act. I find that the defendant at the time was not in law responsible for his actions or his speech, and I am unable to punish him in any way. For his conduct in the chasm, he will be formally convicted of attempted suicide while temporarily insane, but he leaves the course without a stain on his character."

Terrific, I thought. Just the kind of stuff I wanted. It went into the manuscript and I thought nothing more of it after thanking Peter for sending it to me.

The manuscript had gone into type when I had a call from my son, Mike. Peter had sent me the anecdote in typescript and Mike had picked it up one day and read it, chuckling all the while. When he called he reminded me of it and told me of some other cases he had come across in a book by

A. P. Herbert, a British jurist. He told me of one case where a man had written a check on the back of a cow and sued when his creditor had refused to accept it; of a challenge to the legality of marriage, and a few other similarly strange incidents, which included my friend Haddock and the chasm at the Mullion Golf Club.

All of them had appeared first in *Punch*, the British humor magazine, all of them involved Mr. Haddock, and all of them were obviously fish stories. After a hurried call to explain that I'd been duped, the publisher yanked the episode out of the book.

As far as I know, everything else in the following pages actually did happen in some form. Stories change in the telling of course, particularly over a period of years, but as far as I know all grew around a kernel of truth. For example, I wonder about Gene Sarazen's story of why he arrived late for his first round match in the 1922 PGA Championship in Pittsburgh a month after winning the U.S. Open. It may stretch the sense of reason, but Sarazen claims a passer-by saw him at dinner in Columbus, Ohio, and told him he had only a few minutes to catch a train to Pittsburgh, whereupon he grabbed his bag and raced off to the station barely in time to jump onto the end platform of the moving train. This sounds like something from *Punch* as well, but Sarazen says it's true.

On the other hand, there is no dispute that Gene's double eagle actually happened. If he told about it today without corroboration, his story of the 1922 PGA would seem more plausible.

Again, if I had only recently beamed down after an extended tour with Captain Picard and picked up a book on golf, I'd believe the Sarazen tale of 1922 more easily than I'd believe that Bobby Jones could truly win the Grand Slam in 1930 and then quit the game; or than that Ben Hogan could play 90 holes in four days in 1950 on battered legs and win the National Open in a playoff; or that Jack Nicklaus could win the 1986 Masters Tournament after he had turned forty-six. These things did happen, though, and so if they did, why not all the rest?

Stories like these abound in golf if only we can coax them out of those who lived them. Finding sources is the key; I stalked them.

I'd always been interested in history; it seemed to me that knowing what had happened in the past helped immeasurably in relating what was happening today. As a young reporter covering the Masters at a time when there was no thirty-six-hole cut and everyone played all four rounds, I followed a certain ritual. Since I worked for an afternoon newspaper that had no Sunday edition, Saturdays were free. With no pressure to follow the minute-by-minute progress of those actually winning the tournament I chose instead to follow the old heroes, men like Billie Burke, the 1931 Open champion, Sam Parks (who had won in 1935), Ralph Guldahl, Horton Smith, the great amateur Jess Sweetser if he played, and Sarazen, of course. When I could I would take a seat next to Al Laney, of the *New York Herald Tribune* in the press facility, a corrugated metal

Quonset hut that was quite advanced for its time, and walk with him or Herbert Warren Wind when they'd put up with me, hoping to learn more about this wonderful game and its history.

As I headed out on the course one day during the 1950s, Bill Fox called and asked if I'd mind if he came along. A short man, rather slender, with snow white hair and a ruddy complexion, Fox worked for an Indianapolis newspaper. While he wasn't one of my idols, I knew he'd been around for years, and I thought this would be a great opportunity for me. I told him, "Sure come along," all the while gloating at my luck: "What stories he can tell me."

We headed down the path toward the first fairway, cut across through the woods, and gradually worked our way to the lower parts of the Augusta National Golf Club, talking about nothing in particular. As we broke through the trees, I saw that we had arrived near the green of the fifteenth hole. There on the left the ground rose to the crest of the hill, and just a little bit to the right stood the broad pond, and beyond it the wide but shallow green. It was here back in 1935 that Sarazen had holed that remarkable four-wood shot for his double eagle 2, the most famous golf shot ever played. Television was just beginning to catch on in the middle 1950s, and the huge audience the Masters draws each spring hadn't been exposed to the wonders of Amen Corner—the wonderful sweep of holes from the tenth through the thirteenth. No one knew about them. Instead, because of Sarazen and because the Masters telecast began at the fifteenth, it was probably the most famous hole in all of golf.

We stood quietly for a moment, drinking in the scene—the emerald grass and the sparkling, un-naturally blue water (it was dyed)—and I waited impatiently for him to tell me all about this hole; maybe he'd even been there that day twenty years earlier when Sarazen tore the heart from Craig Wood with his double eagle, then beat him in a playoff the following day.

Finally Bill spoke. Looking first to the left and then to the right, he turned his head toward me and said, "Tell me, young man, what hole is this?"

Some sources were better than others.

Port St. Lucie R. S.
January 1995

Acknowledgments

Books of this sort by their nature can't be done without help. I had plenty, much of it from friends working on books of their own who with uncommon generosity sent me material they thought I could use. I am particularly indebted to George Eberl, whose little gem *Golf Is a Good Walk Spoiled* brims with charming and baffling incidents relating to the rules of the game; and to Rhonda Glenn, who sent me batches of manuscript from her own *The Illustrated History of Women's Golf*, a story worth reading. Many incidents from those works are retold here.

Others offered their guidance as well. Dave Marr was particularly helpful, and so was Dick Miller, the author of *Triumphant Journey*, the story of Bobby Jones and the Grand Slam. Ken Bowden pointed me to helpful stories of Jack Nicklaus, Doc Giffin helped with Arnold Palmer anecdotes, and Robert Macdonald, the publisher of "The Classics of Golf," not only sent copies of his books, he told me which pages to study as well.

Nor can I overlook Robert Guenette, who put together a series of films for the United States Golf Association entitled "Heroes Of the Game." I have used some of the jewels he dug up.

Through some assignments he gave me, George Peper, the editor of *Golf* magazine led me to some useful information; Jerry Tarde, the editor of *Golf Digest*, was always ready with encouragement; and I extend everlasting thanks to Gary Van Sickle and Terry Galvin of *Golf World*. It was Gary who first traced the tortuous turns that led to John Daly's winning the 1991 PGA Championship when he started out no closer to playing than ninth alternate. Gary and Terry sent the account along to me. I am indeed indebted.

I must thank John Matheny of the United States Golf Association for a number of the Ben Hogan stories. As a young man John spent time as a shop boy at Shady Oaks and heard so many tales of the great man.

Abroad, Dermott Gilleece, in Dublin, led me to Harry Bradshaw tales; and Bruce Critchley, a former British Walker Cup player and commentator with the BBC, told me some colorful and ribald stories, not all of which could be used here.

So many others helped in so many ways that I'm embarrassed because I know I've failed to mention more of them than I've managed to thank.

I must mention this, however. The collegiality of those who write about golf is one of the joys of this business. Everyone works to become the best he, and lately she, can be, but after more that forty years I've found that nearly every one of them has been willing to share. This spirit has made mine such a happy career.

Golf Anecdotes

Introduction

Everyone's a Critic

As it is with any activity that involves strong personalities, golf is full of bizarre tales.

It can be a perverse game; millions love it and can't play often enough, but others despise it. The American critic H. L. Mencken claimed that anyone who played golf should be barred from holding public office. Mark Twain called it a good walk spoiled, and Winston Churchill claimed it was like chasing a quinine pill around a pasture.

Don't Be Presumptuous

Ben Hogan was capable of biting sarcasm and caustic comment that could peel the hide from a rhinoceros. He could freeze a man with an icy glare or make him feel he was on fire. Those were rare moments, saved for those who asked questions he felt were particularly stupid or made requests he felt were presumptuous. His victim's status in life mattered not at all.

Hogan was appointed captain of the 1967 American Ryder Cup team for the match at the Champions Golf Club, in Houston. Although Arnold Palmer was gradually giving way to Jack Nicklaus in the middle to late 1960s, he was still at the top of the game, the player most people, given the choice, would want to watch play. Aside from all the tournaments he'd won—four Masters, two British Opens, and one U.S. Open in particular—he'd been picked for three Ryder Cup teams, once as both player and captain. He won a place on the 1967 team as well, his first with Hogan as captain, a man who took every competition seriously. There would be no nonsense; Hogan played to win.

Their first meeting under those terms sparked an exchange that caused the other players to cringe and hunker down in their chairs to avoid barbs that flew like shrapnel at Verdun.

Palmer had been given permission to arrive a day late. Before he arrived, the team met and debated which ball they would play, whether the small ball used everywhere else in the world except North America, or the bigger American ball they had all grown up playing. The size difference amounted to only six one- hundredths of an inch, but the smaller ball flew farther, especially into the wind. Assuming the British and Irish side would use the smaller ball, the Americans decided it would be foolish to give up the distance they'd lose by playing the bigger ball. They chose the small ball.

Palmer came the next day. Flying his own plane, he zoomed low over the golf course, announcing he was on his way, and arrived at Champions a short time later. Champions has a very big locker room. While Hogan sat at one end, Palmer strode through the door at the far end, a good distance away, and initiated a conversation that was shouted from both ends. Seeing Hogan, and somewhat annoyed about the choice of balls, Palmer called, "Hey, Ben, what's this about playing the small ball?"

Obviously already fuming because Palmer had not arrived with the rest of the team, Hogan glanced up and called back, "That's what we agreed before you decided to come."

Then Palmer announced, "I don't play the small ball."

Unimpressed and perhaps sensing a challenge to his authority, Hogan snarled, "Who says you're playing, Palmer?"

Palmer did play the first round of foursomes. Paired with Gardner Dickinson he won two matches, worth one point each.

When the pairings for the second day's four-ball matches were announced, Palmer's name was missing, setting off a panic among the press. They wanted an explanation, but not one among them had the nerve to face Hogan alone. In cases like that, the simplest solution is to ask for a press conference, where they would find safety in numbers.

The conference called, the press crept into the room where Hogan sat on a raised platform beside a PGA representative. No one had the nerve to speak first. As the silence grew increasingly uncomfortable, the PGA man spoke up. "Gentlemen," he said, "you asked for this meeting. Doesn't anyone have a question?"

Finally one brave soul sacrificed himself for the common good and said, "Mr. Hogan [no one called him Ben], is it true Arnold isn't playing tomorrow?"

Hogan glared and said, "That's what the pairings say."

"Could you tell us why?"

Hogan glared a stronger glare and admitted, "I could, but I won't."

The conference broke up. Palmer did indeed sit out the morning four-ball, but he teamed with Julius Boros and won the afternoon match, and then won two singles the following day. In the three days of the Ryder Cup, Palmer won every match he played and accounted for five points as the United States won, 23½ to 8½. The Americans wouldn't have dared lose. Hogan wouldn't have liked it.

Squelched

The game is full of squelches, some even unintentional. Sam Snead stood banging balls on the practice tee early in the week of the U.S. Open one year while a lone spectator stood watching him. After a time the spectator called to Sam and asked, "When do the pros start playing?"

A little vain to begin with, Sam was naturally put off. "I don't know," he snarled, "I was just sent out here to break in the course."

Take a Drop

Robert Trent Jones was probably the most widely known of all the golf course architects, but not everyone greeted every one of his courses with wild-eyed raves; some players thought Trent went too far in creating punishing holes. Running into him one day, Jimmy Demaret slapped him on the back and said, "Saw a course you'd really like, Trent. On the first tee you take a penalty drop."

Evolution of the Captain

Horace Hutchinson, who blossomed late in the nineteenth century and won two of the first three British Amateurs, introduced golf to his logic tutor at Oxford. That evening the tutor was asked to describe the game as he saw it. Pausing not a beat he said, "It consists in putting little balls into little holes with instruments ill adapted to the purpose." His description satisfied both sides—those who love the game, and those who hate it.

Hutchinson, of course ranks among those who loved it best. He is also one of those who caused a change in how affairs were run in the golf world. Through the early history of British club operations, it had been traditional that the captain won his office in a tournament held at the end of the competitive year.

Hutchinson was an exceptionally fine golfer. He played as top man for Oxford and reached the British Amateur final in three consecutive years. In his finest moment he beat the great John Ball by one hole at the Royal Liverpool Golf Club, at Hoylake, Ball's own course, and won the 1887 British. He rose to become the first English captain of the Royal and Ancient Golf Club of St. Andrews after turning to writing about the game.

As a member of Royal North Devon, a club known better as Westward Ho!, Hutchinson won the club's Autumn Medal in 1875 and assumed the role of captain. The captain in those days automatically became the arbiter of all things concerning golf within the club. What he said carried the day. There was a problem with that. Hutchinson was only sixteen. Westward Ho! changed the rules; the following year the captain became an elective office; no more young boys running the club. The idea caught on, and the old method died.

A Humbling Game

Golf has been called a humbling game, and indeed it is. The great Irish amateur Joe Carr was playing a casual round at Rosses Point, in County Sligo, in western Ireland, the scene of many a triumphal day earlier in his golf career.

Joe had built a wonderful record. From 1953 through 1967 he had won three British Amateurs, six Irish Amateurs, had played in nine Walker Cups, and had captained the 1965 team, which held the United States to a halved match at Five Farms, in Baltimore. He was the best-known amateur ever to come out of Ireland.

Well past the years when he could still play first-class competitive golf, Joe was having a bad day at Sligo. He'd hit one shot to the left, another to the right, seldom hit a green, and occasionally failed to get his ball airborne.

He had a young caddie who not only didn't know him, but had never even heard of him. Trying to ease Joe's growing agitation, the caddie asked, "Have you ever played Rosses Point before, Sor?"

"Oh, yes," Joe answered. "Many times."

"You know they play the West of Ireland championship here, Sor," the caddie went on.

"I know," Carr said, and then added smugly, "I won it twelve times."

The caddie thought that over for a while, and when Joe squibbed another ball along the ground, he concluded, "It must have been fierce easy to win in those days, Sor."

Language Barrier

In a game that has known many men whose accomplishments impressed themselves at least as much as they impressed others. Charles Blair Macdonald would stand close to the top of any ranking of the egotistical. Convinced he was the best golfer in America in the latter years of the nineteenth century, he refused to accept as worthy of the name championship any tournament he didn't win.

Macdonald eventually did win the U.S. Amateur, the first held under the imprimatur of the United States Golf Association. H. J. Whigham, who succeeded him as champion and later became his son-in-law, went on to become editor of the magazine *Town and Country*. Of his father-in-law, Whigham observed that Macdonald "never learned to master the passive verb."

Bull-headed, domineering, and overbearing, Macdonald accomplished wonderful things nonetheless. He laid out a number of golf courses of lasting quality, including the charming Mid Ocean Club, which ranges along the ocean in Bermuda. Its fifth hole, a 430-yard par 4 calling for a drive across a lake, stands among Macdonald's finest designs. The fairway swings left, allowing the golfer to cut off as much of the lake as he dares. The more he cuts off, the shorter the approach to the green, which sits on a peninsula jutting into the lake. Babe Ruth once lost eleven balls trying to bite off too much.

When the course was finished, Governor General Sir James Wilcox and an Admiral Packenham asked to look it over. Arriving at the fifth tee, the admiral said he didn't believe anyone could clear the lake. Macdonald insisted it could be done, but the admiral was still skeptical. "Let's see you do it," Packenham challenged.

Never known to back away from a challenge, Macdonald reached for his driver. As he teed his ball and took his stance, the admiral asked where he was aiming. At that moment two dogs pranced along the fairway. Macdonald said he would drive where the dogs were. Joking, the admiral asked, "Which one?" Macdonald looked up and said, "The second one."

He hit a beauty. The ball cleared the water easily, bounced once on the fairway, and then everyone heard the dog yelp. The ball had smacked into its hindquarters. Macdonald was one of those infuriating people who could back up bragging with performance.

Immodest to a fault, when he built his golf course in Southampton, on Eastern Long Island, he called it nothing less than the National Golf Links of America. It was a landmark course, the first truly modern American course that followed the strategic concept of design, allowing the player several choices of approach to playing the hole. Oakmont, in the suburbs of Pittsburgh, outdates the National by about ten years, but it is a penal course, extracting severe penalties for the mis-played shot.

With all his character flaws, Macdonald sometimes showed a pixie-ish sense of humor. He ruled, nevertheless, as the absolute monarch at the National. After the course had opened, Daniel E. Pomeroy, a member, suggested a windmill would look nice sitting on a hill overlooking the sixteenth green. Macdonald agreed. He had it built and sent the construction bill to Pomeroy.

Macdonald insisted on an unusual conditions for the clubhouse. Buildings ordinarily are considered obstructions; a ball in a building, on it, or close enough to interfere with a player's stance or swing can be moved without a penalty. Macdonald felt the National's clubhouse should be an "integral part of the course." You play the ball as it lies no matter where that might be, or else call it unplayable, taking the appropriate penalty.

Impromptu Ruling

Each year the National holds a tournament called the Gold Putter. Ken Gordon, a member of the USGA's executive committee and a pretty good golfer, lost his first round match and had gone to his room in the upper floors of the clubhouse to take a nap. Gordon stripped to his underwear, stretched out on the bed, and felt himself dozing off when someone rapped on his door.

Swearing softly at the interruption, Gordon opened the door and found a young man holding his golf shoes in his hand, trailed by his caddie carrying his full complement of clubs. The young man explained that he had pulled his approach to the eighteenth onto the clubhouse roof. "Mr. Gordon," the young man said, "the only way I can get to my ball is through your room."

Ever suave, Gordon said, "Yes, of course," and led the way. There was some sticky business with the screen in Gordon's window, which was solved in the end, and then caddie and player scrambled onto the roof. After a brief search the young man found his ball in a rain gutter, clearly unplayable. The young man appealed to Gordon, hanging out the window watching. "What can I do?" he asked.

In his time Gordon had officiated at the U.S. Open and the Masters, and so he was equal to the challenge. His pipe clenched tightly between his teeth, Gordon told him he could invoke the unplayable ball rule. He could lift the ball from the gutter, clean it if he wished, and drop it within two club-lengths of where he'd found it in the gutter.

The young man tried twice, but each time his ball bounced and hopped back into the gutter. Gordon told him to place the ball where it had struck the roof the second time he'd dropped it. His ball properly placed, the young man pitched off the roof and onto the course, still alive in the match, although lying 4.

The young man and his caddie climbed back inside Gordon's room, thanked him for his help, and went on their way. Thinking this was the only time he could remember giving a ruling in his underwear while he hung out a bedroom window, Gordon went back to his nap.

The Missing Mate

Golf has had its trying moments. Two under par through the thirteenth hole of the 1992 U.S. Open's second round at Pebble Beach, California, Nick Faldo was in good position, but then he ran into trouble on the fourteenth hole, a long par 5 that doglegs right. Two decent shots put him within 9-iron range of the green, but he pushed his shot toward a tree leaning over a bunker. Everyone saw the ball go into the tree, but no one saw it come down. It must still be up there.

Just in case he couldn't find it, Faldo played a second ball and pushed another shot. It splashed into the sand. Since he'd be assessed a two-stroke penalty if he played the second ball, Nick thought it would be a good idea to climb the tree and look for the original. If he found it he could declare it unplayable and save one stroke. With the gallery gathered around, he climbed up and shook the tree. No result. "Higher. Higher," the gallery called. Nick climbed higher and shook again. Still no result.

Now he began to feel silly. With the gallery laughing and the other players wondering when he would ever come down, Nick thought of himself as Tarzan swinging through the jungle. As loudly as he could, Faldo called, "Where the hell's Jane?"

It Comes and Goes

In another, even more critical moment, Craig Wood could have won the 1939 U.S. Open except for one mis-played shot. The regulation seventy-two holes ended with Byron Nelson, Denny Shute, and Wood tied at 284, forcing an eighteen-hole playoff the following day. Shute wasn't the same player he had been and fell out of the hunt early, but Nelson and Wood battled stroke for stroke through the first sixteen holes. When Nelson three-putted the seventeenth, Wood moved a stroke ahead with only the eighteenth to play.

This is where Sam Snead needed only a bogey 6 to win and took his infamous 8 instead. Wood hit a nice drive, but he pulled his second shot into the gallery, hitting a spectator who turned out to be a distant relative of Alfred M. Landon. Three years earlier Landon had lost the presidential election to Franklin D. Roosevelt by epic proportions. He carried only Maine and Vermont; Roosevelt won the other forty-six states. This obviously wasn't a good time for the Landon clan. Nor was it a good time for Wood. He had already lost the 1935 Masters to Gene Sarazen and now he lost to Nelson. Byron birdied the eighteenth and tied Wood, then beat him in a second playoff.

Wood's victim was lucky. Taken to a hospital, he returned the next day to see the final playoff round.

After years of falling short, Wood finally hit the top in 1941, winning both the Masters and the Open. Suffering from a painful back, he played the Open wearing a steel corset. His back caused him so much pain he nearly withdrew, but instead he carried on, shot 284, and won by three strokes over Denny Shute.

After the presentation Wood climbed into a car to drive to his hotel with Fred Corcoran, who ran the PGA Tour, and Bill Cunningham, a sports columnist for the *Boston Herald*. Along the way they approached a driving range. Always alert for publicity, Corcoran said, "Craig, don't you want to hit a few balls?" Going along with it, Wood said, "By all means; pull in." Wood climbed out, bought a bucket of balls from the range owner, banged out the balls, climbed back into the car, and they drove away.

Cunningham found all this a little bizarre. "Why did you do that?" he asked. Without cracking a smile, Wood explained, "I didn't want to lose my touch."

Porky's Practice

In another zany moment, Porky Oliver was warming up for the Canadian Open one year. Arriving at the club, he deliberately picked the smallest caddie he could find, saddled him with his huge golf bag, and marched him to the practice tee. Once the caddie had succeeded in his struggle, Oliver had him unzip one of the pockets and dump what must have been 100 balls onto the ground and arranged them in a neat pile. Porky told the youngster to move out onto the practice ground. When the kid had walked out about 100 yards and turned around. Oliver waved him along. He went another 100 yards, and Oliver waved him along once more. Now at the very end of the practice ground, the caddie stood up against a fence, about 275 yards out. With the caddie in place, Porky took out his wedge. He shanked his first shot. Next he hooked one. He topped his third and hit his fourth sky high. With this he whistled to his caddie—who hadn't come close to fielding one of those balls—called him back in, had him scoop the balls back into the hag, helped him lift it onto his shoulder, and marched back to the clubhouse. End of practice.

Buy American

One year a major golf ball manufacturer gave a dinner party for about one hundred golf pros and their wives. Announcing there would be no speeches, a company representative said he did, however, want to call his guests' attention to a very serious problem. He claimed the country was being flooded with Japanese golf balls that he said were being sold at "ridiculously low prices." He claimed the cheap balls were hurting the pros' business and that unless the government imposed a stiff tariff, the Japanese would corner the golf ball market.

At each setting the company had placed a small, pearl-handed jackknife for repairing ball marks. The bottom of each box bore the inscription "Made in Japan."

Life's Frustrations

The game can be frustrating because of mysterious circumstances. Byron Nelson played a particularly fine drive during the 1941 Hershey Open. He watched it clear a tree at the bend of a dogleg and hit squarely in the fairway. When he arrived at the spot though, he couldn't find it. With no option but to call it a lost ball, he walked back to the tee and played an identical shot. Reaching his second ball, Byron muttered "I don't understand it. My first drive landed in the same spot, but I couldn't find it." The two-stroke penalty for the lost ball cost Nelson $300 in prize money, a significant amount in 1941.

A week passed, and then Byron received a letter from a young man who told him he had taken a girl with him who had never been to a tournament before, and on the way home she told him, "Remember that lost ball they were all looking for?" Reaching inside her purse she took out a golf ball and said proudly, "Here it is. I picked it up when nobody was looking."

The young man apologized and went on to say he read where the penalty cost Byron $300 in prize money. "At least let me reimburse you for your loss," he wrote. The envelope contained three one hundred dollar bills.

The Blind Shall Lead Them

Corcoran once refereed a two-hole match between two blind golfers played at night during the Cleveland Open. The next day Corcoran saw Ky Laffoon on the practice putting green. One of the wackier characters to grace the tour in its early years, Laffoon listened while Corcoran told him how well the blind men putted. Fred demonstrated how one of them never moved his head during the stroke. While Corcoran coached him, Laffoon tried the blind man's method, liked it, and went out and won the Cleveland Open.

The next day a newspaper headline read, "Blind Man Teaches Laffoon How To Putt."

Thanks But No Thanks

The game can be a little stuffy as well. Corcoran lined up a celebrity four-ball to play an exhibition in Norwalk, Connecticut, for the benefit of the local hospital. This was not to be your normal exhibition match; as Corcoran described it, he planned to take golf temporarily out of what he called the public library atmosphere.

He persuaded Babe Ruth, Gene Tunney, Gene Sarazen, and Jimmy Demaret to play together. Corcoran brought out a combo from the Fred Waring band, which was appearing in New York, and Colonel Stoopnagle, a 1930s radio character. The combo, marching to a lively beat, followed the players around and never paused in their music-making, not even in mid-stroke, and the Colonel followed along in a sound truck offering absurd commentary. It was a quite boisterous affair.

Once as Ruth lined up a putt, the gallery grew politely quiet. Ruth turned and yelled, "How about a little noise. How do you expect a man to putt?"

The match turned out to be wildly successful. About 6000 fans turned out, *Time*, *Life*, and *Newsweek* covered it, along with the wire services and one radio network. The often zany carryings-on seemed to have little effect on the players. Demaret shot 70 and Sarazen 71.

The following day Corcoran had a telephone call from Harold Pierce, of Boston, the president of the USGA. Pierce said, "I heard you had quite a bit of excitement in Norwalk the other day." Corcoran admitted this was true and added, "It was quite a match."

"Tell me about it," Pierce said. Corcoran went into exquisite detail while Pierce chuckled. When Corcoran finished, Pierce said, "It sounds like a lot of fun. I hope you never do it again."

Isn't Imagination Wonderful?

Finally, some of the game's more appealing incidents didn't happen at all. Shortly after Max Faulkner won the 1951 British Open, readers of London's *Daily Mail* read that before a blow had been struck in anger, a young man had approached him early in Open week, handed him a ball, and asked for his autograph. Faulkner took the ball and in a supreme seizure of confidence scribbled, "Max Faulkner—1951 Open Champion."

Everyone agreed it must have been true; hadn't they read it in the newspaper? What really happened is this. Ian Wooldrige, the *Daily Mail*'s gifted sports columnist, sat before his flickering fire one evening growing increasingly desperate for an idea. The deadline for his column approached with alarming speed, and poor Ian hadn't a thought in his brain. Suddenly, in a flash, he saw the vision of Faulkner predicting his own victory at Royal Portrush. The championship had been played in Ireland, hadn't it, and didn't Ireland reek with leprechauns, little people who knew everything? Who would question that one of them had pricked Faulkner's brain and given him prescient knowledge of what lay ahead? Who could say it didn't happen? Who could deny it? Certainly Faulkner wouldn't. So Wooldridge wrote it.

Part One

Royal Bans to Tournament Golf

For centuries the search for how, and even where, golf began has both fascinated and perplexed those who care about such things. No one can say for certain; they can offer only theories. One of those theories traces the game to Roman times, when boys raced through the streets batting a leather ball with a stick. Others say the game originated on the frozen canals of Holland, where the Dutch played a game called *het kolven*, and since the heavy, unwieldy clubs they used to roll a ball along the ice toward a goal post was called a *kolb*, or *kolf*, we see the temptation to believe golf came from Holland, brought across the North Sea by Dutch traders.

It seems more likely, however, that golf as we know it originated on the linksland of Scotland, sandy wastes left behind by the receding sea, that shepherds used for grazing sheep. It is easy to picture a shepherd swinging his crook to knock a rounded stone toward a convenient rabbit hole. Heightening the challenge by moving ever farther away, he might have planted another crook to mark the hole's position—the first flagstick.

Whatever its origins, scholars believe golf was played as early as the twelfth century, although not on established courses as we've come to know them. It spread so and had become such a passion by the fifteenth century that those who ruled Scotland agreed it was far too popular and actually undermined the country's defenses.

Scotland had broken free from English rule by 1338 through a series of wars, which had eliminated the chance of cordial relations between the two countries. Consequently, when Henry V invaded Normandy, in 1420, and won the battle of Agincourt, King Charles VI of France asked Scotland for help. The Scots obliged by sending 7000 men to bolster the French army. The English broke an Easter-week truce the following year by plotting to ford a river and spring a surprise attack against the Scots, many of whom were "playing ball and otherwise amusing themselves with other pleasant or devout occupations."

Fortunately for the Scots, some of them saw the English advancing and organized a defense strong enough to turn them back, inflicting the first serious defeat on the English since they had slaughtered the French at Agincourt.

What were the Scots playing? It could have been golf; it was certainly played by then, and since it was simply a cross-country game at that time, it could be played anywhere. Perhaps the memory of that day when the Scots nearly lost yet another battle to the English because their minds were focused on games rather than their military purpose lingered with King James II of Scotland. Thirty-six years later he issued a famous decree that "The futeball and the golfe by utterly cryit downe and not usit." A similar decree was issued by James's successor, and in 1491, a year before Columbus sailed to the new world, the Third Parliament of James IV renewed the ban with an act that stated, "It is statute and ordained that in na place of the realme there be used Fute-ball, Golfe, or uther sik unproffitable sportis contrary to the commoun good of the Realme and defense thereof." James preferred his subjects practice archery—the better to kill Englishmen.

The first notice we read of golf, then, tells us we can't play it. Of course, the first man to ban golf didn't last very long, either. In the thick of the fight, James II stood alongside one of his primitive cannons during the siege of Roxbury Castle. When it blew up, James of the Fiery Face blew up with it.

Banns and the Ban

Although it was never officially lifted, The ban against golf ended for practical reasons on February 11, 1502, at the altar of Glasgow Cathedral, where James IV not only signed the Treaty of Glasgow—guaranteeing perpetual peace with England and, some say even more important, rescinding the ban against golf—but also betrothed himself to Margaret Tudor, the daughter of Henry VII, who had signed the treaty for England.

James and Margaret were married with proper ceremony in August of 1503, at Holyrood House, in Edinburgh, the seat of the Scottish royal family. Since peace lessened the need for archery practice, golf could be played with a clear conscience. On a clear February day in 1504, James took advantage of the truce and played a round of golf with the Earl of Bothwell. This was the first golf match we know of. Unfortunately, no one bothered to say who won.

Expense Accounts

King James IV, who was later killed at Flodden Field after the truce broke down, had his clubs made by a bowmaker. For the year 1502 the Lord High Treasurer entered in his accounts:

Item: The xxi day of September to the bowar of Sanct Johnestown (now Perth) for clubbs, xiiijs.

Then, a year later, these items:

The third day of Februar, to the King to play at the Golfe with the Erle of Bothuile, iij France crowns, sumena, xlijs.

For Golf Clubbis and Ballis to the King that he playit with, lxs.

In Time of Sermonis

Even after the Peace of Glasgow, golf was forbidden while church services were in session. Those who violated the ban weren't beheaded, as some early accounts suggest, they were fined, probably 40 shillings. The ban didn't prevent the clergy from taking up the game, however. During the time of the religious troubles of the seventeenth century, Michael Bruce, a noted Scottish preacher with obvious experience in the game, delivered a sermon at Cambusnethan that he phrased in golf terms, because his congregation might understand this language better than ecclesiastical expressions.

"The soul-confirmed man," he said, "leaves the Devil at the two more," a reference to the scoring method of the time in which the winner was determined by the number of holes won and the scores by hole kept only in relative terms—that is, the godly man had defeated the Devil by two strokes on the critical hole.

A Divided House

Henry VIII played tennis, but his first wife, Catherine of Aragon, daughter of the Spanish monarchs Ferdinand and Isabella, preferred golf, which may or may not account for what happened to their marriage.

Catherine evidently carried on a long correspondence with Thomas Cardinal Wolsey, the Archbishop of York, who managed foreign affairs for the king. In one letter she wrote that Henry's subjects were happy because they were busy playing golf, an obvious slip. In another she wrote that the king's subjects would he happy that she would be busy with golf, "for my heart is very good to it."

A Prince's Debt

When James VI of Scotland packed his bags and headed south to succeed to the English throne as James I after the two countries united, in 1603, he took his golf clubs with him. He not only played the game himself, he passed on his passion to his two sons. As he played one day, Henry, Price of Wales, the older of the two, was accompanied by one Newton, his schoolmaster, who stood uncomfortably close to the Prince as he was about to play a shot. Henry warned Master Newton to move aside and give him room, but evidently the schoolmaster didn't move far enough. As the prince began his backswing, another man warned him, "Beware you hit not Master Newton." Annoyed that he had been interrupted in mid-stroke, the Prince glared and snarled, "Had I done so I had but paid my debts."

First International Match

Arguments over the origins of golf reached the royal line during the seventeenth century. While he served as Commissioner to the Scottish Parliament under his brother, Charles II, the Duke of York, who later became King James II of England, fell into an argument with two English noblemen of the Scottish court over where golf sprang from. Probably because golf was being played at Blackheath, near London, at the time, the Englishmen claimed golf was an English game. The three men decided to settle the dispute by playing a match over the Links of Leith, close by. The noblemen would team against the duke and any Scottish partner he liked.

After a long search for the best golfer he could find, the duke chose John Patersone, a poor shoemaker who came from a long line of first-rate players and who himself was considered the best of his time. The duke and his partner won. As his reward, Patersone was given an equal share of what evidently was a considerable stakes. He built a house near the Canongate, in Edinburgh, with his winnings, and the duke caused an escutcheon to be fixed to the wall bearing the Patersone coat of arms—a hand holding a golf club and the motto "Far and Sure" inscribed above it. The house still stands.

Early Fanatics

Bishop Gavin Hamilton, a Scottish prelate, grew so addicted to the game and played so often on the links at St. Andrews that he apparently became conscious-stricken in mid-round one day, scurried home, went to bed, and died, "not having given any token of repentance for that wicked course he had embraced."

Since golf was forbidden during church hours, the Earl of March assured there would be no sermons to keep him from his game. As Commandator of the Priory of St. Andrews, he held a position of some power. When the church lost its clergyman, the Earl conspired with others in the town to hold the position vacant so there would be no sermons. These charges were made by James Melville, the son of the minister at Montrose, who claimed further that the earl used the money that otherwise would have gone to the church to pay for his "goff, archerie, guid cheer, etc."

The day before his marriage, on November 9, 1629, the young Marquis of Montrose bought golf balls in order to play against Sir John Colquhoun, the Laird of Luss, his imminent brother-in-law. The next day he was married to Magdalene Carnegie, "But scarcely had the minstrels ceased to serenade them when we find Montrose at his clubs and balls again." He was only seventeen at the time, his bride a year younger, and they evidently enjoyed a honeymoon of mixed interests. Nine days after the wedding Montrose sent a messenger to St. Andrews for repairs to damaged clubs, six new clubs, and a further supply of balls.

Montrose was probably among the first to play with the new featherie ball, which had been introduced about 1618. Until then golf was played either with balls made of a hard wood, or leather stuffed with cotton or wool waste.

Not Worth the Price

On his way to the Battle of the Boyne in the Jacobite Rebellion of 1690, King William III sailed to Ireland and his meeting with the forces of the deposed James II from the town of Hoylake. Although today the site of the Royal Liverpool Golf Club, among the most storied in Britain, the value of the land fell as silt from the River Dee piled up along the Hoylake shore.

By the early nineteenth century, the entire town was owned by Samuel Baxter. When his daughter was about to be married, in 1809, he decided to sell a portion of his estate. A fisherman named Eccles offered £90. Since this wasn't quite what Baxter wanted, he offered to throw in the sector of the township where the Royal Liverpool stands today for a further £10. Eccles talked over the proposition with his wife, who was evidently a thrifty woman. She said the town wasn't worth the money, and the sale was canceled. A hundred years later the portion Mrs. Eccles turned down for £10 was sold to Royal Liverpool for £30,000.

The First Tournament

John Rattray, an Edinburgh surgeon, ranked among the finest medical men of his time, so well known he was dragged from his bed by Charles Edward Stuart, Bonnie Prince Charlie, to care for his troops on his march into England in 1745. He was rescued through the efforts of Duncan Forbes of Culloden, himself a member of the Company of Gentlemen Golfers, and for a time the secretary of the club. An accomplished golfer himself, Rattray was mentioned in *The Goff*, a mock epic poem dating to the eighteenth century. He won the first golf tournament we know of.

Until the middle of the eighteenth century, competition was limited to individual matches, either singles or alternate shot foursomes, but in 1744 the Company of Gentlemen Golfers, which evolved into the Honourable Company of Edinburgh Golfers, petitioned the city of Edinburgh to provide a silver golf club as a prize for an annual competition on the Links of Leith.

The city's magistrates agreed, but they added the condition that the competition should be open to "as many Noblemen and Gentlemen, or other golfers" from any part of Great Britain and Ireland that enter during the eight days leading up to the tournament. The winner was to be called Captain of the Golf, and act as arbiter of all disputes relating to the game.

The competition was played on April 2, but it was a flop. Just twelve men entered, all of them local, and only ten actually played. Details have been lost, but we do know that Rattray won and repeated the following year.

First Rules of Golf

In order to play the first tournament, the club had to write and post uniform rules. For many years the original rules of the Society of St. Andrews Golfers, which developed into the Royal and Ancient Golf Club of St. Andrews (R and A) were believed to have been the first written rules, but the code adopted for the 1744 tournament of the Gentlemen Golfers of Edinburgh surfaced just before the beginning of the Second World War. Since with one exception they are identical to the rules adopted by the St. Andrews Society ten years later, it seems obvious St. Andrews copied from Edinburgh. In the lone difference, a ball in "watery filth" at Edinburgh must be teed and at St. Andrews dropped.

The 1744 Code of Rules

I. You must tee your ball within a club's length of the hole.

II. Your tee must be upon the ground.

III. You are not to change the ball you strike off the tee.

IV. You are not to remove stones, bones, or any breakclub for the sake of playing your ball except upon the fair green, and that only within a club's length of your ball.

V. If your ball come among water, or any watery filth, you are at liberty to take out your ball and bringing it behind the hazard, and teeing it, you may play it with any club and allow your adversary a stroke for so getting out your ball.

VI. If your balls be found anywhere touching one another, you are to lift the first ball, till you play the last.

VII. At holing, you are to play your ball honestly for the hole, and not to play on your adversary's ball not lying in your way to the hole.

VIII. If you should lose your ball by its being taken up, or any other way, you are to go back to the spot where you struck last and drop another ball and allow your adversary a stroke for the misfortune.

IX. No man at holing his ball is to be allowed to mark his way to the hole with his club or anything else.

X. If a ball be stop'd by any person, horse, dog, or anything else, the ball so stop'd must be played where it lyes.

XI. If you draw your club in order to stroke, and proceed so far in the stroke as to be bringing down your club—If then your club break in any way, it is to be accounted a stroke.

XII. He whose ball lyes farthest from the hole is obliged to play first.

XIII. Neither trench, ditch or dyke made for the preservation of the links, or the Scholars' holes, or the Soldiers' lines, shall be accounted a hazard, but the ball is to be taken out, teed, and played with any iron club.

The Origin of Stroke Play

Ten years after the Company of Gentlemen Golfers inaugurated its tournament, the Society of St. Andrews Golfers organized a tournament of its own, in 1754, with a feature unique to its time. Matches had always been scored by holes won; the total number of strokes was seldom counted, but one player was either playing the like, the same number as his opponent, or playing the odd, the one more, or the two more. When the hole ended, they began again. This method was certainly good enough for individual matches, but it didn't work well when a large group played.

The Society of St. Andrews solved that problem neatly, laying down the condition that "In order to remove all disputes and inconveniences with regard to the gaining of the Silver Club, it is enacted and agreed by the captain and the gentlemen golfers present that in all time coming whoever puts in the ball at the fewest strokes over the field . . . shall be declared and sustained victor."

Thus stroke play was born.

Dinner with Garrick

While he was visiting London in 1758, the Reverend Dr. Alexander Carlyle, a prelate from the Lothian coast, near Musselburgh, Scotland, was invited to dine with David Garrick, the great English actor, at Garrick's house, in Hampton. Dr. Carlyle was known as a golfer, and Garrick told him and his other guests to bring along their golf clubs. The group played, then returned to the Garrick house. Dinner over, Carlyle noticed an archway and beyond it the River Thames. Calling to Garrick, he said he'd show him a shot that might surprise him. Carlyle then drove a ball through the archway and into the river. Impressed, Garrick asked Carlyle to give him the club, and Carlyle agreed.

Record Scores

In the competition for the Silver Club in 1767, James Durham shot 94 at St. Andrews, an astonishing score, so low it stood until the Autumn Meeting of 1853, a period of 86 years that carried the game into the gutta-percha ball era. Lieutenant John Campbell Stewart lowered the record to 90 in winning the King William IV Gold Medal, and then two years later George Glennie shot 88. His record stood for twenty-four years, until Horace Hutchinson shot 87, in 1884.

While he was at St. Andrews University, Glennie was so much better than the other students they handicapped him by allowing him to play with only one club, a battered old mid-spoon, the rough equivalent of the modern 3-wood. Even then he was too strong for the rest of the field.

Uniforms

No matter how well or poorly eighteenth-century golfers played, they were expected to dress well. Members of the Honourable Company of Edinburgh Golfers were required to wear their uniforms whenever they stepped onto the links at Leith. Minutes of the Honourable Company's meeting of November 16, 1776, tell that Lieutenant James Dalrymple was convicted five times of playing without his uniform. Having confessed to the heinousness of his crime, he was fined six pints of ale. The minutes fail to say who collected.

High Shots

Until the middle 1980s, members of Seminole Golf Club, in North Palm Beach, Florida, often stood on the eleventh tee and tried to hit a water tower rising high above U.S. Route 1, which runs close by. This game ended when the tower was removed. In golf terms, the origin of this particular sport can be traced back to the eighteenth century. In 1798 two members of the Royal Burgess Society of Edinburgh tried to drive a ball over the spire of St. Giles Cathedral, in downtown Edinburgh. A man named Sceales, who came from Leith, and a printer named Smellie were allowed six balls each. They both cleared a weathercock more than 160 feet high. An un-named Edinburgh golfer matched this record in 1835, and then some years later, the writer Donald McLean won a substantial bet by driving a ball over the Melville Monument in St. Andrews Square, in Edinburgh, a height of 154 feet.

Cock o' the Green

Andrew McKellar gave up his position as a butler toward the end of the eighteenth century, bought a tavern in Edinburgh, turned its management over to his wife, and spent the rest of his life playing golf at the

Bruntsfield Links—every day, all day, no matter the weather. He left the house after breakfast and usually returned at dusk, except on those evenings when he lingered to play the short holes by lamplight. Even in winter he played, and when the snow hardened enough, he'd use red balls.

He wasn't a very good player and he wasn't a member of any of the clubs that played Bruntsfield, but he was as enthusiastic as any man. On the rare times he won a match, he was known to dance around the hole.

Because of his passion for the game, his wife was left to run the business, and like any wife abandoned for golf, she didn't like it. She tried to embarrass him late one evening by bringing him his dinner while he was still playing. Apparently a man of narrow vision, McKellar missed the humor and suggested that if she chose, she might wait until the game was over, because he hadn't time for dinner just then.

Sir David Baird

They made them hardy in the old days. When the North Berwick Golf Club in Scotland was organized in 1832, Sir David Baird became the first captain. An unusual man, he often played wearing a tall hat, which might have helped him hold his head still. A dedicated golfer, he drove to Musselburgh on a drenching wet day, played eight rounds over the nine-hole course, and then drove back home, in Willie Dunn's terms, "without changing a stitch."

The Longest Drive

The original hardwood golf balls were superseded about 1618 by balls made with a leather cover stuffed with cotton and wool waste. Later still, ballmakers boiled a hatful of goose feathers to make them pliable and rammed them into the leather cover using instruments resembling awls. These balls were hard and flew great distances.

Monsieur Samuel Messieux, a teacher of French in Madras College, part of St. Andrews University, apparently played excellent golf. He won the gold medal of the Royal and Ancient Golf Club in 1827 and the club's silver cross in 1840. He is best remembered though for playing the longest drive ever recorded with a feather ball. On a frosty day in 1836, with only a slight wind at his back, Monsieur Messieux ripped into a shot from the fourteenth tee at the Old Course and drove his ball 361 yards.

Captain Cook

Johnny Fischer, the 1936 U.S. Amateur champion, collected art in his later years. Toward the end of his life he gave away a number of paintings, among them a picture of Captain James Cook, the great British explorer, holding a golf club as he stood on the sands of Sandwich, in southeastern England. Although this region is rich in golf courses, like Royal St. George's, Prince's, and Royal Cinque Ports, so far as anyone knows, none were operating in Cook's time. Furthermore, no one seems to know if Cook ever played the game. Nevertheless, it is known that when the great explorer stumbled upon the Australian continent, he stepped ashore at Botany Bay and walked about at LaPerouse, a suburb of Sydney. Today it is the site of the New South Wales Golf Club.

Part Two

The Age of the Gutty Ball

Although a new feather ball could be driven great distances, it had flaws. Feather balls were expensive, costing about half a crown apiece. They were fragile; they broke apart easily. Even if they held together they lost their flight characteristics under constant pounding, and they lost their shape as well. The feather ball's shortcomings tended to limit the game's appeal.

Then, in the middle 1840s, a statue of the Hindu god Vishnu arrived at the home of the Reverend Doctor Robert Adams Paterson, in St. Andrews, and the game was changed forever. A wrapping of gutta percha, a gummy, malleable substance made from the sap of a tropical tree, had protected the black marble Vishnu during its shipment from Singapore and lay around the Paterson house for months, until someone found it could be heated, reshaped, and used to resole their shoes. One of the Patersons found still another use.

As his re-soled boots began to fall apart, Robert Paterson, a student for the ministry and an ardent golfer, molded the shoe sole into a rough sphere, slipped onto the links early one April morning in 1845, to play a round with his new golf ball. It lasted only a few holes, but a brother who lived near Edinburgh improved on it, and the Patersons were granted a patent.

"It Won't Flee"

Allan Robertson, a short, blocky Scot, most likely the best golfer of his time, operated a good business making feather balls in St. Andrews; his shop turned out 2,456 during 1844. Four years later, in April of 1848, a man named Tom Peters showed one of the new balls to Robertson and Tom Morris, who worked for Allan at the time. Neither man had seen one. Peters warned that because of this new ball, the days of the feathery were over. Robertson scoffed, but Peters persuaded him to try it. Instead of hitting the ball fairly, Robertson slashed down, causing it to duck in flight, which Peters said no one could do better than Allan. Robertson sneered and cried, "Bah. That thing'll never flee."

As a young man playing with clubs handed down from affluent members of local golf societies, Robertson spent whatever leisure hours he could snatch from a long and hard day's work developing his game on the links at St. Andrews, often at 4 o'clock in the morning, in the gray dawn of a Scottish summer. He was born into golf; his grandfather had been a caddie, and his father, David Robertson, had been senior caddie of those who served the Royal and Ancient. Allan followed his ancestors as a caddie, but he evolved into what is generally considered the world's first professional golfer.

In addition to serving as caddies, his ancestors had been ball-makers for generations. Allan followed them. He had a natural flair for the game, of course, and although he was never formally attached to St. Andrews, he was responsible for early improvements to the condition of the Old Course.

Robertson was described by a contemporary biographer as a "short, little, active man, with a pleasant face and a merry twinkle in his eye. His style was neat and effective, and he held his clubs near the end of the handle, even his putter high up. With him the game was one of head as much as of hand; he always kept cool, and generally pulled through a match, even when he fell behind. He was a natural gentleman."

Championship golf hadn't been born when Robertson was at his peak, but even so he played one of the game's unforgettable shots.

Allan was involved in a big money match at alternate shots over three rounds at the Old Course of St. Andrews in 1853. With a Mr. Erskine, an R and A member, as his partner, he played Willie Park, Sr., of Musselburgh, another great professional of the day, and a Mr. Hastie, another R and A member.

With Park and his partner one hole ahead with two holes to play, someone offered £15 to £5 they would win. Someone else accepted. The bet seemed secure when Park put his side's approach onto the green of the seventeenth, the diabolical Road Hole, where a paved road runs against the right side of a long and narrow green sitting at an angle to the approach. Erskine, meantime, had missed the green to the right, leaving Allan on the road playing "the two more." What happened next is part of golf lore.

Studying the shot with infinite care while all around him spectators quoted odds and exchanged money, Allan lofted a little pitch that hit on a footpath within a fraction of his target, bounced onto the green, crept slowly toward the hole, and finally tumbled into the cup for a 4. Stunned, Hastie rammed his first putt four or five feet past the cup, and Park uncharacteristically left his putt short. Robertson and Erskine had won the hole and evened the match. Park was so upset at losing a hole he had counted on winning, especially one that would have closed out the match, he topped his drive on the eighteenth into the Swilcan Burn and lost that hole and the match as well. In Wellington's terms, it had been a damned close run thing.

The Great Money Match

Robertson and Morris often paired together in money matches. Since their livelihoods depended on the continuing sale of feather balls, both men vowed they'd never play this new ball, but Tom, a mild man of good manners, broke his pledge and teamed with a gentleman golfer in a match using gutta-percha balls. When Allan found out, the two men parted. Morris opened his own shop, making not only feather balls, but clubs and gutta-percha balls as well.

Nineteenth-century matches often lasted for days, because they weren't all played by holes; some were played by rounds. Early in 1843, for example, Robertson won a match of 20 rounds—two rounds a day for ten days—beating Willie Dunn on the last day, two rounds up with one round to play.

Robertson and Morris paired together against Willie and Jamie Dunn in the biggest money game of the century. With £400 as the stake, they played three rounds: the first at Musselburgh, the home course of the Dunns, on the far side of Edinburgh from St. Andrews; the second at St. Andrews, the home of Robertson and Morris; and the third at North Berwick, a neutral site.

Off to a fast start, the Dunns won the first round by 13 and 12, but Robertson and Morris evened the match by winning at St. Andrews. Still, the Dunns seemed to have the match in their pockets when they led at North Berwick by four holes with eight to play. Robertson and Morris, though, were never beaten so long as there were holes to be played. Odds of 20-1 were offered against them at this point, but they fought back, won that round by one hole, and with it the match.

Robertson is said never to have lost a match at even terms, although he did lose in alternate shot foursomes. Paired with Willie Dunn, Robertson was beaten by Tom Morris and Bob Anderson over eighteen holes at St. Andrews, in 1854, but three years later, in 1857, he teamed with Andrew Strath, from another St. Andrews family, and beat Morris and Willie Park over thirty-six holes. It should be pointed out, though, that Robertson rejected repeated challenges from Park, who had beaten Morris twice in £100 matches.

Iron Clubs

Robertson was one of the game's great innovators. The modern set of golf clubs is composed mainly of irons, but in Robertson's day, which saw the transition from feather ball to gutta-percha ball, clubs were made mostly of wood, quite often applewood. There is little question that irons would have worked their way into the game without him, but it is irrefutable that Robertson's great skill and imagination brought them into the game earlier than they might have.

At first, all iron clubs were called cleeks, the Scottish term for a hook. They were used at first only to dig a ball from a lie so poor that it might break a wooden club. Robertson, though, found another use; he used it to run shots onto the green.

Robertson made the cleek famous, and incidentally started a new branch of the club-making business. Until he did, club-makers produced only wooden clubs. Some were works of art (Hugh Philp often spent an afternoon polishing the head of applewood to give it the luster he wanted). Originally blacksmiths, those who made irons called themselves cleek-makers.

•

The First National Championship

George Glennie and Lieutenant John Campbell Stewart, two men who had set scoring records at St. Andrews, played first-rate golf throughout their lives. They turn up again in 1857 when golf took another turn toward the game as we know it today.

Pre-eminent until the middle years of the nineteenth century, private matches lost much of their appeal with the dawn of championships—competitions that drew their fields from players representing different clubs. By then the Gentlemen Golfers had evolved into the Honourable Company of Edinburgh Golfers, and the Society of St. Andrews Golfers had become the Royal and Ancient Golf Club of St. Andrews. They continued to conduct tournaments, but confined their fields to their members. Everything changed in 1857 when the Prestwick Golf Club, on the west coast of Scotland, sent letters to seven other leading clubs inviting them to play a match at either Prestwick's twelve-hole course or at St. Andrews, which by then had eighteen holes. Bowing to the majority's wish, the championship was played at St. Andrews, July 29, 30, and 31, 1857, with a silver claret jug as the prize. The championship was played as alternate shot foursomes: each club could enter one pair.

Invitations were extended to Prestwick, of course, St. Andrews, North Berwick, Perth, Musselburgh, Carnoustie Panmure, Leven, and Royal Blackheath—believed to be the first club formed in England. It was the only English club invited, but the English had their revenge.

Not only did the eight clubs accept, but the Honourable Company, along with Edinburgh Burgess, Bruntsfield, Montrose, and Dirleton Castle—

all of them in the general vicinity of Edinburgh—asked to be included as well.

Neither Carnoustie Panmure nor the Honourable Company could find two players to represent them, which cut the field to eleven teams and created an awkward format for match play. It worked out nicely though when Edinburgh Burgess and Edinburgh Bruntsfield played a halved match in the second round, and each advanced to the third. Blackheath then defeated Bruntsfield by six holes and St. Andrews defeated Burgess by three.

Represented by Glennie and Stewart, Royal Blackheath won the championship, sweeping through by beating Perth by eight holes, Leven by twelve, Bruntsfield by six, and the Royal and Ancient by seven holes in the final match.

This was a team competition; the first individual championship was played a year later, again at St. Andrews. The championship was won by Robert Chambers, who defeated a Mr. Wallace, of Balgrumo, in the final. An extremely slow player, Wallace had driven his opponents mad with his creeping pace. Chambers, however, didn't let the pace disturb him. He said, "Give me a novel and a camp stool, and I'll let the old chap do as he likes."

The British Open

Allan Robertson died in 1859, leaving something of a void. In spite of his refusal to play Willie Park, he had generally been conceded the best golfer of his time, and now, with his death, golf had no recognized leading player. The Prestwick Golf Club then acted on a suggestion by Major J.O. Fairlie that the time seemed right for a professional tournament organized by the leading clubs. The prize was to be a red Morocco leather belt; Prestwick members immediately subscribed five guineas. None of the other clubs participated, and so Prestwick took on the championship by itself.

The original invitation, handwritten in blue ink on light blue notepaper, is encased in a clear plastic envelope and bound into a loose-leaf notebook stuck away in a small and nearly inaccessible room on the upper floor of the Prestwick clubhouse. The invitation reads:

Prestwick Golf Club

Challenge Belt

It is proposed by the Prestwick Golf Club to give a Challenge Belt to be played for by Professional Golfers; and the various clubs in Scotland and England are requested to name and send the best players on their Links not exceeding 3 to compete for it. The game to be 36 holes or three rounds of Prestwick Links, the player who succeeds in holing his ball in the fewest number of strokes to be the winner.

The winner of the belt for three successive years to keep it.

> The game to be played over the Links at Prestwick at 12 o'clock noon on Wednesday the 17 October current. It is understood that the players must be known and respectable "cadies."
>
> All disputes to be settled by the Captain of the Prestwick Golf Club or the person acting for him. The gold medal of the Club will be played for on Thursday the 18th at noon.

The notice was sent to the following clubs: St. Andrews, Musselburgh, Bruntsfield, Leven, Carnoustie, Panmure, Montrose, Perth, Blackheath, Darlington Castle, North Berwick, and Aberdeen.

Although twelve clubs were invited to send players, only eight men entered: Willie Park, from Musselburgh; Tom Morris, Prestwick; Andrew Strath, St. Andrews; Bob Andrew, Perth; Daniel Brown, Blackheath; Charley Hunter, Prestwick St. Nicholas; Alex Smith, Bruntsfield; and William Steel, Bruntsfield.

Prestwick at that time followed a peculiar routing. Holes criss-crossed one another, four of its greens were common to two holes, and players standing on the ninth green had to duck while shots from both the second and tenth tees whistled overhead. (The club held a re-enactment tournament in 1979. It drew 120 entrants, far too many, and several players were hit, although none of them were seriously hurt. The next year several caddies considered it too dangerous and refused to work. The tournament was dropped.)

Willie Park won with 174, or six under 5s. Tom Morris placed second, with 176, and Andrew Strath came in third, at 180. These were the only scores under 190, or ten over 5s, a score any twelve or thirteen handicapper of the late twentieth century could beat. That first Open, however, was not open at all; it was limited to professionals. Major Fairlie offered another proposal the following year, that the Open be open to all the world, professionals and amateurs alike.

Young Tom Morris

At the time of the early British Opens, Prestwick's first hole measured 578 yards, longer than any on the rota of courses of the late twentieth century (at 577 yards, the sixth at Royal Troon is the longest). When he won the 1870 British Open, Young Tom Morris, who was only twenty, began with a 3 on the first hole. No one knows how he did it, but the score is authentic. He played that first twelve holes in 47 strokes, an average of just under four strokes a hole, remarkable scoring more than a century later. Following with two 51s, he shot 149, an average of 74.5 for two eighteen-hole rounds, and won by twelve strokes over Bob Kirk and David Strath. Young Tom's average score for the British Open stood for thirty-four years. It was never beaten with the gutta-percha ball.

Under the terms of the championship, the red morocco leather belt, embellished with silver ornamentation, was to be given permanently to the man who won three Opens in succession. This was Young Tom's third straight, and since he had retired the belt, the Open was canceled.

Young Tom was a genius at the game. He won his first match of consequence when he was thirteen, winning a prize of L5 at Perth by beating William Greig, a local prodigy. Three years later, when he was just sixteen, he entered a tournament that drew all the best players to Carnoustie, perhaps the strongest course in Scotland.

When the Morrises arrived, the great Willie Park asked, "Whit hae ye brocht the laddie for, Tom?" Morris answered, "Ye'll ken whit for soon enough."

Young Tom tied Park and Bob Andrew for first place, then beat them in a playoff. Later he won an Open Professional Tournament at Montrose and followed up by beating Park in a challenge match. In a money match against Bob Ferguson in 1869, he shot 77 at the Old Course, beating the record of 79 Robertson had set some years earlier.

Speaking of his son, Old Tom said, "I could cope wi' Allan [Robertson] myself, but I could na cope wi' Tommy."

Tommy brought scoring to new levels. When he won his first Open, in 1868, succeeding Old Tom, his father, he shot 157, five strokes under the previous record. The next year he shot 154, and then 149 in 1870. He also brought new strategy to the game. Where Robertson has used his cleek to play run-up shots, Young Tom used it for pitching. He is said to have been the first golfer to learn how to impart backspin to a ball.

Young Tom was very strong; he is said to have snapped wooden shafts simply by waggling the club. Nevertheless, his putting was probably the strongest part of his game. He stood erect, with the ball off his right foot and the toe of his left foot pointed toward the cup. He took great pains with short putts, and above all he was consistently good.

Young Tom won the Open once again, when it was revived in 1872 after a lapse of one year, making four in succession. He never won again, although he continued to play and win money matches.

He and Old Tom were playing Willie and Mungo Park, Willie's brother, at North Berwick in September of 1875 when in the midst of the match a telegraph messenger pushed through the crowd and approached Old Tom. The message said his son's young wife had given birth to their baby but lay gravely ill. The match was stopped, and one of the wealthy backers of the match offered his yacht for the journey across the Firth of Forth and up the coast to St. Andrews. Before they could leave a second message arrived; Tommy's wife had died.

Young Tom was crushed. When the party arrived home he cried, "It's not true."

He played twice more, to honor commitments, and then three months later he died, on Christmas day. The romantic version says he died of a broken heart. More likely drink had something to do with it.

Years later Bernard Darwin, a golfer, a Dickensian scholar who also wrote about golf for *The Times*, London's leading newspaper, asked Leslie

Balfour Melville, who had seen them both, how Young Tom compared with Harry Vardon, the great golfer of a later era. After considering the question for a moment, Balfour Melville said, "I can't imagine anyone playing better than Tommy."

Unfinished Open

Even though it had taken the initiative, Prestwick had never planned to run the British Open alone. After Young Tom Morris retired the Open championship belt, Prestwick persuaded the Honourable Company of Edinburgh Golfers and the Royal and Ancient Golf Club of St. Andrews to join with it not only to buy a new trophy but also to conduct the Open. With the agreement, the championship then had an established rota.

While over the years the R and A has established a reputation for impeccable administration of its championship, the 1876 British Open, the second at St. Andrews, turned out an absolute disaster.

To begin with, it was scheduled for September 30, on the same day Prince Leopold, a son of Queen Victoria, was to drive himself in as Captain of the R and A, a colorful ceremony in which the new Captain stands alone on the first tee of the Old Course before a group of members and spectators, and as a cannon booms, drives a ball down the wide fairway.

Although Leopold was to become the second Royal Captain, he would be the first to drive himself in; his older brother Edward, the Prince of Wales (Edward VII), had served as Captain in 1863 but hadn't gone through the ceremony, in fact, hadn't even traveled to St. Andrews.

The prospect of watching a member of the royal family execute the ceremony evidently threw the R and A off its game. Excitement ran so high the club forgot to reserve the course for the Open. Chaos reigned. The Open field had to go off alternately not only with R and A members but with the town's artisan golfers as well.

With the course crowded, the weather bad, and the ground heavy, scores ran unusually high. Only Bob Martin and Davie Strath, with 86, broke 90 for the first eighteen holes, and when Martin, the first of the two to play, shot 90 in the afternoon, four under 5s for thirty-six holes, the way seemed clear for Strath to win, for word was that he was playing well. Then things went badly.

On the fourteenth, the Long Hole In, Davie pulled his drive into a group of artisans playing the fifth hole, which runs parallel, and felled one Hutton, an upholsterer; Davie's ball hit Hutton on the forehead. Although the blow seemed to do no damage to Hutton—he recovered quickly and walked home under his own power—it upset Strath. A local newspaper account said, "This put Davie off a bit, for he took 6." Then he lost another stroke at the fifteenth.

Leaving the sixteenth green, he heard that Martin had finished with 176. Now Davie needed two 5s to win. By then overcrowding had exhausted everyone's patience, the day was growing late, and Strath was playing in semi-darkness. The rules of the competition stated clearly that players in front should be allowed to putt out. This was to cost Strath the championship.

Davie made his 5 at the seventeenth, playing a long shot to the heart of the green that, helped along by a strong following wind, had run farther than Strath had any reason to expect. His ball hit another player's leg and stopped short of the dreaded road that crowds against the green.

A 6 at the eighteenth and Strath had tied Martin (they both stood seven strokes clear of Willie Park, who finished third).

Immediately after Strath had finished, Martin's supporters protested that Strath had violated the rules by playing toward the seventeenth before the players ahead of him had left the green. In keeping with this Open's reputation as the worst managed of all time, the committee couldn't decide whether or not to uphold the protest and decreed a playoff. Feeling the protest should be decided before a playoff, Strath refused to play. Martin walked the course the next day and claimed first prize of £10. Strath was awarded second prize of £5.

J. O. F. Morris, Old Tom's youngest son, seemed to have the 1878 championship locked up. His 161 at Prestwick had weathered every challenger. Only Jamie Anderson, the defender, had a chance, but to tie Morris, he had to play the last stretch of four terrifying holes in one over 4s, a difficult assignment at best. Thinking deeply, Jamie said, "I can dae't wi' a five, a fower, a three, and a five."

Anderson began by holing a full-blooded iron shot for a 3 at the fifteenth, then holed a putt all the way across the sixteenth green for a 4. He overshot the hole of the next, a par 3, but his ball climbed a mound, hesitated for an instant at the top, then trickled back down and into the cup for a hole in one, the first in British Open history.

Anderson had nearly been disqualified. As he prepared to play his tee shot a spectator noticed he had set his ball ahead of the markers and warned him he would be disqualified if he played the shot. Anderson calmly picked up his ball, re-teed it behind the markers, then made his ace.

He closed out with his 5, beating not only his own schedule, but Morris by four strokes as well.

He needed every one of those strokes, though, for Bob Kirk, playing behind him, was tearing up the course, and came to the last hole needing a longish putt for the 4 that would tie Anderson. He was a trifle strong with his stroke; the ball ran true to the hole, hit the back of the cup, bounced up, but missed. Kirk then missed the next putt and finished two strokes behind.

Close, But

Although only Young Tom Morris has won four successive British Opens, Bob Ferguson came close. He had won three from 1880 through 1882, and he looked as if he had won his fourth straight in 1883. After apparently throwing away his opportunity by making a 10 on an early hole,

Ferguson played the last three holes at Musselburgh in 3s and tied Willie Fernie. One stroke up coming to the last hole of the playoff, Ferguson made his 4, but it wasn't good enough. Not only did Fernie drive the green, but he holed an extremely long putt for a 2, winning by one stroke.

Crime and Punishment

Locked in a close fight with Willie Park, Jr., at Musselburgh, in 1889, Andrew Kirkaldy hung a putt on the edge of the cup on the fourteenth hole. Stepping up to the ball, Kirkaldy played a careless backhanded stroke and missed the ball completely.

Not sure just what had happened, an official approached Kirkaldy and asked, "Did you try to putt that ball, Andra." Kirkaldy answered, "Yes, and if the hole was big enough I'd bury myself in it."

With that missed stroke, Kirkaldy dropped into a tie with Park, and lost the playoff. Kirkaldy never did win a British Open.

Hilton at Muirfield

Prize money had been raised from £20 to £110 for the 1892 British Open, but it was won by Harold Hilton, who was given none of it because he was an amateur. He also didn't spend much time at it. The first seventy-two-hole championship, it was played Wednesday and Thursday, two eighteen-hole rounds each day.

Hilton left his home in Liverpool, on England's west coast, on Monday evening and traveled all night, arriving at Muirfield early enough on Tuesday morning to play three practice rounds, played two more rounds on Wednesday, two more on Thursday, then took the train home on Friday.

Course Closed

Maintaining the St. Andrews Old Course was no easy job, particularly when both the British Amateur and Open were scheduled within a month of one another. The course needed a rest between tournaments, but this was a public course, owned by the townspeople of St. Andrews. How can it be closed when nothing is going on.

Old Tom Morris figured it out. After the Amateur had been settled, he filled in all the cups. The Old Course remained closed for two weeks. Evidently it was in fine shape for the Open.

Blackmail

Cheape's Bunker, a sand pit that cuts into the second and sixteenth greens of the Old Course, takes its name from James Cheape, an eccentric Lord of Strathyrm, the adjoining estate, who owned the land, and in 1894 was given the rights to gather shells from the golf course. He ground the shells to dust to maintain paths on his estate.

To gather those shells, Cheape had the right to dig. When the British Open came to St. Andrews in 1895, Cheape threatened that unless he and his family were given viewing rights, he would dig up the entire sixteenth fairway. The R and A bowed to this terrorist threat.

Late in the twentieth century, Cheape's descendants still lived in the old house and held the right to do their washing in the Swilcan Burn.

Slight Delay

Good friends, Harry Vardon and J.H. Taylor tied for first place in the British Open at Muirfield, in 1896, setting up a playoff. They would not play the following day, however, because they were scheduled for another tournament, at North Berwick, a few miles from Muirfield. They played at North Berwick as scheduled, then returned to Muirfield for the playoff the next day. Vardon won by four strokes.

Squeezed Out

Vardon and Willie Park, Jr., were locked in a battle for the 1898 British Open at Prestwick. Park led Vardon by two strokes after fifty-four holes, but when he began the third round with a 6 at the first after Vardon had made his 4, they stood even. They were still even approaching the last hole, but Vardon put himself in danger by throwing everything he had into his drive, hoping to reach the green. His ball squirted off to the right and left him with a difficult pitch over scraggly rough. Nevertheless, he played the shot beautifully, holed from eight feet, and posted a score of 307.

His round completed and his score turned in, Vardon stood alongside the last green waiting for Park to finish. Willie played a marvelous drive that reached the front of the green and followed with a putt that pulled up three feet short. To force a playoff, Park must hole the putt.

The tension had mounted so during those last few minutes the crowd began edging toward the green, finally oozed onto the putting surface, and while Vardon struggled to hold his place, the gallery finally pushed him aside. By then Vardon, the man most concerned with Park's putt, could see nothing.

Considered a deadly putter at twice that distance, Park stroked his putt. Unable to see, Vardon listened to the reaction of the crowd. When he heard a disappointed "Ooo . . . ah," he knew Park had missed and that he had won his second championship.

On Putting

As astute observer of the game, as well as one of its leading players of the late nineteenth century, Park had formulated the doctrine, "The man who can putt is a match for anyone."

Worth Striking For

The 1899 championship was scheduled for the Royal St. George's Golf Club, in Sandwich, England, but it nearly didn't come off. A number of professionals threatened to strike unless they saw a considerable increase not only in the amount of prize money but in the number of prizes as well. The strike failed, partly because the game's leading players—Harry Vardon, J. H. Taylor, and James Braid, who made up the Great Triumvirate—failed to go along with the others. They wouldn't have helped, anyway, because the sponsors had decided they would not yield to a strike.

Nevertheless, the sponsors announced during the championship that first prize would be raised from £40 to £50, and second place from £20 to £25.

Speed Record

While Harry Vardon won the 1899 championship, Jack White stormed from behind to snatch second place with 75 in the last round, setting the course record for Royal St. George's Golf Club. The news of White's great round sped off to London by telegraph, but with no understanding of golf, an editor at a London newspaper wrote a headline claiming White had raced around the course in seven minutes and five seconds.

Braid at Muirfield

James Braid began the 1901 British Open, at Muirfield, by hooking his first drive over a stone wall and out of bounds. Nevertheless, he rallied, and came to the seventy-second hole with strokes in hand. Attempting to reach the green of the eighteenth, which lay some 200 yards off, Braid slashed at his ball and stood horrified as the shaft splintered and the head went flying off toward the clubhouse. Nevertheless, he had made such good contact, the ball carried onto the green and he beat Vardon by three strokes.

Heavy Weather

On the October morning the R and A had scheduled its 1860 Autumn Medal, a raging gale ripped through St. Andrews. At the height of the storm, as rain lashed the old town, word came that a large fishing vessel lay helpless in St. Andrews Bay, drifting toward the rocks. With volunteers needed to man the lifeboat, moored within 200 yards of the first tee, hardly anyone stepped forward. R and A members offered fishermen money to crew the boat, but they felt they would be asking to be drowned.

Captain Maitland Dougal, who was about to tee off in the Medal, led by example. He stepped into the boat and manned the lead oar. Enough volunteers followed, and they set off in the heavy seas.

Through superb seamanship and help from another vessel, which sheltered the lifeboat as it drew alongside the helpless ship, the crew was taken off and the lifeboat returned to St. Andrews.

Dougal had been at the oars for five hours. Back on land again, he changed into dry clothes, drilled a hole in his gutty ball, stuffed in buckshot to add weight and help it bore through the gale, and shot 112, taking second place in the Medal, eight strokes behind William C. Thompson, who won with 104.

Match Lost

Robert Clark, whose book *Golf, a Royal and Ancient Game* is among the earliest on golf (it was published in 1875), was engaged in a foursomes match at Musselburgh in 1870, and played the tee shot on the last hole, a par 3. It was almost dark, and no one was able to follow the ball in the gloom. The match depended on this shot, but they couldn't find it. Eventually they gave up and conceded the hole and the match. Walking off the green, one of the players glanced into the hole, and there sat Clark's ball. Too late, though; under the rules of the day, the match had ended when Clark conceded the hole. This may be the only incident on record when a match was lost because of a perfectly played shot by the losing side.

Old Sutherland

The Sutherland Bunker, behind the Cottage Bunker, on the fifteenth hole of the Old Course at St. Andrews, was named for a man with excessive passion for the game. A member of the R and A, he was not a very good golfer, but he was probably the game's first gallery marshal. While he was watching a match between two of St. Andrews's better golfers, a train from Dundee discharged its passengers, who made their way across the golf course to the broad sandy beach beyond. Incensed at this intrusion, Sutherland raced toward them, frantically waving his cleek and herded them off the line of play, calling to his friends, "It is disgraceful of the railway people bringing a parcel of uneducated brutes down here when they knew a real match was going on."

He thought of little else but golf. Introduced to a lady from London, Sutherland greeted her by saying, "How are you? Mr. Glennie never played better than today. He went round in 106. A grand golfer." The lady, of course, had never heard of George Glennie, nor for that matter of golf itself.

Whose Grip

The Vardon Grip, the most popular method of holding a golf club; was it the invention of Harry Vardon? Actually, no. What is known as the Vardon Grip was developed by an amateur. John Laidlay experimented with a new method of gripping a club after 1878, when he won his first medal. Until then, and well after that time, most golfers gripped the club in their palms with the fingers and thumbs of both hands wrapped around the handle, much as one holds a baseball bat.

In a letter written late in his life. Laidlay said, "Just before 1885, I took to gripping one finger of the one hand over one finger of the other, which I believed to be a good thing to do." Later he wrote, "It was looked on as most foolish for a good young golfer to have done. . . . It must have been several years later when I first played with Harry Vardon, and I am certain that had he used it I should have observed him at once. Personally, I have not the least doubt that I was the first person to use this grip."

Laidlay adopted the grip when he was about twenty. He was ten years older than Vardon and nine years older than J. H. Taylor, who also used the grip. It is obvious then, if Laidlay has laid the facts down accurately, that Vardon and Taylor copied the grip from him, not the other way around.

Laidlay, after all, won the British Amateur championships of 1889 and 1891. Taylor won the first of his five British Open championships in 1894, and Vardon the first of his six in 1896. By then Laidlay's grip had been established.

How did it become known as the Vardon Grip? No one is quite sure. The most logical explanation hinges on Vardon's winning the 1900 U.S. Open. Since golf was relatively new in the United States, and most American golfers had been taught by expatriate Scots brought up with the old palm grip, they hadn't seen such an innovative method of holding the club until then. Of course both Vardon, the winner, and Taylor, the runner-up, used the same method, but since Vardon won, the grip was named for him.

Even though all this took place in the late years of the nineteenth century, feelings still ran high well toward the end of the twentieth. Shortly after the London *Daily Mail*'s golf correspondent Michael McDonnell published an account of how it had been Laidlay rather than either Vardon or Taylor who had developed the grip, he began receiving correspondence from angry relatives. Jack Taylor, the son of J. H., said in McDonnell's report, "They all say Harry Vardon introduced [the grip], but that's not a fact. My father never gripped the club any other way. The rest of the players of the time used the palm grip." Then a distant relative of Vardon's wrote, "You have blackened his name, and all of us Vardons want it put right. J. H. Taylor's son may be the last direct link, but there are still a lot of us Vardons around . . . so beware." A Vardon enthusiast called the article, "An insult to Vardon's reputation."

Park vs. Morris

Willie Park, Sr., and Old Tom Morris played five big money matches, the last, in 1882, at Musselburgh, Park's home course. Willie stood 2 up with six holes to play when Bob Chalmers, an Edinburgh publisher who was acting as referee, stopped play because spectators were interfering with the balls. Chalmers and Morris then went into Foreman's pub, and after waiting for a time, Park sent word that if the two didn't come out and resume the match, he would play the remaining holes by himself and claim the stakes. They didn't, and he did.

Hell Bunker

While it has been relatively stable throughout the twentieth century, the Old Course at St. Andrews evolved over a long period of years. Hell Bunker, a deep and wide depression about 80 yards short of the fourteenth green, began as a much smaller obstacle in 1882, while Old Tom Morris served in the dual positions of professional to the R and A and greenkeeper of the Old Course.

One day an indignant golfer fumed to Morris that the condition of the course was so bad he had had only one decent lie all day, and that was at the bottom of Hell Bunker. His ball lay so well, he was able to play a wooden club from it.

Morris's features, usually dark in their normal state, turned black. Immediately he sent a work crew to the site armed with picks and hoes and had them hack away until no golfer could ever again play from Hell Bunker with a wood.

Years later, Gene Sarazen found himself in that bunker during the British Open. Rather than try to advance his ball toward the green, Sarazen had to play backwards, toward the tee. He made 8, and lost by one stroke.

Night Golf

Professor P. G. Tait, a distinguished mathematician and professor of philosophy at Edinburgh University during the late years of the nineteenth century, spent his summers at St. Andrews. He played golf incessantly, often five rounds in a day, the first beginning at six in the morning. This wasn't enough for him, though. A man with a fine sense of humor in addition to an inventive mind, the professor coated a number of golf balls with phosphorescent paint and arranged a night-time match with Mrs. Tait, Professor Crum Brown, and Thomas Henry Huxley, one of the century's leading scientists and among the first to support Charles Darwin's theories of organic evolution.

The match was going well and the idea proving a success until the flammable paint set Professor Brown's glove afire. Before he could rip it off it burned his hand. An account tells us, "A chastened group returned."

A Disproved Theory

Professor Tait once worked out a theory that the gutta-percha ball could be driven no farther than 191 yards. His son, Freddie, tore that theory to shreds. One of the finest golfers of his time, Freddie drove the green of the thirteenth hole at St. Andrews, a distance of 341 yards, and the shot was said to have *carried* 250 yards.

Amateur Status

Royal Liverpool held an amateur tournament in 1885, which later blossomed into the British Amateur championship. Douglas Rolland, a stonemason from Elie, near St. Andrews, sent in his entry, but Rolland had placed second in the 1884 Open and won prize money. Although no precise definition of who was or wasn't an amateur had been written, Rolland clearly wasn't an amateur. What about John Ball, though, who was among those expected to win? As a boy of fourteen or fifteen, Ball had tied Bob Martin for fourth place in the Open of 1878, at Prestwick, and had been awarded the princely sum of ten shillings. Uncertain whether or not he should accept the money, he asked Jack Morris, his chaperon, what he should do with it. "Put it in your pocket," Morris said.

Was he or wasn't he an amateur? The committee decided that because he was so young when he accepted the prize money, he could remain an amateur. For many years afterward the rules stipulated that an amateur must not have played for a money prize after reaching the age of sixteen.

More Confusion

That 1885 Amateur championship raised confusion to lofty heights. For example, both players in halved matches advanced in the early rounds, but toward the later stages they replayed the entire match. Thus Alan F. Macfie and W. M. de Zoete played fifty-four holes against one another. After they halved their first two, Macfie finally won by one hole in the third.

Furthermore, with forty-eight players, the field was uneven, but that could have been worked out easily by giving some players first-round byes. Not here though. Instead of awarding byes at the beginning, they were held over to the end, so that when the amateur reached the semi-final stage, only three players were left. Macfie drew the bye and rested in the clubhouse, drinking tea and munching cucumber sandwiches while John Ball and Horace Hutchinson were at each other's throats. Hutchinson beat Ball by 3 and 2, but the next day Macfie, a good player but not in the same class with either Hutchinson or Ball, won by 7 and 6. Since this was a rather informal beginning, it took 40 years for the R and A to recognize Macfie as the first Amateur Champion.

A. J. Balfour

In 1886, five years before the British statesman A. J. Balfour was appointed Chief Secretary for Ireland, Lord Frederick Cavendish and T. H.

Burke, two government officials, were murdered in Dublin's Phoenix Park. An enthusiastic golfer, Balfour played an occasional round at Phoenix Park's little golf course, protected only by two plain-clothes detectives. By insisting on his weekly game, Balfour was popularly considered to be challenging potential assassins, but his games didn't last very long. When the Land League, organized to reform land practices, learned Balfour was one of the regular players, League members sneaked into the park in the still of night, and while most of Dublin slept, wielded picks and shovels and dug up the course, ending golf in Dublin for the time being. (The residence of the American Ambassador to Ireland sits in Phoenix Park today.)

Through his enthusiasm for golf, Balfour might have done as much for the game in Britain as President Eisenhower did nearly a century later in the United States. Playing the Old Course at St. Andrews, Balfour once reached the eleventh hole, a demanding par 3, with his tee shot, but he was left with a long, long, difficult putt. Consulting his caddie, Balfour was told, "Hit it a yard to the left." Balfour duly rapped his ball, but it finished a good yard to the right of the flagstick. Fuming, his caddie shouted to a companion, "And these are the bastards running the country."

Freddie Tait

One of the finest amateur golfers ever bred in Scotland, Freddie Tait had a touch of mischief and daring. In 1896, as a subaltern in the historic Black Watch Regiment, he climbed to the top of the Rookery, the highest peak in Edinburgh Castle, which rises 350 feet above the city, and in a match against a fellow officer, drove a ball toward a fountain in Princes Street Gardens, not only 350 feet below but 300 yards off in the distance. After playing their drives, the two soldiers raced down the hill to the Gardens but couldn't find either ball. A passer-by had picked them up and taken off with them.

Tait once drove a ball off line and through a man's hat. The owner happened to be wearing it at the time and insisted Freddie pay him five shillings for a new one. When the round ended, Tait grumbled to Old Tom Morris about the expensive shot he had played, but Morris had no sympathy for him, saying, "You ought to be glad it was only a new hat you had to buy and not an oak coffin."

Tait was a genuine Scottish hero, a dashing figure whose captivating personality and pure power were storybook stuff. Born in 1870, he came along at the same time as the great John Ball and Harold Hilton, two of Britain's finest amateurs. A series of great matches with Ball reached their climax in the final of the 1899 Amateur championship, at Prestwick.

Tait had gone ahead in the morning, but Ball struggled back in the afternoon and led by a hole after the sixteenth. Playing the seventeenth, the Alps, where a high hill blocks the view of the green, both men mis-hit their approaches, but where Ball's ball lay cleanly in the sand bunker across the front of the green, Tait's ball lay in accumulated water within the same bunker. With the match at stake, Tait had to play the shot. Not only must he dig the ball out of the sand in the era before the sand wedge, he had to play it high enough to clear the bunker's front wall, shored up by railroad ties. Standing ankle deep in the water, Tait flung himself into the shot. A plume of water flared up from his club, the ball popped out, cleared the ties, and rolled onto the green. They halved the hole in 5s, and Tait birdied the eighteenth, sending the match to extra holes. Ball won on the thirty-seventh.

Tait and Ball met again at Royal Lytham and St. Annes in October. Four down after twenty-three holes, Freddie fought back and defeated Ball on the thirty-sixth. This was Tait's last match. A graduate of Sandhurst, Britain's West Point, he was shipped off to South Africa three weeks later and fell in battle the following February, killed by a Boer bullet as he pushed ever forward, once again taking the fight to the enemy.

Golf in America

Some forms of golf had been played in North America as early as the seventeenth century—indeed, players had been cautioned against playing through the streets of Albany, New York—but the game seemed to have petered out after the United States and Britain fought the War of 1812. Then, in 1887, Robert Lockhart, a linen merchant from Musselburgh who lived in Yonkers, New York, brought back clubs and balls from a business trip to Scotland. He took them to a field in New York City near the Hudson River that is now Riverside Drive, and nearly hit an iceman with a wild shot. A mounted policeman patrolling nearby smiled, and after watching Lockhart and his sons bang away, asked if me might try a shot or two himself. Lockhart agreed.

The policeman climbed from his horse, took his stance, waggled a few times, as he had seen Lockhart do, then lashed into the ball, sending a beauty screaming down the field, as far as any Lockhart had hit. Pleased with himself, the policeman teed up another, then flung himself at the ball, obviously trying to hit it even farther. He missed completely, tried again, missed another, then missed twice more. Disgusted, he climbed onto his horse and rode away.

Lockhart's Legacy

Lockhart was friendly with John Reid, another Yonkers resident and native of Scotland. On one of Lockhart's frequent trips abroad, he had brought tennis equipment home, and played on Reid's front lawn, believing he was introducing lawn tennis to the Americas. He found he was too late; the game had arrived in 1874. Later he took his clubs and balls to Reid's house, an act that had a lasting effect on golf in the United States. On February 22, 1888, Reid invited six friends to his house. They laid out three rude holes in his cow pasture, across the street from the house, and while Alexander Kinnan, Kingman N. Putnam, Henry O. Tallmadge, Harry Holbrook, and John C. Ten Eyck watched, Reid and John B. Upham played golf. Smitten by the game, the group met again at Reid's house on November 14 and formed the St. Andrew's Golf Club. In one of their first acts, they elected Lockhart their first member.

Willie Dunn's Influence

Willie Dunn was a small, neat man, about five-feet-six inches tall and weighing 125 to 130 pounds. He wore a trim mustache that curled up at the ends, had bright, clear eyes, a fresh, eager face, and when the occasion called for it, wore a winged collar with a four-in-hand tie, stirrup trousers (so that too much shoe didn't show), a coat of just the right length, and a fedora set squarely atop his head.

While living the good life in France, Dunn was said to have taught golf to more earls, lords, and duchesses than any other professional. In the custom of the day he had overseen construction of the golf course at Biarritz, the ultra-exclusive resort on the Bay of Biscay, just north of the Spanish border, while giving lessons to aristocrats, noblemen, and other men and women of enormous wealth who wintered there. He taught Queen Natalie of Serbia and impressed her so much she hired him to lay out a golf course on her Biarritz estate.

Dunn also met W. K. Vanderbilt, the American socialite who visited Biarritz frequently. When Vanderbilt showed an interest in golf, Dunn took him to the famous Chasm Hole, which called for a pitch of about 125 yards across a deep ravine. Demonstrating how the game is played. Dunn teed his ball on a small mound of sand, and hit a crisp shot into the cool, clear air. The ball flew directly at the flagstick, landed on the green, and braked itself within a few feet of the hole. He tried another, and still another, laying each ball within holing distance of the cup.

Dunn's demonstration led to the golf course at Shinnecock Hills, on eastern Long Island. Vanderbilt was so impressed, he contacted some friends when he returned home, and with Edward S. Mead, of the Dodd, Mead Co. publishing house, and Duncan Cryder, organized Shinnecock Hills.

The Feminine Influence

The first golf holes in Boston were lined by flower pots. While golf is assuredly of Scottish origin, it arrived in two important American centers through France, at Shinnecock Hills and at Boston, both at about the same time. Frances Boit, a young American girl who had returned home from the Continent, had played the game at Pau, another French spa, and brought her clubs with her when she spent the summer of 1892 with her aunt and uncle, the Arthur Hunnewells of Wellesley. Frances had assumed she would find several golf courses around Boston, but when she didn't, she demonstrated the game on the Hunnewells' front lawn.

The Hunnewells had their nephew and Mrs. Hunnewell's brother as neighbors. Among them their lawns covered enough ground for Miss Boit to lay out seven short holes. Casting about for something to line the holes, Frances, with the feminine touch, settled on unused flower pots. Frances Boit thus became the first American-born golf course architect.

Auspicious Beginning

Arthur Hunnewell tried the game and was quickly won over. Laurence Curtis, however, was a more important recruit. He was a guest of the Hunnewells that day, and he, too, had been invited to play. An influential member of The Country Club in Brookline, he was so captivated by the game he wrote to the executive committee proposing the club add golf and suggesting a course could be laid out for about $50. The committee agreed.

A six-hole course opened in April of 1893 with an exhibition staged by Curtis, Hunnewell, and George E. Cabot. Hunnewell, by an incredible stroke of luck, made a hole-in-one from about 90 yards. The gallery, though, didn't seem impressed. Made up of largely of horsemen, archers, and riflemen, the spectators reasoned he had done no more than he had intended, and since he was supposed to be an expert showing them how the game should be played, he should have expected to hit the ball into the hole. Indeed, disappointed when no one else holed in one, they walked off with little hope the game would catch on.

The Long Drive

Samuel Parrish had been the first secretary of Shinnecock Hills. Playing golf one wintry day he unleashed a drive that won him the reputation as a phenomenally long hitter. Recalling the incident, Parrish wrote:

> We were at the north end of Lake Agawam, Southampton, New York, looking south toward the ocean. There was a strong north wind blowing down the lake at the time, and as I was able to steady myself on a patch of snow, the drive was a fair success, so that the ball went sailing down the lake until it struck the ice, and then kept on with but little diminution in its velocity. Had the ice and wind held out, the ball would doubtless still be going, but it finally struck a snow bank on the shore of the lake and stopped. Morton then solemnly paced off the drive and reported its length to have been 489 and a half yards. . . . I enjoyed a tremendous reputation

as a driver, my fame having penetrated to Boston, and I was the recipient of many congratulation.

In Your Eye

Walker Breeze Smith, president of the Tuxedo Club, in Tuxedo, New York, wore a glass eye. Showing a bizarre flair for gamesmanship, he won the pivotal point against John C. Ten Eyck, of a socially prominent New York family, in one of the early team matches. Smith led Ten Eyck by four holes when the match ended, but he had Ten Eyck beaten as early as the third hole. Standing at the tee, Smith told Ten Eyck that golf was a matter of courage and keeping your eye on the ball. He pulled a flask of Scotch from his pocket and invited Ten Eyck to have a drink, explaining that whiskey supplied the courage. Then, with his ball sitting on the tee, he plucked out his glass eye and set it atop his ball, saying, "That means keeping your eye on the ball."

The he drove his ball, shattering the glass eye—no great loss; Smith had a boxful. He went through the same routine on every hole. "You can imagine what this would do to a fellow's game," Ten Eyck complained.

First U.S. Amateur

Newport Golf Club, in Newport, Rhode Island, conducted the first United States Amateur championship in September of 1894. The Amateur was to be played at thirty-six holes of stroke play over a course with a number of unusual features—a rock that rose thirty feet high a few yards in front of the seventh tee, a stone quarry in front of the sixth tee, a swamp along the left of the second hole and an open ditch six feet wide about 280 yards from the tee, and a series of stone walls that not only bordered some holes but cut directly across fairways. One of those walls helped decide the championship.

Charles Blair Macdonald, a wealthy Chicagoan who had learned the game as a student at St. Andrews University, led the first round with 89, but his game collapsed the second day and he shot 100, losing by one stroke to W.G. Lawrence, a Newport member who had learned to play at Pau. Macdonald lost when he topped a shot that rolled against a stone wall crossing the property. It cost him a two-stroke penalty. Macdonald complained bitterly that stone walls were not proper hazards.

A month later St. Andrew's held another Amateur championship in Yonkers, and once again Macdonald was in the thick of it. Even with the semi-final and final rounds scheduled on the same day, Macdonald went to a party the previous evening thrown by Stanford White—the architect who was later shot by Henry Thaw because of his liaison with his wife, Evelyn Nesbit Thaw, a chorus girl from the cast of the hit musical *Florodora*, and the subject of the movie *The Girl in the Red Velvet Swing*. Macdonald hung around until five o'clock in the morning.

On White's advice, Macdonald took some strychnine tablets after a short nap. Stimulated by the drug, Macdonald defeated Lawrence in the morning, but he complained he didn't feel at all himself at the luncheon break. Ever helpful, White suggested a good steak and a bottle of champagne. Macdonald followed the prescription. In the afternoon he played Laurence Stoddard, a member of St. Andrew's, hole-for-hole through the scheduled eighteen holes of the match, but once again he lost because of one bad shot. He sliced his drive into a plowed field on the first extra hole and took three shots to work back to the fairway.

The USGA

Macdonald complained bitterly once again, insisting no single club could conduct a national championship. The rivalry between Newport and St. Andrew's, along with the controversy over the two Amateur championships, caused so much turmoil that something clearly had to be done. Shortly after the St. Andrew's tournament, Henry O. Tallmadge, the secretary of St. Andrew's, Theodore Havemeyer, of Newport, and Laurence Curtis, of The Country Club, invited representatives of those three clubs, along with delegates from Shinnecock Hills and the Chicago Golf Club (Macdonald) to a dinner in New York. They met on December 22, 1894, and formed what is now the United States Golf Association.

Have a Nice Day

Willie Park, Jr., the son of the first British Open champion and champion himself in 1887 and 1889, had been almost invincible in a series of money matches in the United States during 1895. Among his victims he counted Willie Campbell, Willie Norton, and Joe Lloyd, and he won two of a series of three matches against Willie Dunn, at $200 a match. With the first U.S. Open conducted by the infant United States Golf Association coming up, he would have been the logical favorite, but for some now forgotten reason he sailed for home in July, shortly after his last match with Dunn. As the gangplank was being drawn away from the steamship *Etruria*, taking Park back to Scotland, Willie was handed a message from Dunn, wishing him a happy voyage home.

Shippen at Shinnecock

A black man nearly won the second U.S. Open conducted by the United States Golf Association. John Shippen, whose father was a Baptist missionary to the Shinnecock Indian reservation, on eastern Long Island, entered the 1896 Open, played at the Shinnecock Hills Golf Club, where he caddied. Shippen was sixteen.

When the rest of the field, mainly immigrant British professionals, heard of Shippen's entry, they threatened to withdraw rather than play with a black man. Theodore Havemeyer, who ran the American Sugar Company and financed much of the United States Golf Association's activities, acted quickly. He told them they could play or not play as they wished, but John Shippen would be in the field.

The Open was played at thirty-six holes then. Shippen shot 78 in the first round, tying five other men for first place, then hung on, still a contender, through much of the second round. His game fell apart on the thirteenth hole, a rather short par 4. Recalling the incident some years later, Shippen remembered he had played Shinnecock many times before, and, "I knew I had to drive up the right side, but I played it so far right my ball landed in a sand road. That was bad trouble in those days. I kept hitting the ball along the road, but I couldn't get it out. I wound up with 11. Unbelievable." Shippen shot 81 and tied for fifth place, at 159, seven strokes behind Jim Foulis, the winner.

Joe Lloyd at Chicago

Joe Lloyd was among the longest hitters of his day. Willie Anderson, a seventeen-year-old Scottish immigrant, led the 1897 Open with 163 when Lloyd reached the eighteenth tee of the last round. The eighteenth measured 465 yards, quite a long hole in the days of the gutta-percha ball, but Lloyd ripped a long drive, then scorched a brassie that hit short of the green, bounced on, and rolled within eight feet of the pin. He holed the putt for an eagle 3 and won by a stroke. No one since has won the Open by scoring an eagle on the last hole.

Collateral

Golf professionals of the late nineteenth century had such bad reputations that when Fred Herd won the 1898 Open, the USGA insisted he put up security for the trophy, fearing he'd pawn it for drinking money.

Willie, Come Home

Willie Davis, the man who laid out the original course at Shinnecock Hills and re-designed Newport for the first U.S. Open, remained at Newport as the club professional. He took time off during September of 1899 to play in the Open, held that year at the Baltimore Country Club. Before he played a shot, Willie received a message. A peeved member wanting a game ordered him home. Willie went back to Newport and stayed long enough to quit.

Part Three

From the Rubber-Core Ball to the First World War

No one had ever been totally happy with the gutta-percha ball. Like baseballs of the day, it was dead. A good feather ball could be driven farther, and man, with his restless nature, constantly experimented. A ball called the Eclipse was introduced during the 1880s. A mixture of cork and rubber, it bounded along the ground, leaping over bunkers and flying totally out of control on pitches to greens. It soon disappeared. Someone else developed a ball called the Putty, equally lively and uncontrollable. It failed too.

Both balls, along with many others, were based on molding a solid mass into a sphere. Coburn Haskell took a different tack. A non-descript player from Cleveland, Haskell revolutionized the game during an idle moment in 1898. Sitting on the porch at the old Cleveland Country Club chatting with Joe Mitchell, the club's professional, Haskell held a tangled bundle of rubber bands, twisting them, squeezing them together, and bouncing them off the floor. Suddenly an idea struck. If he could wind these bands under tension around a solid core and cover them with a layer of gutta percha, he could produce a much livelier ball than the gutty. Surely such balls could fly farther than any before them.

Haskell contacted Bertram Work, a friend at the B. F. Goodrich plant, in Akron, some twenty miles south of Cleveland. Work agreed Haskell might have the germ of an idea, and offered to turn loose the bright young men in the factory to experiment and work out the manufacturing process. If anything came of it, they would split the profits 50-50. They took out their patent in 1898. The advantage became obvious early, the ball flew a good twenty yards farther than the old gutty.

J. H. Taylor was given a supply just before he played in the 1900 U.S. Open; he finished second to Harry Vardon. Accustomed to the gutty ball, Taylor refused to play the new ball in the Open, but a week or so later he tried one during a casual round at the Rockaway Hunting Club, on Long Island. Taylor was not a long hitter—his usual drive covered about 175 yards. Another group stood on the first green as Taylor stepped up to his ball, but since the hole measured 240 yards, J.H. assumed they were well out of range. Giving it a good rap, Taylor met the ball solidly. It shot off like a rocket and rolled to the feet of a man holing out. Surprised and stunned, Taylor rushed ahead and apologized, but at that moment he realized the old gutty ball was doomed. The new ball simply flew too far for the gutty to survive.

Jack Jolly, an American professional who helped popularize the Haskell ball, was especially proud of how he engineered its acceptance in Scotland. Born in St. Andrews, he knew how frugal the Scot could be, and used this knowledge to apply a touch of salesmanship. Traveling through Scotland by train, he would stand on the back platform, and whenever he passed a golf course, he'd throw a dozen or so balls onto a fairway. "There isn't a Scot alive who would refuse a free ball," he laughed.

Jolly also invented the liquid center for the wound Haskell ball. He solved the technical problem of centering the core by filling baby bottle nipples with water, freezing them with dry ice, then winding the rubber thread around them.

Herd Convinced Them

For the most part British golfers reacted strongly against the Haskell ball. The magazine *Golf Illustrated* invited comments from leading professionals before the 1902 British Open with similar results. Most of them condemned the ball. James Braid, the 1901 champion, claimed the Haskell balls split too easily, and he added the ball did *not* carry farther than the old gutty, although it might run farther. He claimed further that a bold putt was likely to jump out of the hole. Harry Vardon ended his comments with the prediction that nobody playing the Haskell would win the championship.

In an ironic twist, Sandy Herd wrote:

> In answer to your letter in regard to the Haskell ball, I think it is a very difficult ball to play with. It drives all right but that is about all I can say about it. As regards putting, especially on hard bumpy greens, it is simply off altogether, it hasn't got the click to guide you on the putting green such as the gutta ball has. What one thinks he can gain in the drive he will very soon lose on the put-

ting green. It is too fiery altogether on the green. I hope all the professionals play with it at Hoylake except myself. So much for the Haskell.

On the day before the opening round, Herd played a practice round with John Ball, the great amateur. Herd had reached thirty-five by then without having won an Open championship, and he saw the years slipping away. He felt unusually frustrated because of his friendship with Vardon, Taylor, and also with James Braid, who had won in 1901.

Ball played the practice round with a Haskell ball and offered one to Herd. Sandy declined the offer, but he changed his mind quickly. Stunned as Ball outdrove him through the first few holes, he asked if he could borrow one after all. He became an immediate convert, bought a few from the limited stock in the professional's shop at the end of the day, shot 307 in the Open, and won by one stroke over Vardon and Braid.

Herd's victory gave the gutty its final death blow. Realizing something new had been added to the game, golfers wanted the new ball, but the balls were hard to find. Prices shot to exorbitant levels. Some golfers were known to have paid 30 shillings for one ball, a price beyond reason. Once bought, the balls were guarded like jewels.

Pneumatic Ball

To combat the relatively new Haskell ball, made by B. F. Goodrich, the Goodyear Company introduced a pneumatic ball for the 1906 season. With a rubber shell filled with compressed air, the pneumatic ball had an unfortunate tendency to explode in mid-flight.

Using a Goodyear ball in an exhibition match at St. Andrew's, in New York, Willie Dunn sliced a shot into the gallery. The ball exploded and injured a spectator.

It had other problems as well. Alex Campbell, the professional at The Country Club, near Boston, was in contention for the 1907 U.S. Open, at the Philadelphia Cricket Club when his ball was punctured and all the air bled out. Campbell four-putted with the deflated ball, replaced it when the hole ended, and lost by three strokes.

Realizing its mistake, Goodyear retired the air-filled ball.

Still the battle for the gutty raged on. The British PGA ruled in 1902 that the gutta-percha ball must be used in tournaments held under its banner. Reacting to its apparent dictatorial stance, some amateurs believed the selfish interest lay behind the ruling. Some amateurs claimed the professionals favored the gutty because the softer Haskell ball wouldn't damage clubs as easily, cutting off a source of income. They also maintained professionals wanted everyone to play the gutty because they earned money remolding those that broke apart.

Horace Hutchinson, a pioneer among those who wrote about the game, may have been the most important early convert to the Haskell. An influential R and A member, he had won the British Amateur championships of 1886 and 1887, and he was to become captain of the R and A in 1908. His opinions carried weight. Hutchinson favored the Haskell because he found it easier to play in the sense that it flew farther than the gutty and, of course, made the holes play shorter.

No man is without his detractors, though. Those against the Haskell claimed Hutchinson's reasoning represented the point of view of a weakening man, the whimpers of one past his prime. In fact, he was only forty-three.

Prisoner of War

Born in England, H. J. Whigham moved to the United States after graduating from Oxford and joined the faculty of Lake Forest College, north of Chicago. A first-class golfer, Whigham was brought to America by members of the Onwentsia Club, of Chicago, so they might boast of a member who could play on equal terms with Charles Blair Macdonald of the Chicago Golf Club.

Macdonald had won the first U.S. Amateur championship run by the USGA, in 1895. Whigham succeeded him in 1896, won again in 1897, then married Macdonald's daughter. Whigham became a newspaperman, equally adroit as a drama critic and golf writer, and also as a war correspondent. At the outbreak of the Spanish-American War he was sent to Cuba, reported on the battle of San Juan Hill, and eventually was captured by the Spanish at Matanzas. The Spaniards suspected he was a spy. Released, at the war's end, he returned to the United States and resumed his career.

President McKinley

William McKinley was the first United States President to show an interest in golf. While he recuperated from an illness in White Sulphur Springs, West Virginia, in 1899, stories telling of his playing golf caused quite a bit of comment. A cartoonist for the *Boston Evening Record* depicted him in knickers, and an accompanying article asked, "Shall the President or shall he not become a golfer? Shall he allow the thought of a possible lack of dignity to interfere with his restoration to health?"

Vardon at Chicago

Although he was considered one of the greatest golfers who ever lived, Harry Vardon was a terrible putter. He often froze over the ball while a muscle in his right arm twitched visibly. He showed just how bad he could be on the last green of the Chicago Golf Club during the second round of the 1900 U.S. Open, during his first visit to the United States. He missed one short putt, leaving his ball less than two feet from the hole. Then he missed again, snagging the clubhead in the wiry grass. He still won, though, beating J. H. Taylor, his friend and fellow Englishman, by two strokes.

Golf and the Olympics

Golf had a short life in the Olympic movement; the game was played twice, in 1900 and in 1904, but never returned. Because of the Dreyfus Affair and its charges of anti-Semitism, the Olympics were played almost secretly. Baron Pierre de Coubertin, who had revived the Olympic Games in 1896 after they had been abandoned centuries earlier, was shoved aside for the 1900 revival, in Paris, and the organization of the Games turned over to the Paris Exposition Company.

Peggy Abbot, a member of the Chicago Golf Club, had been visiting Paris with her mother, an ardent golfer. Seeing an advertisement, they entered what they thought was simply an international golf tournament at Compiegne, a Paris suburb that was to become famous in later years as the site of surrender in the two world wars.

The women's championship was to be played over nine holes. Peggy Abbot shot 47 and won. Her mother placed seventh.

They believed they had won the Paris City championship. Neither ever knew they had won Olympic medals.

With a score of 82–85—167, Charles Sands won the men's championship. A member of the St. Andrew's Golf Club in Yonkers, New York, Sands had been beaten 12 and 11 by Macdonald in the final of the first United States Amateur championship conducted by the United States Golf Association, in 1895.

Peggy Abbot later married Finley Peter Dunne, the creator of Mr. Dooley, the political satirist.

Four years later, when the Olympics shifted to St. Louis, in conjunction with the St. Louis World's Fair, golf played a part once again. Nearly 100 men entered, among them H. Chandler Egan, a Harvard student from Chicago. Egan would win two consecutive U.S. Amateur championships and develop into one of the finest golf course architects.

Moving easily through the early rounds, Egan met George Lyon, a 46-year-old Canadian, in the final. A superb athlete who excelled at cricket, baseball, tennis, and track, Lyon had taken up golf at the age of thirty-eight and had become quite good. By then he had won three Canadian Amateur championships. He beat Egan by 3 and 2. When he was called to receive his gold medal at the presentation ceremony, Lyon walked the length of the room on his hands.

Lyon sailed for England to defend his championship in the 1908 Olympics, but when he arrived he found that he had no competition. No one else had entered. He was offered another gold medal but refused it.

Golf a Menace

In their triennial convention in San Francisco, in 1901, Episcopal bishops denounced Sunday golf. One wrote that while the Anglican Church is not puritanical, and the member should be left to do as he wished once he'd done his religious duty on Sunday, the surroundings of a golf course were often bad.

"They drink too much Scotch whiskey out on the links. They sometimes bet heavily. It is this side of Sunday golf playing that I most disapprove of. This style of playing has increased so that it is now a menace and should be stopped."

The Birth of U.S. Steel

At the time the biggest deal in the history of American business, the arrangement that created United States Steel was swung at the St. Andrew's Golf Club, in Yonkers, New York. A native of Scotland and an enthusiastic golfer, Andrew Carnegie was locked in a match with the financier Charles Schwab, who had been trying to convince Carnegie to sell Carnegie Steel; Carnegie had been holding out. During the course of the round, Schwab finally talked Carnegie into selling out to a new combine, provided they could agree on the price. Carnegie calculated the amount he wanted and jotted $480 million on the back of the scorecard. Schwab agreed, U.S. Steel was born, and Carnegie became immensely rich.

While Carnegie didn't exactly hoard his money, neither did he throw it away. When he was put up for membership in the American St. Andrew's, he was sent a bill for $140 by John Reid, the founder of the club and a close friend. The bill evidently staggered Carnegie so badly he wrote back, "I cannot consider this a proper use of money—no, Sir. I will never be there, and besides, it is a rich club. For the St. Andrew's Society, anything—for a swell club, nothing."

Carnegie relented after a time and became a member. Later he referred to Reid as the "first president of a golf club in all America, and such a club as St. Andrew's must ever be remembered in the history of that unequaled game."

Willie Anderson

While they might have enjoyed watching the skilled professionals play the game, club members didn't invite them to dinner. Or lunch. On the morning of the first round of the 1901 U.S. Open, at the Myopia Hunt Club, north of Boston, a member stood on the clubhouse steps and told the assembled professionals they were to take their lunch in the kitchen. Willie Anderson, a twenty-one-year-old Scottish immigrant about to win the first of his four Open championships later in the week, stood on the fringe of the group idly swinging his mashie. As the substance of the message became clear, Willie's swing picked up speed, and for a moment he lapsed into the dialect of his youth.

Furious, he slashed at the ground with all his power, gouged a huge divot from the velvety lawn, and snapped, "Na, na, We're na goin t' eat in the kitchen." Astonished, the club member ducked back inside, and sometime later the professionals were told they could have their lunch in a tent raised especially for them.

Anderson had a melancholy personality. He won the U.S. Open in 1901, 1903, 1904, and 1905, four championships within five years. No one has ever won four so quickly, and no one else has ever won three in succession (Ben Hogan later won four in six years). Anderson was essentially a modest man who got along extremely well with the other players. Even though he beat them regularly they admired him, respected him, and liked him. On the other hand the public seemed indifferent to him.

Anderson had grown up in North Berwick. Tom Mercer, the professional at the Sound Beach Golf Club, in Old Greenwich, Connecticut, had come over from Edinburgh. Although Mercer was ten years older, he and Anderson became close friends. In moments of gloomy introspection, Willie often muttered to Mercer, "They don't know me. They don't know me."

Anderson died young, possibly the result of heavy drinking. Willie played his last Open in 1910. He shot 303, finished five strokes behind Alex Smith, and placed eleventh, beaten even by his younger brother, Tom.

By then a good player could earn decent money in exhibition matches. In the fall Anderson set out on a series of thirty-six-hole exhibitions around Pittsburgh. In his last match he teamed with Gil Nicholls against William C. Fownes, the current U.S. Amateur champion, and Eben Byers, who had won the 1906 Amateur. The match was a classic. Playing terrific golf the amateurs shot a better-ball score of 69 in the morning against a par of 72. They stood two holes up at lunch. Anderson and Nicholls went to work in the afternoon, playing at the same levels the amateurs had played in the morning. They had pulled even after thirty-five holes, but with the match at stake Byers chipped in from off the edge of the eighteenth green and saved the day for the amateurs.

Anderson seemed noticeably tired over the last few holes. He labored into the golf shop, sprawled in a chair and groaned he wouldn't play another round of golf that year. Returning to his home in Philadelphia, Anderson died two days later, on October 25, 1910. He was thirty.

Walter Travis

Walter J. Travis, the first overpowering American amateur, had been born in Australia and hadn't taken up the game until he reached thirty-four. In 1900 he became the first to win the U.S. Amateur with the rubber-cored ball, but he is remembered best for causing the R and A to ban center-shafted clubs, those with shafts that enter the clubhead forward of the heel.

Travis had already won his three U.S. Amateur championships by 1904 when he embarked for England and a try at the British Amateur. He had been playing so badly when he arrived that when a Mr. Phillips, a member of the Apawamis Club, in Rye, New York, offered to lend him his putter the day before his first match, Walter accepted.

Called the Schenectady for the city where it was made, this strange-looking mallet-headed club had its shaft set in the center of the head. Travis putted like a demon. Holing putts from every range, he defeated the best the British could throw against him. In the fifth round he beat Harold Hilton, who had already won two British Opens and two British Amateurs, and in the semi-finals he beat Horace Hutchinson, who had won two British Amateurs.

He met Edward Blackwell for the championship. A member of the Crosse and Blackwell family, Blackwell had once driven a ball 358 yards, and another time had driven the eighteenth green at St. Andrews, an even longer shot. The match was played in bitter silence. Talking about it later, Travis called Blackwell the most remarkably silent man he'd ever met, "thereby proving himself an unconscious humorist of the first water."

Travis admitted he didn't talk much during a serious match, but after Blackwell had outdriven him by miles on the first two holes, to be sociable Travis complimented him on his driving while they stood on the third tee. "He merely murmured some sort of acknowledgment," Travis recalled. "That was the end of any attempt at conversation."

Although Blackwell consistently outdrove him by fifty to seventy yards, Travis, his cigar clenched between his teeth, holed a series of twenty- to thirty-five-foot putts and won the match by 4 and 3, becoming the first American winner of the British Amateur.

After Travis had holed two or three longish putts, Waldo Burton, another American, offered to bet an English friend that Travis would hole the next long putt. The Englishman accepted, and Burton collected. He made the same offer on the next four holes when Travis's ball lay from twenty to thirty-five feet from the hole. Travis holed every one; the Englishman finally stopped accepting bets.

About a week after the final, Travis was playing a casual round at Sandwich when two men approached and asked if he would attend a dinner to be given the following Saturday evening to commemorate his winning the championship. Hurt by what he perceived as rude treatment by the British, who so obviously didn't want a foreigner to win their sacred championship, Travis turned them down. He confessed the thought instantly flashed through his mind that had Blackwell won, the dinner would have been held that same evening, and Travis, the victim, would have been seated at Blackwell's right hand.

Each day after he completed his match, Travis bought at least one and occasionally two clubs in the professional's shop. By the time he played Blackwell, Walter had only two clubs left from his original set—a mashie and, of course, his putter.

Travis's putting had so impressed the British public that St. Andrews clubmakers threw themselves into full scale production to fill the flood of orders for this new putter. It was a mistake. They had made thousands when the R and A banned the center-shafted design. The heads were dumped on the scrapheap. The R and A maintained the ban until 1952.

Travis's love affair with the Schenectady became an on-and-off thing. Writing about it later he admitted, "I have never been able to do anything with it since. I have tried it repeatedly, but it seems to have lost all its virtue."

He continued to use the Schenectady occasionally until finally it was mounted on the wall of the Garden City Golf Club, in Garden City, New York, where it remained for many years. It disappeared during a party one night and was never returned.

Over the years, Travis played a number of important matches against Jerry Travers, a much younger man, without either man's gaining a substantial edge. By the time they played one another for the last time on the national level, Travis had won three U.S. and one British Amateur, and Travers was the defending U.S. Amateur champion.

They met in the semi-finals of the 1908 Amateur at the Garden City Golf Club, where Travis not only held a membership but had redesigned the course. With Travers outdriving Travis by many yards, and Travis holing everything in sight, they finished the morning round with Travis one hole ahead. Travers clearly outplayed Travis as the afternoon round began, but Travis fought back, and held a two-hole lead with four holes to play. But Travis, 46 years old by then, had tired and had no more to give.

They reached the eighteenth tee with Travis needing a win to take the match to extra holes. The eighteenth at Garden City is a nice par 3 across a lake formed by excavations during the construction of the course. In addition to the pond, the green is protected by bunkers, one a pit six feet deep with steep walls. Travis's tee shot found it. When Walter climbed in he dropped from sight. The gallery stood quiet while the glint of a clubhead flashed in the sun and a shower of sand flew from bunker. But no ball. Again the gallery saw a clubhead and heard a thud as it dug into the sand. Once again no ball. Two shots gone, Walter climbed from the bunker, and with a sad smile shook hands, conceding the hole and the match. Travis himself had designed those bunkers and fought to keep them in place against the objections of other members. They particularly despised the bunker that had defeated him.

Travis had a son, Bart. At the beginning of the First World War Bart slipped off to Canada, joined the Royal Canadian Air Service, and shot down six German planes. But he couldn't play golf, despite the everlasting hopes of the old man. Playing in the qualifying round for the club's spring invitational tournament, Bart hit a particularly fine shot into the eighteenth that settled about five feet from the cup.

When the old man saw it, family pride awoke and he squirmed through the crowd for a front-row view. He beamed when the putt fell for a very nice 2, then, chest out, he strode up to Bart and asked what he had shot, of course expecting a very good score. Bart frowned and said, "Let's see, he said, mentally adding up the strokes, "That gave me 101, Sir." Stunned, Travis grumbled , "Oh, hell," and, shoulders sagging, pushed his way back through the crowd and into the clubhouse bar.

Remodeling some holes at the Essex County Club, in Massachusetts, Travis revealed the secret of his phenomenal success as a golfer. He and some others were discussing the possibility of turning a stretch of land into a new short hole. Asked how far he thought they were standing from a certain tree, Travis said, "Between 155 and 157 yards." One of the others asked 'Why don't you say between 155 and 160 yards, Walter." Travis said, "Because it isn't. It's between 155 and 157 yards." He seemed so certain, the group decided to measure the distance. Travis was off. The tree stood 157½ yards away.

Origin of Birdie

In nineteenth-century America, the term "bird" described a person or thing of excellence. For example, "He's a perfect bird of a man." Eventually the word worked its way into golf. Abe Smith, a rather ordinary golfer with no accomplishments to speak of, was playing with friends at the Atlantic City Country Club, in New Jersey, during the fall of 1903. During the course of his round he hit an unusually good shot to a difficult par 4. When his ball came down stiff to the pin, Smith spun around to his friends, beamed, and cried, "That's a real bird of a shot." Although at first "bird" meant no more than a good shot, through popular use the term evolved into a designation for one under par, hence "birdie."

Change in Scoring

Team matches like the Walker Cup and the Ryder Cup award a single point for each individual match. The system is relatively new. The British universities of Oxford and Cambridge have played against one another since 1878. In its original form the match was decided by the total number of holes won by each side. For example, a match ending 3 and 2 awarded three points to the winning side.

The method changed after the 1907 match. Oxford had led by a 12-hole aggregate, but in a match between the two captains, Cambridge's W. T. Allen overwhelmed Oxford's C. N. Bruce by thirteen holes, swinging the match to Cambridge. Bruce had been ill, illustrating that a single team member either playing off form or under the weather can lose a match that seems to have been won handily. For better or worse, Oxford and Cambridge adopted the system of one point per match, setting the precedent for every other team match.

Harry Vardon

Early in Harry Vardon's career, his father considered his younger brother Tom the better golfer, even though Harry had won more prizes. Tom held the professional's position at Royal St. George's, in Sandwich, England, and later moved to the United States, where he played with but little distinction. His father often said, "Harry may win the prizes, but it is Tom who plays the golf."

Harry Vardon had been fond of all sports. At Ripon, a club in England where he had taken his first job as a professional, he played cricket and also tried boxing at the Boffy, the gardeners' sleeping, eating, and resting place, where they engaged in all kinds of games. When Harry arrived one evening, a gardener called to him, "Harry, we've got someone who will put you through your paces tonight." He was astonished to find he was to box a rather buxom young girl, the daughter of one of the gamekeepers. The girl landed a blow on the nose that Vardon said would have done credit to the heavyweight champion. It also ended whatever aspirations he might have had held of becoming a boxer.

The 1896 British Open was scheduled to end on a Thursday. Vardon and J. H. Taylor tied for first place, which called for a playoff. The championship was played at Muirfield, but since a tournament had been arranged for Friday at North Berwick, perhaps a mile away, the outcome of the British Open, the most important competition in golf, was delayed a day.

Vardon hadn't putted as well as he would have liked during the Open championship. Between rounds of the tournament at North Berwick, he noticed an old cleek that had apparently been thrown away, liked the look of it, had it re-shafted, beat Taylor by four strokes over thirty-six holes the next day, and won the first of his six British Open championships.

As a young man, Vardon had formed the Ganton Football Club and played center forward, the key position in the attacking line. Returning home tired of golf after his first American trip, in 1900, he found the club unbeaten so far. He was named captain and played goaltender. He felt he wasn't competent to return to center forward.

A few years after he left Ripon he found himself in the neighborhood once again and decided to call on the old couple who owned the house where he had lived. Neither the man nor his wife recognized him, and when Vardon insisted, the women said, "No you're not Harry Vardon. Harry Vardon is too famous and he 'as made too much money now to come and see folks like us." Nothing Vardon could say could convince her he was Harry Vardon.

Vardon stood at the peak of his game early in the twentieth century. By 1903 he had won three British Opens and was generally considered the game's leading player. Shortly after winning the 1903 British Open, at Prestwick, Vardon was taken ill. He hemorrhaged on the Totteridge golf course, where he was the professional. Taken home, he rested overnight and felt so well he returned to the club, but as he bent over to tie his shoes he began to hemorrhage again. Diagnosed as having tuberculosis, he spent months recovering in a sanatorium in Norfolk.

Vardon was so much straighter off the tee than any other golfer of his time that around the turn of the century a group of scholars arranged tests hoping to learn the secret of his accuracy. In one test Vardon hit a series of drives from a dirt tee. The scholars charted the position of his feet in relation to the ball on every drive and then erased his footprints so he had nothing to guide him when he addressed the next ball. Vardon hit about two dozen drives; each drive split the fairway.

Studying the charts, the scholars found Vardon varied his stance slightly on each shot, his left foot angled a shade more open on one, his heels closer together on another, his right foot nearer the ball on still another, and so on. They concluded the variations in Vardon's stance had no effect at all on how he hit the ball.

Vardon met Andrew Kirkaldy in the final match of a tournament at Montrose, in Scotland. The successor to Old Tom Morris as the honorary professional at St. Andrews, Kirkaldy had nicknamed Vardon the Greyhound because the others had so much trouble catching him. Vardon won on the last hole when Kirkaldy three-putted. The match over, Jack White, another golfer, sympathized with Andrew. Kirkaldy snapped back, "Never you mind about that. I beat all the blighters in my own class."

Vardon and Taylor played an international match against Braid and Sandy Herd in 1905, thirty-six holes at St. Andrews and Troon in Scotland, and St. Anne's and Deal in England. Vardon and Taylor had opened a lead of twelve holes before reaching St. Annes, but Braid and Herd had fought back and won five holes in the morning portion. All four men, who were good friends, went to lunch together at a nearby hotel. Returning with about fifteen minutes to spare before the afternoon round was to begin, they were stopped at the gate by a policeman who asked them for their admission badges. Told they didn't have badges because they were the players in the game, the policeman didn't budge; either they show their badges or they wouldn't be admitted. The argument went on for several minutes until someone in authority convinced the gendarme that unless he stepped aside, there would be no afternoon match.

Even though he won six British Opens, Vardon had enough candor to admit he ranked among the mediocre putters. "There are many ways of performing the operations successfully," he stated. "I can claim, however, to be in a position to explain how not to putt. I think I know as well as anybody how not to do it."

Playing a thirty-six-hole stroke-play competition at Hyères, in France, in 1908, Taylor played what he believed was a pretty good shot to the eighth, a par-3 hole, but he couldn't see the ball as he approached the green. Wandering about looking for his ball, Taylor was approached by a spectator, who said, "If you're looking for your ball, it's in the hole."

Playing at Northwood, in Middlesex, England, Vardon pushed his approach to the eighteenth so badly his ball was stymied by the clubhouse. Not one to waste strokes, Vardon ordinarily would have played a safe shot to the left and pitched to the green, but he must have felt particularly daring on this day. With his ball lying only two or three yards from the clubhouse and the wall rising thirty feet, Vardon had to play the shot almost straight up and at the same time put such spin on the ball that once it cleared the top of the building it would move forward and land on the green. He had actually to play a shot that would change direction in mid-flight. He pulled it off, laying the ball within a few feet of the hole.

Approaching the eighteenth at the end of his second round of the day, Vardon missed the green again, hitting his ball into a rocky stream. He decided to play the shot rather than drop from the hazard, and nearly holed it.

It's Called Confidence

Showing a decided lack of faith, some of Vardon's friends felt apprehensive for him when he scheduled a match against an extremely long hitter on the other man's home grounds. Approaching him, one of his friends said, "Harry, we feel sorry for you today because you're up against a very long hitter."

Vardon looked at him coolly for a moment, then asked, "How long is the first hole?"

"350 yards," they answered.

"Can he drive the green?" Vardon asked.

"No," they answered.

Whereupon Vardon snipped, "Well I can get there in two."

Vardon had a sense of his place in the game's history. When he returned to the United States for his 1913 tour, he reflected on his first visit in 1900 and noted that the game had grown enormously. In later years he said,

> "When I look back on this tour, it is with a feeling of utmost satisfaction that I realize I was actually the means, through the medium of my visit there, of starting that which was to become in later years the great golf craze of America. As one gets older and has finished participating in the big events, it is satisfactory to think that one was the cause of advancing the game of golf throughout such a vast country as the United States."

During his 1913 visit the manager of Jordan, Marsh, the Boston department store, invited Vardon to give an exhibition of driving balls into a net. Vardon accepted. He was supposed to perform at half-hour intervals, but as each half-hour ended, the applause from the assembled shoppers grew so enthusiastic that he kept on hitting balls. At one point the manager told him his performance was going over so well, the store had sold out all its golf equipment. Still, Vardon kept hitting balls, but the job became so boring that after a time he varied his routine by playing mashies toward water taps on the ceiling. Vardon had just hit one when the manager raced in telling him to stop, because he might set off the fire extinguishing system and flood the store.

Toward the end of the First World War, Vardon was awakened one night by an explosion. A German Zeppelin had dropped a bomb that landed in his garden, destroying part of his house and killing a neighbor.

Vardon's brother, Tom, eventually settled in St. Paul, Minnesota. Harry and Ted Ray played a match against him and Jack Burke during their 1920 tour of the United States. Before the match, Tom told everyone who would listen that, "Although I always play my best, my damn brother always goes one better than me." Tom, however, had some revenge on this day. Tom holed a good long putt on one hole, leaving Harry a shorter putt to halve the hole. As Harry's ball rolled toward the cup and seemed ready to drop, a frog leaped out of the hole and knocked the ball away.

Harry broke the course record in the afternoon portion of the round, to which Tom remarked, "I expected you would."

Biting Tongue

After a match against Vardon and Ray at the Capital City Country Club, in Atlanta, Howard Beechert, the club professional, asked Vardon what he thought of Capital City's bermudagrass greens. "Fine," Vardon exclaimed. "They're the best grape vines I've ever played on."

Requited Romance

Since Vardon's wife, Jessie, had no interest in golf, he missed having someone to share the joys and disappointments in his career until later in his life he met a young dancer performing at the Royal Hotel, in Hoylake, England. Although she didn't play golf, Tilly Howell recognized the name of Harry Vardon and asked for his autograph. Twenty-two years younger than Vardon, she found him an attractive and charming man, and evidently he found her charming as well. Within a few days they became seriously involved.

A year later he returned, and their affair picked up again. With Tillie genuinely interested in Vardon's career, he suggested she move to London, where they could be close together. In 1925, realizing she was pregnant, she moved to Liverpool. Their son, Peter, was born in January 1926. Realizing she couldn't raise the child, Tillie turned him over to her married sister, and she eventually moved to Birmingham, where she could be near him. Vardon visited often, bringing the boy presents, but eventually Tillie felt she had to break with Harry when Peter began asking questions.

Vardon continued to send Tillie money and he paid for Peter's education. Even though Tillie remained intensely loyal to him and never married, she waited until after Vardon's death in 1937 to tell her son of his father.

Taylor's Tribute

Harry Vardon died in London in 1937. J. H. Taylor, his great rival in so many tournaments and partner in many others, spoke at his funeral:

> By the death of Harry Vardon, the game of golf has lost one its greatest players, and I, and many hundreds of others, a dear and valued friend. We had known each other for more than 40 years. We both made our debuts in the Open Championship in the same year, in 1893, at Prestwick. That is a long time ago, and Vardon and I have been associated as servants of the game for nearly the whole of our lives.
>
> Vardon, as is known, learned his golf in Jersey, a favorite spot for the development of skillful players. I am often asked whom I consider to be the best golfer ever I saw, and with a life experience behind me of having seen all the great players of the last 50 years, I give it as my mature and considered judgment that Vardon was the greatest of them all. His style was so apparently simple that it was apt to mislead. He got his effects with that delightful, effortless ease that was tantalizing. It was a legend of the game that Vardon was never off the center of any fairway in two years of play. I can scarcely subscribe to this, but I do say without fear

of contradiction that Vardon played fewer shots out of the rough than anyone who has ever swung a golf club. If the test of a player be that he makes fewer bad shots than the remainder, then I give Vardon the palm. He hit the ball with the center of every club with greater frequency than any other player, and in this most difficult feat lay his great strength as a player.

In addition to his wonderful skill, Harry Vardon will be remembered as long as the game lasts as one of the most courteous and delightful opponents that could ever be. I have good reason to appreciate this because Vardon and I, in the pursuit of our calling, met some hundreds of times, and although I was generally unsuccessful, I give it to him that when I was fortunate enough to win, he gave me the fullest possible credit.

Another tribute I should like to pay to my old friend. Throughout the years I knew him, I never heard him utter one disparaging remark about any player. He was at all times most anxious with his help and advice. Allied to his magnificent skill, Harry Vardon will be always remembered as one of the most kindly souls that ever existed. And to know him was to love him.

Arnaud Massy

Shooting 312 at Hoylake, England, in 1907, Arnaud Massy beat J.H. Taylor by two strokes and became the only French winner of the British Open. Four years later, in 1911, he tied Vardon at Sandwich and forced a thirty-six-hole playoff. Playing masterful golf in the first round, Vardon opened a five-stroke lead, then gained two more through the early holes of the afternoon. Massy managed to win some back and stepped up to the tee at the seventeenth hole, a par 3, holding the honor. There the Frenchman hit a wonderful shot within twelve feet of the pin, but the merciless Vardon put his shot easily inside him. Throwing up his hands, Massy conceded defeat. It is the only incident of a player's giving up during a playoff. Massy had played 148 strokes through 34 holes. When Vardon holed out of the 35th he had taken 143. On his way back to the clubhouse, Massy muttered, "I cannot play zis damn game."

Margaret Curtis

The 1907 United States Women's Amateur championship ended with a match between two sisters, the Misses Harriot and Margaret Curtis, the only time relatives of any description played against each other for a national championship. Harriot Curtis won the 1906 championship but lost the 1907 final to her younger sister.

In the year when Harriot Curtis held the United States Women's championship, Margaret Curtis entered a stroke-play tournament at Walton Heath, in the suburbs of London. Paired with May Hezlett, the British champion at the time, Margaret Curtis was having a good round and was in fact leading by five strokes standing on the eighteenth tee.

Both Miss Curtis and Miss Hezlett hit fine drives, but Miss Curtis overclubbed her approach shot. Her ball not only flew over the green, it dived into a gorse bush and sat inches above the ground, enmeshed among the brambly branches. With no previous experience with gorse, Miss Curtis tried to hack her way out. She couldn't, but she kept trying. She finally holed out in 13 and lost the tournament by four strokes.

Thirty years later Margaret Curtis returned and arranged for a lesson from James Braid, who had won five British Opens early in the century and spent many years as the Walton Heath professional. Braid greeted her by asking, "How are you, and how is your sister?"

Surprised that Braid knew not only of her but that she had a sister as well, Margaret Curtis asked how he could have remembered. Braid answered that he not only knew of her sister but "I even remember that day thirty years ago when you ended up in that gorse bush. As a matter of fact, the bush is still there, and I would be delighted to take you out there right now and show you how to play the shot."

Two years after the outbreak of the First World War, before the United States was drawn in, Margaret Curtis sailed for France, where for two years she held the position of Chief of the American Red Cross Bureau for Refugees in Paris. Occasionally she took time off to play a few rounds of golf at the Saint-Cloud Golf Club, on the outskirts of Paris. One of her close contacts, an executive with American Express, was an enthusiastic golfer as well. Without her knowledge he sent a van to the Curtis home in Manchester, Massachusetts. Astonishing everyone at home, the driver announced he had come to pick up Miss Curtis's clubs. They were then dispatched to Paris, where Miss Curtis used them the remainder of her time there.

An outstanding administrator, Miss Curtis took three months' leave once the war ended and allied herself with a Quaker organization working in the devastated areas of France. Soon, though she was drawn back to the Red Cross. In 1921 the Red Cross asked her to take charge of the social workers on their staff for the Child Health Program in Central Europe. This meant going to Poland, Czechoslovakia, Austria, and France. In the winter of 1923 she was sent to Greece to help resettle refugees from Smyrna.

Don't Give Up

George Duncan, the 1920 British Open champion, won his first tournament in 1907 in spite of one horrendous hole. Playing at the Worseley Golf Club, in Lancashire, on England's west coast, Duncan hooked his approach shot out of bounds. He dropped another ball and hooked it out of bounds as well. He continued hooking shots out of bounds until his eighth shot, including penalties, landed on the green. Duncan holed the putt to avoid soaring into double figures on a single hole. Not one of the game's great putters, Duncan offered sound advice nevertheless. He said, "If you're going to miss 'em, miss 'em quick."

President Taft

A mid-90s golfer who always played in striped morning trousers, William Howard Taft played almost every day, usually with Todd Lincoln, Abraham Lincoln's oldest son. Playing in the dead of winter at the Chevy Chase Club, just outside Washington, they reached the fourth hole, a par 3 of about 150 yards with a stream immediately in front of the green. It was a cold day and the stream was frozen solid—or so it seemed. Lincoln's tee shot fell short, hit the ground in front of the stream, then rolled onto the ice. Always game, Lincoln stepped gingerly onto the ice to play a recovery. Just as he hit the ball the ice gave way and down he went, about a foot deep in the chilling water.

President Taft didn't crack a smile—he just glared. Lincoln's caddie, though, burst out laughing. A conceited man who despised looking foolish, Lincoln fired the caddie on the spot and sent him back to the clubhouse.

Back to School

Most Presidents played their golf at either Chevy Chase or, later, Burning Tree, both in the suburbs of Washington. An indifferent although enthusiastic golfer, President Wilson preferred either the Washington Golf and Country Club, across the Potomac in Arlington, Virginia, or Bannockburn, in the Maryland suburbs. Playing Bannockburn, the President was shocked one day to see a man rush onto the course, grab his caddie by the ear, and haul him away. Angry, the President demanded an explanation. The man explained he taught school nearby and the caddie was playing hooky. After a brief conference, the teacher allowed the caddie to continue the round with the promise he would report to school as soon as the President finished. Young Al Houghton, the caddie, grew up to become a vice president of the PGA of America.

On the days Wilson played at Washington Golf, the White House would call at about seven in the morning, because his caddie would have to go to school. He rarely completed an entire round, rather he would play for an hour or an hour and a half and then return to the White House.

Wilson always carried a rabbit's foot, marked with a gold band bearing the initials WW. It had been given to him by a congressman. He missed it one day after he returned to the White House. One of his aides called James Scott, the club steward, and said, "Mr. Scott, the President wants to talk to you." Scott took the phone. "Scott," the President said, "in playing golf today I lost my rabbit's foot. Please have Johnny [his caddie] go over the course. I played only a few holes. If he finds it, let me know. I would hate to lose it." Scott went to the school and explained the problem. The principal excused Johnny, he found the rabbit's foot, and the President passed along $5 to the boy.

Fred McLeod

Willie Smith, the 1899 champion, and Freddie McLeod each shot 322 at the Myopia Hunt Club, north of Boston, and tied for first place in the 1908 U.S. Open championship, setting up a playoff the following day.

McLeod didn't sleep at all the night before the playoff, and eventually gave up trying. Dressing, he decided against wearing the shirt he had worn for good luck on Thursday and Friday, the days of the regulation seventy-two holes, but he took it with him to the course just in case. Arriving at Myopia, he went to the practice tee to warm up, hit two dozen balls without getting one of them more than ankle high, dashed off to the clubhouse and changed shirts, then shot 77 in the playoff against Smith's 83.

A native of Scotland, McLeod held the professional's position at the Midlothian Club, near Chicago, at the time of the 1908 Open. When he left Chicago for the thirty-six-hour train ride to Boston and the Open, he weighed a strapping 118 pounds. After four rounds in two days of the Open proper and then an eighteen-hole playoff the following day, McLeod lost ten pounds and weighed in at 108 pounds, the lightest Open champion ever.

Early in the twentieth century the Open champion put his prize money in his pocket and headed for home, where he'd most likely hole up in his golf shop and sell balls and the occasional club to members of the club where he was employed. He had no manager, no ghost-written books, and no endorsements for equipment. His picture may or most likely not appear in newspapers, and hardly anyone would recognize him.

The day after he won the 1908 Open, McLeod played in another tournament at Van Cortlandt Park, in New York City, believed to be the first public course in the United States. Because it was such a loosely organized affair, competing players took their turns with the casual golfers, out for a day of fun, and bags stood in an unending line while the golfers waited to play. As McLeod's group arrived someone whispered to the starter that the Open champion had arrived, and the starter waved them onto the tee ahead of the waiting mob.

After McLeod drove, about fifty or seventy-five spectators followed along with him. The course had backed up at a par-3 hole on the first nine with three or four groups waiting their turn, but as McLeod and his partners arrived word spread once again that the Open champion was here, and so those in front gave way and McLeod played through. Hole by hole the gallery grew as McLeod and his companions sped around the crowded course helped by those in front who stepped aside. When the last putt fell on the final green, the gallery applauded politely and the players walked off. As McLeod walked through the crowd feeling pretty good about himself, he heard one spectator ask another, "Which was the Open champion?"

Braid at St. Andrews

Rain canceled a day's play of the British Open for the first time in 1910. Since St. Andrews had been through a hot and dry summer and rain hadn't been expected, the holes were cut in low areas. When the deluge hit midway through the round, the holes acted as collecting points for water, and the course became unplayable. Word that the round was canceled reached James Braid at the fourteenth tee, but a cautious man, Braid continued on. Playing with a ball heavy enough to sink to the bottom of the hole while others used those so light they floated, he shot 76. The round didn't count, but Braid shot another 76 the next day, and with 299, beat Alex Herd by four strokes. By then he had won five British Opens within ten years, four of them within six. He didn't win another.

Bernard Darwin

Considered the best of all writers who concentrated on golf, Bernard Darwin was in addition a fine player and administrator. He served for a time as chairman of the rules committee of the Royal and Ancient Golf Club of St. Andrews, and he scored for Francis Ouimet when Ouimet beat Harry Vardon and Ted Ray in the playoff for the 1913 United States Open championship.

Darwin played Horace Hutchinson, another Englishman, in the fifth round of the 1910 British Amateur. Darwin had won nothing of real moment, but Hutchinson had by then won the British Amateur twice and had been runner-up twice. Like Darwin, Hutchinson dabbled in golf writing.

The two men struggled through the regulation eighteen holes all square, then moved on to the nineteenth. Up first, Hutchinson pushed his first drive out of bounds, teed another ball, and drove it out of bounds as well. His third shot remained on the golf course. Now Darwin had only to put his drive into play and the match was his. Instead he drove his first ball out of bounds, teed another, drove it out of bounds, teed another and drove it out of bounds as well. He conceded the match.

Johnny McDermott

Johnny McDermott, a quiet, mannerly young man who neither smoked nor drank, became the first great home-bred American professional. Growing up in New Jersey, he took jobs as the club professional first at the Merchantville Field Club, then switched to the Atlantic City Country Club.

McDermott drove himself to improve. He began practice at dawn, often before 5:00 a.m. in the summertime, and continued until eight o'clock, when he opened the shop. After closing late in the afternoon, he played until dark, then practiced putting by lamplight. His mashie, the equivalent of a modern 5-iron, became the stuff of legend. He practiced by hitting shots to a large tarpaulin spread out perhaps 150 yards away. As he improved he gradually reduced the tarp's size, until eventually he played to spread out newspapers.

As a seventeen-year-old, McDermott entered the 1909 U.S. Open, shot 322, and placed forty-ninth. The following year he shot 298 and tied Alex and Macdonald Smith, immigrants from Carnoustie, Scotland. Alex Smith won the playoff. McDermott entered again in 1911, at the Chicago Golf Club, shot 307, and tied Mike Brady and George Simpson, setting up another playoff.

McDermott usually played a Rawlings Black Circle golf ball. A manufacturer offered a $300 bonus if the winner used a brand called Colonel. McDermott switched, then stood at the first tee and drove two Colonels out of bounds. He persisted, though, shot 80, and won. He was nineteen. He is the youngest player ever to win the Open. Then he won again the following year.

On a tour of the United States in 1913, Harry Vardon and Ted Ray entered a tournament at Shawnee-on-Delaware, a mountain resort on the upper Delaware River, in Pennsylvania. The tournament attracted about the same field as the Open, coming up a few weeks later. Playing error-proof golf, McDermott shot 293, eight strokes better than Alex Smith, the runner-up, and thirteen strokes better than Vardon. Ray finished another stroke behind.

As the gallery called for a victory speech, McDermott was boosted onto a chair at the presentation ceremony. With very little grace he said, "We hope our foreign visitors had a good time but we don't think they did, and we are sure they won't win the National Open."

McDermott's speech stunned the crowd. The Englishmen's faces flushed but they said nothing. American players responded angrily. They felt the remarks were particularly ungracious coming from McDermott since the British had received him so cordially on his two visits to the British Open. McDermott apologized to Vardon and Ray. Both men accepted, realizing that Johnny was still young and flushed with his victory. Others weren't so understanding. Johnny received a letter from a USGA official who wrote that because of McDermott's "extreme discourtesy" his entry for the 1913 Open might be rejected. It wasn't, and he missed tieing for first place by four strokes.

McDermott's career ended quickly. He entered the 1914 British Open but didn't even tee off. On the first day of qualifying he missed a ferry that would have connected with a train to Prestwick, the championship site. The round was already under way when he arrived, but understanding officials offered to allow him to play when he explained why he arrived late. He refused, saying it wouldn't be fair to the other players. Dejected, he booked passage home on the *Kaiser Wilhelm II,* the fastest ship on the cross-Atlantic run. Its foghorn blaring, the liner crept to sea through a thick fog covering the English Channel. Suddenly out of the mist the grain carrier *Incemore* appeared dead ahead. The ships rammed; the *Kaiser Wilhelm's* hull was ripped open below the waterline. The ship was doomed.

In the barber shop when the ships crashed together, McDermott was led to a lifeboat. Picked up a few hours later, he returned to England apparently unharmed. The experience had affected him more than anyone knew, though. He had lost money in the stock market a year earlier and the incident at Shawnee had preyed on him. Finally, the shipwreck added to his distress. He entered the 1914 U.S. Open but his spirit had been shattered; he was never in a position to win.

Later that season he blacked out as he entered his shop at Atlantic City. Only twenty-three, his career was finished. He was taken to his parents' home in Philadelphia and spent the rest of his life in and out of rest homes taking an endless series of treatments. He never played in another golf tournament.

Hilton at Apawamis

Harold Horace Hilton, a Liverpudlian who always played with white sneakers on his feet and a cigarette dangling from his lips—some said he smoked fifty on the days he played golf—ranks among the greatest amateur golfers Britain ever produced. He won three British Amateurs from 1900 through 1911 after stunning the world by winning two British Opens. He came to the United States in 1911 to play in the U.S. Amateur.

Racing through the early rounds (he beat Robert C. Watson, the secretary of the USGA, by 11 and 10, and C. W. Inslee by 8 and 6, for example), he met Fred Herreshoff, the nephew of the yacht designer, in the final. Six down after three holes of the afternoon round, Herreshoff fought back and sent the match to the thirty-seventh, a testing par 4 of 377 yards with the green set high above the fairway and bordered on the right by a mound that might have offered a convenient means of banking a shot off it, but it was studded with rocks.

Outdriven by the powerful Herreshoff, Hilton played a spoon for his second shot, but he pushed it far too much. It streaked toward the rocky hill, looking as if this one loose shot had cost him the championship. Instead, the ball rocketed into the bank, hit a soft spot, and dribbled down the hill onto the green. Shaken, Herreshoff half topped his approach and followed with a timid recovery twenty feet short of the hole. Hilton became the first non American to win the U.S. Amateur and the first player to win both the U.S. Amateur and British Amateur in the same year.

Francis Ouimet

Francis Ouimet became the first abiding hero of American golf. When he won the 1913 Open as a young man of twenty, he changed the perception of the game in the United States by showing that the best British golfers could be beaten. In a tension-filled playoff, Ouimet not only won the championship, he defeated Harry Vardon and Ted Ray, two of Britain's leading golfers. Vardon by then had won five British Opens, and Ray had won the 1912 British championship.

Ouimet grew up in Brookline, Massachusetts, across Clyde Street from The Country Club, the site of the 1913 Open. As a young boy he found he could save time on his way to school by cutting through the club's grounds. Occasionally he'd pick up a lost golf ball. By the time he was seven he'd built up a collection of Silvertowns, Ocobos, and Henleys, and a few of the new Vardon Flyers, but they weren't much use without a club.

About two years after Francis began his collection, someone gave a club to his older brother, Wilfred, who had become a caddie. When Wilfred was away, Francis would knock balls around the back yard. Eventually he began to caddie at The Country Club. One day he caddied for Samuel Carr, an enthusiastic golfer who usually acted kindly toward the boys who carried his bag. Playing the eighteenth hole, Carr asked Francis if he played golf. Francis told him he did. Then Carr asked if Francis had any clubs. Francis said he had two, a brassie and a mashie. Carr smiled and said, "When we finish I wish you would come to the locker room with me. I may have a few clubs for you. After putting Carr's clubs away, Francis rushed back. Carr gave him a driver with a leather face, a lofter (8-iron), mid-iron (2-iron),and a putter.

Early morning, about four-thirty or five o'clock, Francis would sneak onto The Country Club's course and play a few holes until a greenkeeper shooed him away. Rainy days, when he was sure no one else would be playing, he'd sneak on again. Word of Francis's antics reached home, of course. His mother warned him to stay off The Country Club's grounds and ended by saying golf was bound to get him into trouble.

One summer he talked Frank Mahan, a friend, into going to the Franklin Park public course with him. To get to Franklin Park he had to walk a mile and a half with his clubs slung over his shoulder, to the end of the street-car line, ride into Brookline, transfer to another line, then change once more. Leaving the last street car he would walk nearly a mile to the course, play six full rounds of the nine-hole course—fifty-four holes in all—then go back the way he had come and arrive home exhausted. He was thirteen.

Francis won the 1909 Boston Interscholastic championship at sixteen and entered the 1910 U.S. Amateur. He failed to qualify, failed again in 1911 and 1912, but he qualified in 1913 and played Jerry Travers in the second round at Garden City. Travers by then had won the Amateur three times, including the 1912 champioship. Ouimet gave him all he could handle.

One hole down after the morning eighteen, Francis fought back and had moved a hole ahead after the seventh of the afternoon, the twenty-fifth of the match. Outdriven on the eighth, Francis played a wonderful approach that pulled up within eight feet of the cup. It looked like a winning shot, but Travers played an even better shot, a glorious iron ten inches from the hole. Ouimet lost a hole he thought he had won and never recovered. Travers won, 3 and 2, and went on to beat John G. Anderson in the final.

As Francis was leaving Garden City, he was approached by Robert Watson, a member of the Garden City club and president of the USGA. Wanting to assure a big turnout for the Open, he asked Francis to enter. Ouimet, who held a job as a salesman in a Wright and Ditson sporting goods store, had taken vacation time to play in the Amateur and didn't intend to ask for more to play in the Open—even though it was to be played across the street from his house. Nevertheless, he entered. When pairings for the qualifying rounds appeared in the newspaper, John Morrill, the store manager, saw Ouimet's name and said, "I see you're now going to play in the Open championship." Embarrassed, Francis explained how he had been pressured into entering, but, he added he didn't intend to play. He would be grateful, though, if he could take some time off to watch Vardon and Ray. "Well, Morrill said with a gleam in his eye, "as long as you've entered, you'd better plan to play."

Knowing Francis was getting ready for the Open, a friend invited him to a friendly round at the Wellesley Country Club the Sunday before the championship was to begin. On a short and relatively easy course, Ouimet shot two rounds of 88, two strokes higher than his previous worst score at Wellesley. His friend thought he and the others in the group had ruined Ouimet's game. Francis told them they certainly hadn't, and that he'd probably worked a lot of bad golf out of his system.

After the scheduled seventy-two holes, Ouimet, Vardon, and Ray had tied at 304. To tie the two Englishmen, Ouimet holed a breaking fifteen-foot downhill putt on the seventeenth. As Francis stepped up to putt, the road behind the green was jammed with cars, blocked by the crowds straining to catch a glimpse of the golf. Just as Francis was about to stroke his putt, another car came along, and the driver, seeing the road blocked, blew his horn steadily.

The round over, friends hustled Francis into the locker room out of the chaos. One asked, "Were you bothered by that car horn on the seventeenth?" Ouimet explained he was so centered on holing the putt he hadn't heard a sound.

On the morning of the playoff, Ouimet warmed up on the sodden polo field at The Country Club until he was told Vardon and Ray were waiting for him on the first tee. Johnny McDermott, who had won the previous two Opens, took him by the arm and told him, "You're hitting the ball well. Now pay no attention to Vardon or Ray; go out and play your own game."

Little Eddie Lowery, Ouimet's ten-year-old caddie, followed behind, but before they reached the tee, Frank Hoyt, one of Ouimet's friends, drew Francis aside and asked if he agreed it might be better if he carried the clubs rather than the small boy who had struggled through the previous three days of competition. Francis told Hoyt he would have to take it up with Eddie. Lowery didn't want to give up the bag. Hoyt offered money, but still Lowery refused. Hoyt appealed to Francis once again, but when Francis looked at the game little youngster, Eddie's eyes filled with tears. Ouimet didn't have the heart to turn him down. Francis told Hoyt that Eddie Lowery would be his caddie.

The three players drew straws to see who would play first. Ouimet drew the longest and took the honor. As he stepped onto the tee, little Eddie told him, "Be sure you keep your eye on the ball." Francis shot 72 in the playoff, Vardon 77, and Ray 78.

Strained Relations

Nerves on edge after a disappointing first day of the 1913 U.S. Open, and perhaps because of McDermott's challenge at Shawnee-on-Delaware, the British party went to war among themselves that night.

At dinner together in the Copley-Plaza Hotel, in downtown Boston, Ted Ray and Wilfrid Reid argued over the British system of taxation. Soon the argument turned ugly as they directed insults at each other's place of birth. Losing control of his temper, Ray, a very big man, reached across the table and landed two powerful blows on Reid's nose, driving him to the floor. His nose bloodied, Reid, a much smaller man, leaped to his feet and went after Ray. The headwaiter sprang between them and the others in the party pulled them apart.

Even though rain had fallen throughout the week, professionals weren't allowed inside the clubhouse of The Country Club to change their soggy clothes. Instead, they were asked to use the stables. When rain continued overnight and throughout the morning round of the second day, however, the membership showed a measure of compassion. Hoping to ward off pneumonia, influenza, or the plague, each starter in the fourth round was handed a small bottle of Scotch whiskey.

Reid, who lived in the United States, shot 147 over the first two rounds and tied Vardon for the thirty-six-hole lead. Beginning the third round at 7 a.m. he holed a pitch for a birdie 3 at the first, then laid his approach to the second within a foot of the cup for another 3. The overnight rain had flooded some of the low-lying ground. Reid played what he thought was another outstanding approach to the third, but the green lay under water and he couldn't find his ball. Forced to play another, he scored a double-bogey 6 and dropped steadily from contention. Later, as the water level fell, a greenkeeper found Reid's ball lying within eight feet of the hole.

USGA Bars Ouimet

After he returned to Boston from Detroit following the 1915 Amateur, where, as the defending champion, he won just one match, Ouimet went into partnership with Jack Sullivan, a boyhood fried, and opened a sporting-goods store. When he won the Open, Ouimet had been a clerk in a Wright and Ditson store in Boston. Essentially he would perform the same services, except now the store would be his own.

The United States Golf Association, which rules on such matters, took a grim view. Calling on a draconian interpretation of a regulation that affected everyone who either directly or indirectly received compensation for services rendered in connection with the game, the USGA stripped Ouimet of his amateur status, ruling that by selling golf equipment in a sporting goods store he forfeited his amateur status.

Nothing had changed except the interpretation of the regulation. Repeated appeals, one by the Woodland Country Club, where Francis was a member, were rejected. Defying the USGA, the Western Golf Association invited Ouimet to the Western Amateur, but still the USGA refused reinstatement. Through it all, Francis maintained his dignity. When the First World War began, he played exhibition matches for the Red Cross, and then reported to Camp Devens, in Massachusetts, to begin his army career.

Once Francis had gone into the military service, the USGA quietly reinstated him, and he went on to a long and glorious career in USGA affairs. He served on the executive committee, rising to vice president during 1946 and 1947, served as captain of six Walker Cup teams, and in 1955 he was given the first Bob Jones Award for distinguished sportsmanship in golf, the USGA's highest honor.

Gold Vase

Sponsored by the magazine *Golf Illustrated*, perhaps the oldest golf magazine in continuous publication, the *Golf Illustrated* Gold Vase, a thirty-six-hole stroke-play competition, began in England in 1909.

The year after Francis Quimet won the 1913 U.S. Open championship, he went abroad hoping to win the British Amateur. The Gold Vase was considered a warm-up, so Ouimet entered. It was left to Harold Hilton, the editor of the magazine, to assign him a place in the pairings. Hilton confessed he had a thousand willing volunteers, but figuring rank had its privileges, he paired himself with Ouimet. Suitably inspired, Hilton shot 151 at Mid-Surrey and won the Vase.

Douglas Fairbanks, an accomplished golfer himself, played in the 1930 Gold Vase. At lunch after the first round, he was asked how he had played. "Like a motorboat," he answered. "Putt putt putt putt."

Hagen Breaks Through

Walter Hagen had made his debut in the U.S. Open by finishing fourth in 1913 behind Ouimet, Vardon, and Ray. Since he hadn't won, Walter felt he wasn't suited to golf and decided to pursue baseball. He had pitched semi-pro ball in Rochester, New York, his home town, and while he vacationed in Tarpon Springs, Florida, in the spring of 1914, Pat Moran, the manager of the Philadelphia Nationals, allowed him to pitch a few practice games. He left Florida for another season as professional at the Rochester Country Club with the distinct impression that Moran would give him a thorough trial the following spring.

Since he intended to give up golf, he told club members he wouldn't enter the 1914 Open, scheduled for the Midlothian Club, near Chicago. Hagen's decision not to play disappointed at least one member. Ernest Willard, the editor of Rochester's *Democrat and Chronicle*, thought Walter had done so surprisingly well in his first attempt, he offered to pay his expenses. Walter smiled and accepted the offer.

Even though, at twenty-two, Hagen was barely old enough to vote, he liked the taste of first class, and since someone else was paying the bills, he felt he could afford it. On the evening before the first round he searched out the best restaurant in Chicago and ordered his first lobster dinner. It didn't agree with him. He awoke in the middle of the night with stomach pains, and by dawn he could hardly stand. The hotel's physician gave him some pills, recommended he eat only milk toast, and told him to take two aspirins before he played, if indeed he felt well enough to start.

Riding the South Shore Railroad to Midlothian on a sweltering August day while the steam engine belched smoke and cinders through the open windows, Hagen suffered miserably. Nevertheless, he insisted on playing, even though every swing caused him pain. Following loose shots with remarkable recoveries, Hagen went out in 35, spurted home with birdies on four of the last five holes, and shot 68, breaking the course record. He went on to win the championship, and never again talked of playing baseball.

Bob Gardner

An accomplished pole vaulter as well as a first-class golfer, Bob Gardner won both the 1909 and 1915 U.S. Amateur championships. Driving from Detroit to his home in Hinsdale, Illinois, after taking the 1915 Amateur, he was tooling along at a fast clip with the silver Havemeyer Trophy gleaming on the seat beside him when a motorcycle cop pulled him to the side of the road. When the policeman approached ready to write a traffic citation, Gardner pleaded, "You shouldn't give me a ticket; I just won the National Amateur." To prove it, he held up the trophy. Dumbfounded, the policeman tucked away his book and drove off.

In his later years, Gardner played much of his golf with his family. Once in an alternate shot tournament at Onwentsia Club, near Chicago, where he had held a membership for many years, he left his wife facing a shot across a ravine. Not sure what to do, Mrs. Gardner asked Bob, "What club should I use." Knowing he had a better chance to clear the ravine than his wife, Gardner told her, "Whiff it." Indignant that he had so little confidence in her, Mrs. Gardner walked off the course.

Andra Kirkaldy

A rugged, rough-hewn, occasionally coarse and crude professional, Andrew Kirkaldy had served in the Black Watch Regiment during Britain's 1880s Egyptian War. After a period in England, he returned to his native St. Andrews, where eventually he became honorary professional to the R. and A. A man of powerful physique, he was a beautiful player to watch. Kirkaldy reached the peak of his career in 1889, tying for the British Open championship but losing the playoff to Willie Park, Jr. He excelled in money matches.

Paired in a foursomes match with Ben Sayers, from North Berwick, he faced a nearly impossible shot when Sayers drove their ball onto the railway tracks that ran alongside the Old Course at St. Andrews then. After inspecting a very bad lie, Kirkaldy called, "Lend me your wee mashie, Ben." Sizing up the situation, Sayers shouted back, "Aw, na, Andra. Break your ain clubs."

Chick Evans

Chick Evans may have been the best ball-striker of the early years of golf in the United States, perhaps until Bobby Jones's time. Jerry Travers recalled watching Chick hit drive after drive down the middle of a practice range, then play twenty-five consecutive iron shots within six feet of his caddie downrange.

Trying to catch Walter Hagen in the 1914 Open, Chick came to the last hole needing an eagle 2 to tie. Since the eighteenth hole measured only

277 yards, a 2 was not impossible. Hearing that Evans had a chance to catch him, Hagen rushed to the home green and joined the crowd standing four and five deep, cheering Evans on. Chick hit a powerful drive that ran almost to the green's edge and left him a good chance to hole a chip for the eagle. The crowd grew so quiet they could hear the *click* as Evans stroked the ball. He had gauged the shot just right, but pulled it slightly. It stopped about a foot to the left of the cup.

A superb golfer who played wonderful irons and drove with the best of them, Evans dominated Western Golf Association tournaments, but he'd been a persistent and inconsolable flop in national competition. He'd been a semi-finalist in four U.S. Amateurs, but he'd reached the final match only once, in 1912. Jerry Travers thrashed him by 7 and 6. He was twenty-six in 1916; time was slipping away.

The 1916 Open was played at Minikahda, in Minneapolis. Evans entered, shot 32 on the first nine but slipped to 38 coming home, then shot 69 in the afternoon. With 139, he led at the halfway point. A 74 in the third round kept him in front, but Jock Hutchison had moved into serious contention with 72 in the morning, and then shot a wonderful 68 in the afternoon. His 288 broke the Open's seventy-two-hole record by two strokes.

Evans was on the thirteenth hole when he heard of Hutchison's finish. Chick felt he needed a birdie here, but at 515 yards the thirteenth was Minikahda's longest hole, and it played even longer because of a creek cutting across the fairway about 100 yards short of the green. Very few players tried to clear the creek with their second shot. Feeling he could take a chance, Evans lashed into the shot with his brassie. His ball cleared the creek and ran onto the green. Two putts and he had his birdie. Chick shot 73 and beat Hutchison by two strokes, lowering the Open record to 286.

It became one of the Open's enduring records, lasting twenty years. Its life must be measured by more than time, though. Shot with seven wooden-shafted clubs, it lasted through an era when golfers regularly carried three times as many, through the early years of the steel-shafted clubs that improved shot-making enormously, and into the time of the standard ball. Evans set his record in a period when golf balls were of no standard size or weight. No other scoring record survived such radical changes in equipment.

Later in 1916, Evans went on to the Merion Cricket Club near Philadelphia for the Amateur. He qualified easily, then raced through the early rounds, beating one W. P. Smith by 10 and 9 and John G. Anderson, quite a good player, by 9 and 8.

Chick met Bob Gardner in the final. Gardner had won the 1915 Amateur; the meeting with Evans was, therefore, the first time, and the only time, the Open and Amateur champions had met in the Amateur final.

Evans moved away to a three-hole lead at the end of the morning round, but Gardner fought back in the afternoon. Only one down after the first nine, Gardner looked as if he might pull even when Evans bunkered his approach to the tenth and played an indifferent recovery perhaps seventy or eighty feet from the hole. So uncertain on the greens he often carried four putters, Evans rolled the putt home, halved the tenth, won the twelfth and thirteenth, and closed out the match on the fifteenth, 4 and 3.

It had been an eventful year for Evans. Aside from setting a lasting Open scoring record, he had beaten the defending champion for the Amateur championship and had become the first man to hold both the Open and Amateur championships of the same year. Only Bobby Jones has matched him, in 1930, the year of the Grand Slam.

Playing a friendly round at the Exmoor Country Club, in Lake Forest, Illinois, in about 1920, Chick drove his ball straight down the middle of the eighth fairway about forty yards short of a great tree that stood directly in his path to the green. When he reached his ball, Chick saw he was stymied. Nonplussed, he drew an iron club from his bag and played a wide slice around the tree. The shot landed on the green. Seeing his friends' eyes bulge, Chick smiled, dropped another ball, and hooked it around the tree and onto the green. Then he dropped another ball and hit it over the tree and onto the green, dropped one more ball and drilled a low-flying shot that flew under the spreading branches and onto the green.

Chick played his last game, a nine-hole round, in 1978 with Ken and Harle Montgomery, two close friends from Glen View, Illinois. He was eighty-eight at the time. His second shot to the ninth, his final hole, pulled up about twenty-five yards short of the green. Harle called to him, "Chick, let's see one of your old pitch shots." Evans smiled, drew out his pitching wedge, waggled it briefly, and lofted a lovely shot that bounced twice and rolled into the cup.

Part Four

From Bobby Jones to the Second World War

In golf, the center of power shifted remarkably after the First World War. Where earlier the great players had all been British, they were American when the war ended. A generation of British youth had been devastated by the war's slaughter. Compared with British and European losses, American casualties had been light, and perhaps because the war had been fought so far away, young Americans had been free to develop. Whatever the reason, Americans now ruled athletics.

When the First World War began, the British Open was the only golf competition of lasting significance. As Americans began to take charge, the United States Open rose to pre-eminence. Power shifted because Americans began to win the British Open regularly.

Although he was a naturalized citizen, Jock Hutchison was the first American to win the British Open, beating the amateur Roger Wethered in 1921. The following year Walter Hagen won at Sandwich. Unlike Hutchison, Hagen was a native son. From then until 1934, when Henry Cotton played such remarkable golf at Sandwich, only George Duncan had been able to break the American siege. And in 1920 Ted Ray became the last foreigner to win the U.S. Open for forty-five years.

The 1920s became the Golden Age of Sport, a decade of enormously popular heroes, like Babe Ruth, Jack Dempsey, Bill Tilden, Red Grange, Knute Rockne. In golf we produced a series of equally compelling heroes—Walter Hagen and Gene Sarazen were two of them. One man—a boy, really—reigned above them all. He was Bobby Jones.

Bobby Jones

Jones built an astonishing record. Within a period of eight years, beginning in 1923 and ending in 1930, he won four U.S. Opens, three British Opens, five U.S. Amateurs, and one British Amateur—thirteen national championships. He accomplished all this even though he played less formal golf than any of the great players. He kept his game in tune with weekend golf, mostly friendly rounds with his father and some friends. At his peak he played only an occasional tournament.

Sometimes he entered only two or three tournaments during a season. In 1923 and 1924, for example, he entered only the U.S. Open and the Amateur. In 1924 he also played in the Walker Cup, and in 1925 the West Coast of Florida Open. Pointing toward the Grand Slam in 1930, he played in seven; in 1922 he played in eight.

From 1923, when he won his first national championship, until he gave it all up, he won 62 percent of the national championships he entered. Of his last twelve Opens, nine of them in the United States and the other three in Britain, he won seven and placed worse than second in only one. In six of his last nine U.S. Opens, he won four and lost playoffs in two others. Twice he placed second without a playoff, and once he placed eleventh.

He was a scoring machine, setting records in both the U.S. Open and the British Open. Leading up to the 1928 Walker Cup Match, at the Chicago Golf Club, he shot twelve rounds in 69, 71, 69, 68, 68, 68, 67, 68, 67, 70, 69, and 67. He finished one round in 3–3–3–3–3–3–3–4–3–4–4, eleven holes in 36 strokes.

While he seemed more comfortable in stroke play, he was at the same time a superb match player. No one ever beat him twice in a match-play championship.

His image has come down to us as the model for all our heroes—modest, charming, and unassuming, gracious in defeat, even more gracious in victory, without a trace of the arrogance, false pride, and the overbearing bravado associated with so many athletes of later times. When the time comes to rank the greatest of all golfers, we must start with him.

Bobby Jones first appeared in a national magazine at the age of nine. He had won the junior championship of the Atlanta Athletic Club, beating Howard Thorne, 5 and 4, over thirty-six holes. Thorne, bigger and stronger than Bobby, at sixteen was also seven years older.

Bobby entered his first national championship in 1916, when he was just fourteen years old. Along with Perry Adair, a young friend from Atlanta, he went to the National Amateur at the Merion Cricket Club, a great course in the Philadelphia suburbs. Bobby qualified easily and met Eben Byers, the 1906 champion, in the first round. They played a strange sort of match. Both struggling to control flaming tempers, they threw so many clubs the players behind said their match looked like a vaudeville juggling act. After a bad shot on the twelfth hole, Byers flung the offending club out of bounds and wouldn't let his caddie go after it.

Even then Bobby showed not only a blistering temper but a fiery competitive spirit as well. Reaching the fifth hole, Byers played a wonderful drive that split the fairway, whereas Bobby pulled his shot into heavy rough. Byers watched Jones thrash around looking for his ball for a time, hit his second shot, and then began walking toward the green. Just then he heard Jones call, "Fore, Mr. Byers." Surprised, Byers called back, "I'm sorry, I thought you had picked up." Never saintly with his language, Bobby snapped, "Picked up, hell. Watch this one."

With that Jones ripped into the grass and flew his ball about three feet from the cup. He birdied, won the hole, and took the match by 3 and 1. Bobby claimed later he had won "because Mr. Byers ran out of clubs first."

Jones lost in the third round to Bob Gardner, the 1909 champion, a man who once held the collegiate pole-vaulting championship. Chick Evans, who had already won the Open, beat Gardner in the final, the last Amateur championship before the United States entered the First World War.

After Jones had been eliminated, someone asked Walter Travis if he thought Bobby might improve to any significant degree. Travis pondered the question and answered, "He can never improve his shots, if that's what you mean, but he'll learn a great deal more about playing them."

With the national championship suspended during the war, J. A. Scott, of the Wright and Ditson Sporting Goods Company, arranged a series of matches featuring Jones and Perry Adair, along with Alexa Stirling and Elaine Rosenthal, all teen-aged golf prodigies, and all except Miss Rosenthal—a Chicagoan—from Atlanta. Together they toured through Massachusetts, Maine, New Hampshire, and Connecticut. Later Bobby and Perry teamed in matches against prominent amateurs like Evans and Gardner, and Jones was invited to play against professionals for war relief.

In one memorable match Bobby lost the first three holes to Cyril Walker, an English immigrant. Furious at his loose play, Jones said he fell into a humor described by Robert Louis Stevenson in *The Master of Ballantrae* as "a contained and glowing fury," then played the next six holes in 4–3–4–4–4–2, won five of them, went out in 32, and won the match by one hole.

The Red Cross and War Relief funds realized more than $150,000 from Jones's matches. He became famous.

When national championships were revived in 1919, Jones reached the Amateur final, where he played twenty-year-old Davey Herron at the Oakmont Country Club, near Pittsburgh, where Herron was a member. They finished the first eighteen holes all square, but Herron pulled ahead in the afternoon and stood 3-up with seven holes to play as they approached the twelfth, a long par 5 of 600 yards. Herron drove into a bunker and played an indifferent recovery. With Bobby in the fairway, he stood a good chance of winning one hole back and perhaps turning the match around. Now fate played its role.

Jones planned to play a full brassie for his second shot. As he reached the top of his backswing, a marshal saw the gallery begin to move, raised his megaphone and shouted, "Fore!" The sudden noise jolted Bobby. He flinched coming into the ball, knocked it into a bunker, failed to get out with his recovery, and picked up, conceding the hole to Herron. Instead of just 2 down with six to play, he stood 4 down. Herron closed out the match two holes later, 5 and 4, but he had been forced to play sensational golf to do it. He played the thirty-two holes in four under level 4s.

Jones played in his first U.S. Open in 1920 at the Inverness Club, in Toledo, where, in one of the two or three greatest pairings of golfing talent ever, fifty-year-old Harry Vardon was grouped with eighteen-year-old Bobby in the qualifying rounds. Vardon had been a hero to Jones ever since, as an eleven-year-old, Bobby had watched him play an exhibition match at East Lake, in Atlanta.

Both men opened the thirty-six-hole qualifier with 76s, and Bobby was playing a little better than that in the afternoon. The seventh hole, which then bent sharply from right to left, called for a drive across a deep chasm and a carry over a line of tall trees. Vardon and Jones easily cleared the trouble and lay safely just short of the green, leaving each of them nothing more than a simple pitch or run-up shot to the flagstick. Away, Vardon played a run-up within birdie range of the pin. Like most Americans even then, Jones preferred the pitch. He bladed the ball. It took off like a rocket, scuttled over the green, and dived into a bunker. So humiliated he wanted to crawl into a hole, Bobby turned red. He finished out the hole with a bogey 5, then walked onto the tee of the eighth, a par 3.

So far Vardon hadn't said a word throughout the round, so Bobby, trying to ease his own embarrassment and at the same time encourage Harry to talk, said, "Mr. Vardon, have you ever seen a worse shot?" As economical with words as with his golf shots, Vardon said simply, "No." The discussion was closed.

With the war over, Americans suddenly felt the fever for international competition. In the spring of 1921, a group was organized to make an assault on the British Amateur, and so long as they were there, but just to play an informal match with a group of British golfers. Recognized by now as a player of remarkable talent, Jones was selected for the American team. The Americans won and then went on to the British Amateur. Jones won three matches, one against a man who shot 87 and still took the match to the eighteenth green. Despite his wretched golf, E. A. Hamlet, his opponent, held Bobby one down with two holes to play and then missed from four feet on the seventeenth. Jones won the hole to even the match, and won on the eighteenth under strange circumstances. Hamlet's first putt had laid Jones a partial stymie, but although Bobby's putt hit Hamlet's ball, his own ball toppled into the cup.

Allan Graham, a Hoylake member, thrashed Bobby soundly the next day, beating him 6 and 5, and then lost to Willie Hunter in the final.

Jones stayed in Britain for the British Open and his introduction to St. Andrews, where he committed a sin that embarrassed him throughout his life. After two reasonably good rounds, he shot 46 on the first nine of the third, made 6 on the tenth, a par 4, and then, seething because he had left a putt short and was about to make another 6 on the eleventh, a par 3, he said to himself, "What's the use?" and picked up his ball.

Rumors spread in later years that he had torn up his scorecard and stormed off the course. He did not. It is true that he quit, but he continued playing and indeed went out in the afternoon and shot a respectable although unofficial 72.

Jones played dashing, all or nothing golf. Back home again, for the 1921 U.S. Open, at the Columbia Country Club, outside Washington, D.C., he stood nine strokes behind Jim Barnes after three rounds. Hopelessly out of the running with just eighteen holes to play, Bobby suddenly began playing the shots everyone knew he could play. He birdied two of the first four holes and came to the fifth with a little hope in his heart.

The fifth was a par 5, measuring 560 yards, with an intimidating bunker slicing across the fairway well beyond the drive zone. The bunker presented no threat to players of Open caliber—they would certainly clear it with their second shots—but they had to be careful of a fence running close along the left.

After a drive of perhaps 275 yards, Jones felt he could reach the green with a good brassie and cut at least one more stroke and perhaps two from Barnes's lead. Bobby laid into the shot with all he had, but he came over the top and yanked the ball over the fence and out of bounds. Never flinching from a challenge, Bobby dropped another ball and tried once more. Again he pulled it over the fence. He eventually finished the hole in nine strokes, shot 77 for the round, and placed fifteenth.

Two months later he had Willie Hunter, the British Amateur champion, two down in the National Amateur, at the St. Louis Country Club. Trying to win another hole, Jones attempted to cut the corner of a dogleg and drive the green. In a stroke of bad luck, his ball hit the tallest shoot of the tallest tree and dropped straight down into a stony depression covered by underbrush. Hunter beat him, 2 and 1.

When he was told that that gambling shot had cost him the match, Bobby answered, "I can play this game only one way; I must play every shot for all there is in it. I can't play safe."

Jones's temper nearly led to his being barred from further USGA competitions. Hunter held Bobby one hole down at the thirty-fifth and played a nice pitch onto the green. Playing next, Bobby skulled his approach ball over the green, and in a flaring rage flung his niblick. It skipped along the ground, shot into the gallery, and hit a woman spectator's leg. She wasn't hurt, but Jones became so upset he lost the hole and the match. Incensed at Bobby's boorish behavior, George Walker, the USGA president, warned in a letter, "You will never play in a USGA event again unless you learn to control your temper."

Bobby had been playing golf at the highest levels for five years by then. With the 1921 Open finished, O. B. Keeler, who wrote about golf for the Atlanta *Journal*, talked about Jones with Walter Hagen. Hagen told Keeler, "Bobby has everything he needs to win any championship except experience and maybe philosophy. He's still impetuous. I'll tip you off to something. Bobby will win an Open before he wins an Amateur." Two years later he did just that.

By 1923, Jones had played golf at the national level for seven years and had won nothing other than local or sectional tournaments. Although he was only twenty-one, he began to wonder about himself and question whether he had the game or the temperament to win a national championship. He entered the 1923 Open, at Inwood, just outside New York City, on Long Island. Playing his usual brand of marvelous golf, Bobby had the championship in hand until the last four holes of the final round, all par 4s. Jones pulled his approach to the sixteenth out of bounds, dropped another ball, and pulled it again. The second ball had better luck; it hit a mound rising on the left of the green, kicked off, and rolled within seven feet of the cup. He holed it and saved a bogey 5. Then he missed the green of the seventeenth and bogeyed again. Now for the eighteenth, where a par 4 would win for him.

Inwood's eighteenth, still another par 4, measured 425 yards with the fairway running level for the first 200 yards, then gradually easing down a gentle incline toward a broad pond Inwood members called a lagoon. The green sat on the far side. Feeling he had the Open won, Jones eased up. He hit a short drive, then went for the green with a spoon. Once again he pulled his shot. His ball cleared the pond but missed the green to the left. It sat down behind a pot bunker and under a chain strung around the twelfth tee.

While officials removed the chain, Bobby sat on the ground and brooded over his last few holes of rotten golf. The chain taken down, Bobby took a little too much turf, dumped his third shot into the bunker and finished with a grim 6. He could have won by shooting 74; instead he shot 76 and 296. Still, only Bobby Cruickshank had a chance to catch him.

As Jones, grim-faced, walked off the green, Keeler approached him and said, "Bob, I think you're champion. Cruickshank will never catch you." Refusing to be consoled, Jones said what he felt in his heart longer than he cared to admit: "I didn't finish like a champion. I finished like a yellow dog."

In a stunning burst of spectacular golf, Cruickshank played the sixth through the twelfth in 2–3–3–4–3–4–3, one over level 3s, lost four strokes to par over the next five holes, and came to the eighteenth needing a birdie 3 to tie Jones. After drilling a drive down the middle, Cruickshank ripped a mid-iron straight at the flagstick. It flew over the lagoon, dug into the green, and pulled up within six feet of the cup. Cruickshank holed the putt, birdied, and forced a playoff.

Jones and Cruickshank struggled through seventeen holes of the playoff, neither man yielding to the other. They came to the eighteenth even; once more the championship would hinge on the final hole. Cruickshank cracked. He pulled his drive behind a tree and couldn't go for the green, but Jones pushed his drive onto bare ground in the right rough about 200 yards from the green.

After Cruickshank laid up short of the water, Jones played a landmark shot, one that has become part of the golf legend. From this dangerous lie he hit a crisp 2-iron that rose from the scratchy ground, bored through the dank overcast, cleared the pond, and braked about six feet from the cup. With the championship in hand now, Jones coaxed his first putt close, tapped in his second, and won. A champion at last, Jones relished the moment and told himself, "I don't care what happens now."

About a year later Jones received a letter from J. S. Worthington, a friend who was going home to England to die. Worthington asked if Bobby could send him one of his discarded clubs as a keepsake. Jones felt he had only one he treasured enough to send to his old friend. He sent him the 2-iron he had used at Inwood.

Sixty-three years later, in 1986, the members of Inwood erected a plaque on the spot from where Jones had played his immortal shot.

Jones had qualified easily for the 1925 Open, but he played some ruinous golf at the Worcester Country Club, in Worcester, Massachusetts, in the first round. Paired with Walter Hagen and starting at 10 o'clock, the best possible time, he drove superbly, putted reasonably well, and played excellent pitches. His irons caused him continuous trouble, though.

Worcester's eleventh hole measured just 400 yards, a drive and a medium iron for players of that day. Bobby drove well enough, but then hit a sour iron that nestled in the rough near the left side of the green. He stood a few strokes over par then, but he didn't seem to be in serious danger of missing the thirty-six-hole cut.

Stepping up to his ball to play a little lob shot, Bobby settled his feet in place and set his clubhead behind his ball. Suddenly the ball moved. Bobby stepped away, set himself again, and played his pitch. No one had seen the ball move, not Hagen, neither of the caddies, and no official on the spot, but before going any further he called a penalty on himself. With his two putts he scored a double-bogey 6.

Feeling he had been too harsh on himself, others argued against the penalty saying they did not believe Jones had caused his ball to move. Bobby insisted, though, and asked for a meeting with the USGA. There he explained that he had grounded his club and therefore must be penalized.

After listening to those against the penalty, William C. Fownes, a USGA vice president, said, "Well, Bobby, it seems a matter for you to decide.

Do you think you caused that ball to move?" Jones answered, "I *know* I did." The penalty stuck, and Bobby was given a score of 77, putting him in peril of missing the cut. A 70 the following day saved him, and he went on to tie Willie Macfarlane for the championship. Over a thirty-six-hole playoff, Macfarlane beat him by one stroke.

Later, when O. B. Keeler mentioned the incident in an interview for the Associated Press, Jones was outraged. He saw nothing unusually commendable in what he had done. Confronting Keeler, Jones told him, "You might as well praise me for not robbing banks."

Every match of the 1925 U.S. Amateur, at Oakmont, was played over thirty-six holes. Winning each of his matches by enormous margins, Jones worked his way to the most unusual final match the amateur had ever seen. He met twenty-year-old Watts Gunn, like Jones a member of East Lake, in the only final ever between members of the same club.

When they played at home, Jones had always given Gunn strokes. Striding up to Bobby as they stood on the tee waiting for the signal to start, Gunn grinned and asked, "Are you giving me my usual three strokes."

Jones said no, "But I am going to give you hell today."

Gunn went off to a fast start, but Bobby rallied, played the thirteenth through the eighteenth in 3–3–4–3–3–4, two under par, and beat Gunn 8 and 7.

Since Walter Hagen had won the 1925 PGA Championship and Jones the 1925 Amateur, a showdown match over seventy-two holes was arranged at two courses in Florida, at Sarasota, where Jones was spending the winter, and at Pasadena, Hagen's base. Jones worked as a salesman for the Adair Real Estate Company at the time; since the company had vast holdings around Sarasota, Bobby felt obligated to play the match because of his job. Playing first-class golf through the winter, he had teamed with Tommy Armour and won every one of a series of seven four-ball matches against the game's top professionals.

The Hagen match had a different ending. Jones's irons suddenly turned sour, but that couldn't excuse what happened. Playing first-class golf, Hagen was never over par in any of the four rounds and stood three under on the sixty-first green. Both men ranked among the game's greatest clutch putters. Well behind by then, Jones rolled in a putt from at least 45 feet, hoping to carry the match a little further. Hagen responded by running in a putt from only a foot or two inside Bobby's. The match ended there; Hagen won, 12 and 11.

Chosen for the 1926 Walker Cup team, Bobby had planned to stay for the British Amateur and then sail for home immediately afterward. The British Amateur changed those plans. He won five matches but lost in the sixth round, the quarter-finals, to young Andrew Jamieson, a British Walker Cupper. Not wanting to go home on a losing note, and again trying to atone for picking up his ball in 1921, Bobby canceled his transatlantic passage and entered the British Open. It must have been fate.

Qualifying for the Open, Jones played two remarkable rounds at Sunningdale, a wonderful golf course in the suburbs of London, and qualified easily, shooting 66-68—134. His first round was a model of symmetry; he used 33 putts and 33 shots with woods and irons. Not only had he gone out in 33 and come back in 33, his scorecard showed nothing but 3s and 4s; not a 2 or a 5. He had missed only the thirteenth green, but saved his par with a snappy bunker shot, and he holed only one long putt, running one in from twenty-five feet on the fifth.

As it is with most rounds of this sort, it could have been better. He missed putts from five feet on the third, a short par 4, and another from ten feet on the seventh, a par 4 of more than 400 yards. At the ninth, a par 4 of only 270 yards, Bobby overdrove the green, chipped back to five feet, and missed another birdie opening.

His 134 led qualifying by seven strokes. Reporting on the day in *The Times*, Bernard Darwin described Bobby's round as "incredible and indecent." His golf that day impressed the Sunningdale membership so strongly a stroke-by-stroke account of it hangs on the clubhouse wall.

With a score of 215, Al Watrous, a professional from Detroit, led Jones by two strokes after the third round of the Open proper, at Royal Lytham and St. Annes. Since they were paired, Bobby suggested that instead of lunching where they would face a horde of well-wishers, they go to his hotel for a quiet lunch and a short nap before the final round later in the afternoon. Watrous agreed. They slipped away to the Majestic where Bobby shared a room with O. B. Keeler. They took off their shoes, Bobby stretched out on his own bed and Watrous on Keeler's, and talked until tea, toast, and cold ham arrived. Keyed up, Watrous managed to eat a slice of ham.

When they returned to the players' entrance, the guard turned Jones away. He had put his player's badge on the dresser at his hotel and forgotten to bring it with him. Neither Jones's arguments nor Watrous's confirmation that Bobby was indeed a player had any effect. Finally Jones told Watrous to go ahead, then trotted off to another entrance, paid the admission fee of seven shillings, and walked in.

Jones struggled through the last round, never once using fewer than two putts on a green and not playing well enough to make up ground on Watrous. Al still led by two strokes with five holes to play. Then Jones came alive and caught Watrous with a par on the 445-yard fourteenth and a birdie 3 on the 468-yard fifteenth, one of the strongest holes in British championship golf.

They were still even at the seventeenth, a par-4 hole of 413 yards that bends slightly left. It became the critical hole of the championship. The drive should be played to the right to open the green to the approach and to avoid barren wasteland dotted with bunkers running along the left. Watrous strung a long straight drive to just the right spot, but Jones pulled his drive into the sandy mess on the left next to the dunes. While the dunes were rather low, they still blocked Bobby's view of the green. When Watrous drilled a long and straight iron onto the green, Bobby's troubles multiplied.

Now Jones was in trouble. To avoid losing a stroke he would have to play a shot off bare, dry sand, carry the ball 175 yards to a green he couldn't see, and have it stop quickly once it hit the green. After walking nearly to the other side of the fairway for a good look, Jones ripped into a 4-iron and caught the ball perfectly. It rose quickly over the dunes, flew dead on line, and pulled up closer to the hole than Watrous's ball.

Shaken, Watrous three-putted, Jones got down in two and won the championship. He is the only American to have won a British Open at Royal Lytham.

No one had ever won both the British Open and United States Open in the same year. Home from winning at Royal Lytham and St. Annes, Jones took the train to Columbus, Ohio, for the U.S. Open, at the Scioto Country Club. He played indifferent golf through the first three rounds and well into the fourth, but in spite of his loose shots he hung on. With nine holes to play he stood four strokes behind Joe Turnesa.

Then Jones turned his game around just at the time Turnesa began throwing strokes away. Turnesa rallied at the last hole and birdied, but Jones still had two strokes in hand by then. Bobby lashed into his drive on the eighteenth and hit it so far he had only a 4-iron left. He rifled his ball within twenty feet of the cup, took two putts, and won his second U.S. Open. In a strong and unpredictable wind, he had played the last twelve holes at Scioto in two 3s and ten 4s.

Later that year Bobby played George Von Elm, a blond, teutonic Californian, in the final match of the U.S. Amateur at Baltusrol. The match went thirty-five holes; Jones stood one under par and beaten, 2 and 1.

In one year of superb golf, Jones had reached the sixth round of the British Amateur, won both the British Open and U.S. Open, and lost in the final of the U.S. Amateur. Would it be possible, he thought, to win all four in the same year? His next opportunity would come in 1930 with the next Walker Cup Match in Britain. He'd think about it.

With four holes to play, Jones led Al Espinosa by four strokes and had the 1929 U.S. Open wrapped up. Then he had to struggle for his life. The fifteenth hole at Winged Foot in Mamaroneck, New York, is an uncomplicated par four of little more than 400 yards with no intimidating hazards or obstacles. Jones overclubbed his approach, misjudged a soft lob that failed to clear a small knob, and before he knew it had made a 7. With three strokes of his comfortable cushion gone now, Bobby needed three tough pars to beat Espinosa.

When he took three putts from twenty feet on the sixteenth, all his strokes were gone, and to tie now he must make 4s on two holes measuring 450 and 419 yards. He made his par on the seventeenth with no trouble, but he pulled his approach to the eighteenth into the rough on the bank of a bunker bordering the left side of the green. Bobby popped the ball from the grass with his niblick and crouched watching as the ball ran toward the hole. It pulled up twelve feet short. Now to tie Espinosa he would have to hole a difficult side-hill putt that would break a foot or so from left to right on a slippery green, with an Open at stake.

With the crowd hushed, not quite believing that the great Bobby Jones had thrown away the Open, Bobby studied his line to the hole. Satisfied, he stood up to the ball, soled his putter, then rapped it sharply. The ball started off left of the cup, took the break, and tumbled into the hole. He had made his par and set up a playoff for the next morning.

Throughout his career, Jones continually showed his consideration for others. Since the concluding two rounds of the Open were played on Saturday, the USGA set the playoff for Sunday morning. Seeing the announced time, Jones suggested the association delay the start so that Espinosa might go to church.

His concern for others, though, had no effect on Jones's competitive fire. Church over, Jones ripped around Winged Foot in 72–69—141 against Espinosa's 84–80—164. Bobby had won his third U.S. Open.

The twelve-foot putt Jones holed to tie Espinosa became legendary, especially around Winged Foot. It was the most famous stroke ever played there, and since it had been made by Jones, its legend grew ever larger as the years passed. Eventually the membership decided to do something about it. In 1954 Winged Foot invited Jones back for a twenty-fifth anniversary celebration of the 1929 Open. It was turned into a big day. Tommy Armour, Gene Sarazen, Johnny Farrell, and Craig Wood, the club's professional at the time, played an exhibition match, in which all four players, along with a few others, had a little fun.

Partly crippled at the time by the disease that would kill him, Jones pointed to the spot where the hole had been cut that day in 1929, and a greenkeeper cut a new hole. Then Jones pointed to the spot where his ball had lain. All four players were to try to hole that slippery putt. First Armour missed, then Wood missed, then Farrell, and finally Sarazen missed. All four tried again, and again they all missed. Others tried. Findlay Douglas, the 1898 Amateur champion, who later became president of the USGA, tried and missed. Joe Dey, USGA executive director missed. No one holed the putt.

By the time of the 1929 U.S. Amateur, Jones had already won four Amateurs, three U.S. Opens, and two British Opens, and so it was hardly surprising he was expected to win once again. He hadn't missed playing in a final since 1923.

It was no wonder then that when he arrived at Pebble Beach, California, for the 1929 Amateur a movie company sent a crew to capture the Jones swing on film. They took Jones to the Del Monte Golf Club, just a few miles away, the director positioned his cameraman high in a tree, and told Jones, "Shoot for the camera."

Knowing how good he was and, therefore, a little uncertain about the director's order, Jones asked, "You mean aim right for him?" The director agreed that was exactly what Bobby should do, because he wanted a picture of the ball coming right at the camera. Shaking his head, Jones agreed he would do just that. Drawing back his driver with his lovely legato swing, Jones ripped into his ball, sent it flying directly at the cameraman, hit the camera, and knocked both it and the cameraman out of the tree.

Evidently Jones lost his touch after that. A few days later he lost in the first round of the Amateur to Johnny Goodman, an unknown golfer from Omaha.

The Grand Slam

Bobby Jones hadn't played four tournaments in one year since 1927, and he'd entered only five altogether through 1928 and 1929, but since he planned to go after all four national championships in 1930, he warmed up by playing in four others before the British Amateur, the first of the Grand Slam events. He would play in the Savannah and Southeastern Opens before sailing for Britain. Once there he would play in the *Golf Illustrated* Gold Vase and the Walker Cup match.

In a terrific struggle, Horton Smith, himself not quite twenty-two but the winner of eight professional tournaments in 1929, beat Jones by one stroke at Savannah. It was a classic battle. Bobby shot 67 and set the course record in the first round, Smith lowered it to 66 in the second, then Jones struck back with 65 in the third. They were tied going into the fourth round. Smith shot 71 and Jones 72.

A few weeks later Jones ran away from the field in the Southeastern Open, at the Augusta Country Club, in Augusta, Georgia, shooting 72–72–69–71—282. With three holes to play he stood eighteen strokes ahead of the field, but he grew careless, threw away five strokes on the last two holes, and won by thirteen.

Bobby Cruickshank was counted among the victims of the carnage. He followed Jones for nine holes and claimed he had never seen such a display of golf. When the tournament ended he had a drink in the locker room with O. B. Keeler and Grantland Rice, the syndicated sportswriter who had named the 1924 Notre Dame backfield the Four Horsemen.

A Scot who had come to live in the United States following the First World War, Cruickshank asked Keeler how many tournaments Jones planned to enter in 1930. Keeler said Bobby planned to play only in the

British and United States Opens and Amateurs. "He's going to win all four of them," Cruickshank said, "and I'm going to bet on it."

Cruickshank sent $100 to his father-in-law, in Edinburgh, Scotland, where betting was legal. His father-in-law bet $100 of his own money as well. They won $5000 each, but the *New York Times* mistakenly reported that instead of $10,000 they split $108,000. Cruickshank had a difficult time persuading the Internal Revenue Service the figure was wrong. Before the difficulty ended, Cruickshank's father-in-law had to send an affidavit from the Edinburgh bookmaker naming the correct figure.

Five unidentified Atlantans won $500 each on $10 investments with underwriters under the Lloyds of London umbrella, betting that Jones would win the Grand Slam. Said to be close personal friends of Jones, they bet $50 with Lloyds at odds of 50 to one. When they tried to bet more, and in fact cabled $2000, the underwriter said it could pay no more than $2500 if it lost, and so returned all but the $50 that covered the bet.

British Amateur

Jones very nearly didn't survive the first match of the British Amateur at St. Andrews. He drew Henry Roper, a coal miner from Nottingham, England. Opening with a birdie 3 on the first, Jones parred the second, birdied the third, holed a full 6-iron from a bunker and eagled the fourth, and followed with a birdie on the fifth. He had played the first five holes in five under par, but he had won only three holes.

Jones finally won the match on the sixteenth by 3 and 2. He had played those sixteen in sixty strokes, with six birdies and an eagle, four under level 4s.

Jones's drive for the Grand Slam had a number of turning points, but none, perhaps, more important than what happened on the seventeenth hole during his fourth-round match with Cyril Tolley, the defending champion. With the match all square, Tolley left his second shot alongside the Road Bunker, just beside the green. Jones then selected an iron longer than might have been expected and drilled a low-flying shot directly at the flagstick.

Spectators were allowed to roam freely about the course in those days and crowd close along the edges of the greens. Bobby's ball landed on the green, bounced once, struck a spectator on the chest, then rebounded about ten feet from the hole. Had it not hit the spectator, Jones's ball could have run onto the road, creating a particularly difficult recovery.

Tolley pitched close enough to make his 4, and Jones missed the putt. They halved the eighteenth, sending the match to extra holes. Faced with a dead stymie on the nineteenth, Tolley had to loft his ball into the hole. He very nearly made it; the ball hit the side of the cup and teetered on the rim. Jones won.

Many spectators felt Jones had deliberately hit that approach into the crowd, knowing his ball would not go through. No one could ever be sure if he had, but Tolley always felt the claim might hold grain of truth.

Jones and Tolley drew a gallery estimated at 20,000 fans. The crowd was so large and spread so far that when they were playing the fourteenth hole, one of the two par 5s, Tony Torrance and Alec Hill couldn't play their approaches to the second, a hole and a half away. They had to wait an hour, until Jones and Tolley had reached the eighteenth tee.

Jones met Roger Wethered in the thirty-six-hole final match. Bobby had become such a favorite with the British galleries that hardly a citizen was left in the town. An estimated 15,000 fans crowded onto the golf course.

Before the match began, the starter announced to the crowd that no one had ever played the Old Course without a 5 on his scorecard. Scrambling for his par 4 on the first after a terrible second shot that didn't quite reach the Swilcan Burn, Jones played impeccable golf through the sixteenth. With nothing higher than 4, he stood five holes up on Wethered. Going for the green with his approach to the seventeenth, Bobby pulled his shot into the Road Bunker. Barely hesitating, he flicked a niblick that eased off the edge of the green and coasted within four feet of the cup. Feeling he couldn't miss, he stroked an overconfident putt that slipped past the cup. Wethered won the hole with a 4, and they went to lunch with Jones four holes up.

Temper flaring, Jones paced back and forth in his room at the Grand Hotel, just across the street from the eighteenth green, furious with himself for having missed the putt and the opportunity to be the first to play the Old Course without a 5.

Starting the second round with a bogey on the first, he ended speculation he might play the afternoon without a 5. He made no other mistakes, though, and won the match, 7 and 6. He'd won the first trick of the Grand Slam.

British Open

Jones moved on next to Royal Liverpool, at Hoylake, for the British Open. He led at the halfway point, but he had a sloppy finish in the third round, shot 74, and Archie Compston passed him with an inspired 68. One stroke behind going into the last round, Jones hit a sickening drive on the second. His ball shot off to the right, hit a gallery marshal on the top of his head, and bounced into a bunker guarding the fourteenth green just behind him. It was a terrible shot; coming just after his loose golf of the morning, when he finished with four 5s and a 4, it looked as if Jones might crack.

But good luck had followed him throughout the Amateur, and it was still with him here. His ball had bounded into a clean lie in the sand. From 120 yards out, Bobby played a precise pitch twenty feet from the cup and holed the putt. A birdie 3 where a bogey seemed likely.

Jones wasn't home safely just yet. Even par through seven holes, he went for the green with his second to the long eighth, a par 5. His ball tumbled down a slope and settled fifteen yards from the green. A weak pitch and then a timid chip left him ten feet short of the cup. He three-putted from there, taking a double-bogey 7. Watching Bobby three-putt, Bernard Darwin wrote, "A nice old lady with a croquet mallet could have saved Jones two strokes."

With three holes to play, Jones figured he needed one more birdie. He also thought the sixteenth, a par 5 of 530 yards, offered his best chance. Once again, though, he pulled his second shot slightly, into a greenside bunker. Sizing up the problem, he realized that because his ball lay so close to the back bank, he couldn't play a normal stroke; he would have to lift his club abruptly on the backstroke and play a sharply descending stroke. His niblick, the club he'd normally have played, wouldn't do.

Earlier in the year, after he had beaten Bobby in the Savannah Open, Horton Smith had shown Jones a new club he'd used to play from sand. It was a heavy instrument with a flanged sole designed to ride through the sand rather than dig in, and a concave face that cradled the ball and almost flung it rather than hit it. He gave an identical club to Bobby, who had never used it in a competition. But he used it here. He caught the ball perfectly; it shot up quickly, cleared the bunker's high front wall, then ghosted toward the cup and stopped within two inches of falling for an eagle 3. He made his birdie, parred the next two holes, shot 75, and finished with 291.

Compston, his game collapsing in the final round, shot 82 and fell back, Leo Diegel lost when he bogeyed the sixteenth, and Macdonald Smith came to the eighteenth needing an eagle 2 to tie. After a perfectly placed drive, Mac sent his caddie ahead to attend the flagstick. He played a gorgeous shot that floated toward the pin, hit short, and slipped past the cup. Jones had won the second leg of the Slam.

Speaking to the press shortly after leaving Hoylake, Jones dropped the first hint that this might be his last year. He told the Associated Press, "This is my last shot at the British Open. It's quite too thick for me. I feel I'm not strong enough to play in another."

Tolley, who had won the British Amateur twice by then and had entered the U.S. Open, traveled to Minneapolis with the Jones party. On the train from London to Southampton, where they would board ship, Cyril asked Bobby how long he'd been in Britain.

"Six weeks," Jones answered.

"Have you ever played so badly for so long?"

Jones shook his head slowly and said, "No."

Riding the train from Liverpool to London following Jones's victory in the 1930 British Open, an American businessman shared a compartment with three Englishmen. One of them asked, "Do you know Mr. Jones personally?" The American said he didn't. The Englishman paused and then said, "I say, what a pity it is that Bobby Jones isn't English."

U.S. Open

The U.S. Open was played at the Interlachen Country Club, in Minneapolis, during a stifling heat wave that drove temperatures into the 100s and the humidity almost as high. Jones opened with a sound 71 and needed a par 5 on the ninth for an outgoing 35. At 485 yards, the ninth could be brought within reach of the second shot by playing across a clear blue lake dotted with water lilies. Jones had birdied the hole consistently in practice, but now he pushed his drive into a tight lie near the lake's near bank.

Regardless of the lie, he decided he would go for the green, but as he reached the top of his backswing, two little girls darted from the crowd up ahead. Seeing them from the corner of his eye, Jones flinched coming into the shot and topped the ball. It squirted off in so low a trajectory he knew it wouldn't clear the water; he felt he was heading for a bogey rather than a birdie.

Once again luck—along with the laws of physics—saved him. The ball flew so low it hit the water like a flat stone, skipped once, then again, and climbed up the lake's far shore and rolled within thirty yards of the green. A terrific pitch dead to the hole and Jones had his birdie. He turned for home in 34.

Rumors spread that his ball had bounced off a lily pad, but it had hit nothing more than the water. He shot 39 on the second nine, and with 144 for the first thirty-six holes, tied Harry Cooper and Charles Lacy for first place.

Leading after sixteen holes of the final round, Jones stepped onto the seventeenth tee to play a monstrous par 3 of 262 yards. Throwing every ounce of his power into his swing, Bobby pushed his tee shot badly; the ball cleared a bunker to the right of the green, hit a tree, and was never seen again. It may have fallen into a dried-up water hole covered with tall brown reeds and swamp grass, but no one saw it. Turning to Prescott Bush, the referee, Jones asked, "What shall I do?" Bush answered, "The ball went into the parallel water hazard. You are permitted to drop a ball in the fairway opposite the point where the ball crossed the margin of the hazard." Bobby took his penalty shot, dropped another ball, pitched onto the green, and made 5.

The decision was the most controversial of Jones's career. First, the status of that area as a hazard hadn't been clarified. Furthermore, since no one actually saw the ball fall into the hazard, many people felt the ball should have been declared lost and Jones should have been told to play another ball from the tee. They insisted he should not have been allowed to drop a ball near the hazard with only a one-stroke penalty.

In a newspaper column written under his name, Gene Sarazen stated in part:

> I am sorry that his [Jones's] victory, great as it was, should be tainted by the decision made by the United States Golf Association in Jones's favor on the seventeenth hole. It seems to me that Jones, who in the eyes of the public is a good sport, should have gone back to the tee and made another drive. There is no doubt in the minds of many who witnessed the unfortunate incident that the ball was lost.
>
> The ball in its mad flight from the tee tore through the trees and disappeared. Nobody saw where it dropped. It was presumed that it went into the swampy land that was ruled to be a parallel hazard.
>
> Rules of golf weren't made to be interpreted on presumption. It was just another of those regrettable incidents that have slightly tarnished Bobby's record.

Among other papers, the Sarazen column ran in the Los Angeles *Examiner*, where it was seen by Everett Seaver, an officer with the Western National Bank and a member of the USGA's executive committee, its ruling arm. Incensed over Sarazen's slur, Seaver wrote to Bush, the USGA secretary, that: "While the association will of course not take the matter up officially, I want to go on record as stating that I feel his remarks about Bobby's victory being tainted are so uncalled for and unjustified that I would just as soon see him barred from further competition."

Prescott Bush later became a United States Senator. He had three sons. George, his middle son, became President of the United States. Seaver had a son, Charlie, who was an accomplished golfer. Later in 1930 he reached the semi-finals of the Amateur, where he lost to Gene Homans by one hole. On the strength of his record, he was named to the 1932 Walker Cup team. Charlie had a son who played the game quite well, but did better at baseball. Tom Seaver pitched for the New York Mets, the Cincinnati Reds, and the Chicago White Sox, won three Cy Young Awards, and was elected to the Baseball Hall of Fame.

Jones came to the eighteenth tee only one stroke ahead of Mac Smith once again. Feeling he needed one more birdie, he lashed into his drive and placed it in perfect position, then followed with a 3-iron that stopped forty feet short of the hole barely on the front of the green. To reach the cup his putt would have to run a short distance up a gentle grade and then climb a steeper slope.

His breath was coming fast now, but over the years Jones had learned not to putt when he was breathing hard. He took lots of time apparently sizing up the putt but in reality calming down. Satisfied, he stepped up to his ball and stroked the putt perfectly. The ball climbed both inclines, broke gently left, and dropped into the hole. A birdie 3, a round of 75, and a seventy-two-hole score of 287.

An hour later Smith reached the eighteenth tee still two strokes behind and found about six groups lined up waiting to play, backed up because of the delay caused by the Jones ruling on the seventeenth. Without a word, all twelve men stepped aside and let Mac play through. Once again Smith needed an eagle to tie. He made his par, and Jones had won the third leg.

U.S. Amateur

The Grand Slam very nearly didn't happen. Between the end of the U.S. Open and the beginning of the U.S. Amateur Jones was very nearly killed twice. His first close call happened as he played with friends at East Lake. A storm began building as Bobby and his friends crossed between the ninth green and tenth tee, but they decided to take a chance and continue playing. Before they were aware the storm had moved in more quickly than they expected, a bolt of lightning struck the ground within forty yards of them. Bobby felt a tingle through his feet. Too close, they thought, and bolted for the clubhouse. As they dashed for shelter another bolt struck a small tree. With lightning popping all around as they raced for the clubhouse, Jones opened his umbrella for shelter from the driving rain. As they were about to duck through the locker-room door they heard a monstrous explosion; a lightning bolt hit a big double chimney just above them.

Jones didn't feel anything, but he realized his umbrella had collapsed and hung draped over his head. Once inside, one of Bobby's friends noticed the back of Jones's shirt ripped down to his waist and a six-inch scratch on his shoulder just deep enough to break the skin. After the storm had blown over they stepped outside and found the chimney had literally exploded and that bits of it had been flung as far as the eighteenth green, 300 feet away. Fragments lay strewn around the ground where they had slipped through the locker-room door. Any one of them could have killed a man had it hit his head. Evidently one of them had brushed across Bobby's back as he raced through the door. Had it fallen an instant sooner, the history of the game might have been different.

A few weeks later Jones had a luncheon appointment at the Town House of the Atlanta Athletic Club, the formal name of East Lake. He decided to walk the seven or eight blocks from his law office to the club. It was a such quiet day in Atlanta that as Bobby turned the corner on the last leg of his walk, he could see the club's entrance up ahead but not another pedestrian on the street. Suddenly, about halfway from the corner to the club, he heard a voice behind him shout, "Look out, mister."

Jones spun around in time to see a runaway car swerve toward the sidewalk, bounce over the curb, and head directly toward him. Bobby leaped clear as the car careered toward the buildings lining the street. It passed directly over the spot where he was standing when he heard the warning cry. There was no doubt he would have been hit and possibly crushed against the wall had that lone pedestrian not shouted his warning.

Jones had begun his career in national championships at Merion in 1916 and had won his first Amateur championship at Merion in 1924. Now he was returning to Merion for the Amateur once again. He arrived ready to play, shot 69–73 in the qualifying rounds, and shared the qualifying medal.

He played Sandy Somerville in the first round. An all-around athlete in Canada, Somerville held Jones even through the first six holes, but Bobby coaxed home a breaking eight-foot putt on a slick green at the seventh, won the hole, shot 33 on the first nine, and won the match by 5 and 4.

In succession he polished off Fred Hoblitzel by 5 and 4 and then Fay Coleman by 6 and 5, bringing up Jess Sweetser, the 1922 Amateur champion who in 1926 had become the first native-born American to win the British Amateur. In winning the 1922 U.S. Amateur, Sweetser had beaten Jones by 8 and 7 in the semi-final round. Now they were to meet in another semi-final. Solidly on his game, Jones beat Sweetser 9 and 8.

In the final once again, Jones played Eugene Homans, a twenty-two-year-old Princetonian. Off his form at the start, Jones played the morning portion of the thirty-six-hole match in 72, Homans in 79. Bobby held a seven-hole lead as the afternoon round began, needed only to halve the eleventh hole to win.

When Homan's putt for a birdie slipped past the cup, Jones had won the match and the Grand Slam.

As Bobby set himself to putt on the fourth green during the final match, a man sitting on the clubhouse porch prepared to sip a spoonful of soup. Just then a waiter who had attended Jones all week tapped the man's shoulder and said, "You cannot eat your soup now; Mr. Jones is about to putt." Startled and more than a little annoyed, the man answered with no small degree of truth: "He's on the fourth green; he can't possibly hear me from there, especially with a high wind blowing toward us." Persisting with his unreasonable demands, the waiter warned, "Nevertheless, you must not eat your soup while Mr. Jones is addressing his ball in preparation for putting. That's final. If you persist I must remove your soup."

Although Jerry Travers had won four Amateur championships, no one seemed to recognize him at Merion. He paid his way in to root for Jones. When Jerry stated, "There never was any player near him," a fan suggested that Jerry himself might have given Bobby a bitter battle. Travers left little question of how he felt about it by saying, "He could give me strokes and beat me on the best day I ever saw."

The day he won the 1930 Amateur, Jones told the Associated Press, "I expect to continue to play golf, but just when and where I cannot say now. I have no definite plans either to retire or as to when and where I may continue in competition. I might play next year and lay off in 1932. I might stay out of the battle next season and feel like another tournament the following year. That's all I can say about it now."

He made his statement on September 27. On November 17 he changed his mind and announced he had ended his career and he would play no more competitive golf. When he won the Amateur he had felt an enormous sense of relief; the long hard year had ended, and he felt nothing remained to be done. Ahead lay nothing but rest and the end of worry. Bobby Jones had no more worlds to conquer. He was twenty-eight.

Although he retired from competition in 1930, Jones continued his keen interest in golf and often attended important tournaments. He went to Portland, Oregon, in 1937 to see the final of the Amateur between Johnny Goodman and Ray Billows, at the Alderwood Country Club, a match Goodman won by two holes.

Billows found his ball partially stymied by a huge tree on the sixth hole. If he hoped to reach the green in two he would have to draw the ball around the tree and at the same time keep it relatively low. He hit a gorgeous iron that stayed under the tree, curled around the branches, ran onto the green, and stopped only a few feet from the hole. Standing with a friend and watching Billows play the shot, Bobby said that was the most magnificent shot he'd ever seen, and that he doubted that anyone else could have hit one quite so good.

A man in the gallery overheard Bobby's remark, turned to him and said, "Jones could do it." Bobby looked at him and smiled, "I'm not so sure he could." The man pierced Jones with an angry look, clenched his teeth, and challenged, "Have you ever seen that son-of-a-bitch play?!"

The Masters

Jones attempted a comeback in 1934. His career ended, he began planning the Augusta National Golf Course, an undertaking he had dreamed of for years. He wanted to play golf for fun again and he wanted a course with wide fairways and very few bunkers that the average golfer could enjoy. Through common friends he had met Clifford Roberts, a New York financier. Roberts agreed to work with Jones on his dream: Jones would take on the course design and Roberts would arrange the financing. Roberts located a 365-acre tract of land in Augusta called Fruitlands, the first commercial plant nursery in the South.

Impressed with the design of Cypress Point, which he had seen when he played in the 1929 Amateur at Pebble Beach, Jones engaged Alister MacKenzie, its architect, to work with him at Augusta. With MacKenzie doing the design and Jones not only offering insights from his years of competitive golf but also playing shot after shot to roughed-out greens and tees, they created a masterpiece. With fairways of perhaps twice the acreage of the normal course, it can be a joy for the bogey shooter but a terror for those who would challenge its par.

Construction was completed in 1932. Proud of their new course, Jones and Roberts wanted to show it off by staging a big tournament. It occurred to them that a tournament might attract members as well. They aimed big. Contacting the USGA, they asked for the National Open. They held preliminary discussions; Augusta felt a tournament would have to be held in either April or May rather than June or July, the usual Open dates, because of the hot summer weather. There was another problem as well. The USGA had always demanded fast greens for the Open, but Augusta's greens were slow because they were planted with common bermudagrass. To bring the greens up to Open speed, Augusta National would need bentgrass, but the strains available then couldn't stand up to Georgia's hot and humid summers. The discussions were dropped.

Instead, Jones and Roberts settled on their own invitational tournament designed principally around those golfers Jones had played against. They called it the Augusta National Invitational at first, but even before a ball had been struck, Roberts called it the Masters, a name Jones rejected at first. He joked that the idea was born of a certain immodesty, but the name stuck nevertheless.

The first Masters was played in 1934. Jones entered, attracting more attention than any other player. Since he hadn't played competitive golf since his retirement four years earlier, he felt apprehensive about how he would handle himself in a field that included Walter Hagen, Tommy Armour, Paul Runyan, Lawson Little, Denny Shute, Craig Wood, Henry Picard, and Horton Smith.

With no such reservations, his fans backed him all the way down to a 6–1 favorite. Jones wasn't so sure. As the first round began, he knew within a few holes he could no longer compete at this level. When it opened, Augusta National began at what later became the tenth hole. Bobby reached the first green well enough, leaving himself a twenty-five-foot putt, but he felt his hands and arms jerk as he stroked the ball. Instead of stopping close to the hole, his ball ran six or seven feet past. He saved his par, but he felt the same jerk with the second putt. His smooth assured putting stroke was gone. Three more pars brought him to the fifth, now the thirteenth hole. As he began the round he felt a touch of nervousness, but that had been normal during his prime and went away after a hole or two. Now it lingered, and he sensed the feeling had grown stronger.

As he stepped up to drive he heard a motion-picture camera whirring behind him. He stepped away. In the best of times this would have caused nothing more than a momentary distraction, but now he couldn't take control of himself. Playing the shot before he was ready, he pushed his ball into the trees lining the right side of the fairway, well off the proper line of play. With that stroke he realized he no longer had the desire, the will, or the willingness to take the punishment necessary to play at that level. There would be no comeback. Jones played the round in 76, shot 294 for the full 72 holes, and tied Walter Hagen, whose career was over, and Denny Shute, the current British Open champion, for thirteenth place. No longer a threat, in later years Jones settled easily into the role of host.

Jones returned to St. Andrews quietly in 1936 hoping to play a quiet round over a course he had come to love. He and his wife, Mary, had been on their way to the 1936 Olympic Games, in Berlin, with the sportswriter Grantland Rice and his wife, Kit. They had spent a few nights at Gleneagles, a luxurious hotel about an hour's drive from St. Andrews. Jones told Rice he couldn't stay this close to St. Andrews without playing the Old Course.

Rice called ahead to arrange an informal luncheon and set up a round with Laurie Auchterlonie, the R and A's honorary professional. If Jones expected a quiet round played in comparative solitude, he must have been disappointed. When he arrived at the Old Course, he found he had to worm his way through 5000 or so spectators waiting to see him once more. Word had spread throughout the town that Jones had returned, and the citizens of St. Andrews, who had been so charmed by his personality and the surreal quality of his golf, swarmed over those ancient links, claiming Jones as their own once again.

Shopkeepers closed their doors and hung signs in their windows announcing "Our Bobby is back." As each moment passed, more St. Andreans poured through the town's narrow lanes and onto the links. Perhaps inspired by the affection shown him by the townspeople, Jones attacked the course. Birdie followed birdie on the outward nine until Bobby reached the outermost limits. He turned for home with a score of 32, about as low as he'd ever shot there. Now, though, the magic left him and he limped back in 39. Still, he had shot a highly creditable 71. He had played his last round at St. Andrews. He returned once again, in 1958, but he was crippled by then with the disease that would kill him.

Even though he could no longer threaten to win a tournament, Jones continued to play in the Masters. After two rounds of the 1947 tournament he withdrew because of a painful soreness in his neck and shoulders. It was nothing new; he'd had cricks in his neck off and on since 1926.

He told his friends, the golfers, "I'll join you on the first tee next year." He was forty-five; he never played in another Masters.

He developed double vision for a period of six weeks in the spring of 1948. In May he had trouble using his right hand. At the same time he began stubbing his right foot as he walked, and his right side burned whenever he swung a club. On August 15 a small group of spectators followed him around the Wahconah Country Club, in western Massachusetts. He played poorly. He never played again.

In May of 1950 he was admitted to the Lahey Clinic, in Boston, for surgery to relieve a damaged spinal disk that doctors believed might be pressing on a nerve. It didn't help. In July of 1956 he visited Dr. H. Houston Merritt at the Neurological Institute of the Columbia-Presbyterian Hospital, in New York. After a series of tests, Dr. Merritt filed a two-and-a-half-page report that concluded Jones had syringomyelia, a rare disease that attacks the spinal column, withers the muscles, dulls the sense of feel, while at the same time causing persistent pain, and eventually kills. There was no cure.

Nevertheless, Jones continued to attend the Masters, patrolling the course and watching the golf from a cart. Seeing his friend Al Laney, a gifted reporter for the *New York Herald Tribune*, Jones invited him to sit with him. Laney asked about his condition. Jones answered honestly by saying, "I've known you longer than anyone in golf. I can tell you there is no help. I can only get worse. But you are not to keep thinking of it. You know that in golf we play the ball as it lies. Now we will not speak of this again. Ever."

One morning, in the strange semi-conscious state of first awakening, Jones tried to step out of bed and fell to the floor. At first he lay still, not understanding, but in the next moment he realized he couldn't stand, couldn't walk, and never would again. The rage boiled over. He balled his fists and pounded the floor, blow following blow, and swore and cursed the chance that had changed him from a vibrant and vital man into a cripple.

The disease usually kills within ten to twelve years. Jones suffered for twenty-three years. At the end he was bedridden and his once stocky body had withered so he weighed only 90 pounds.

While he was doing commentary on a match in Atlanta during the television series "Shell's Wonderful World Of Golf" Gene Sarazen asked Jones for an interview. Jones had always admired Sarazen, and even though he was nearly bedridden by then, he allowed himself to be carried to the course where he sat in a golf cart. Jones reminisced about the parallels in their lives. They were born within weeks of one another, Sarazen on February 27, 1902, Jones on March 17, and they were both married in June of 1924 to girls named Mary.

Remembering the day, Sarazen said, "Bob looked at me and said, 'We both did quite well, didn't we?'

"I just sat there and cried."

Some years before he died, Jones talked about his condition and himself. "When it first happened to me," he said, "I was pretty bitter, and there were times I didn't want to go on living. But I did go on living, so I had to face the problem of *how* I was to live. I decided I'd just do the very best I could."

Jones died December 18, 1971, at sixty-nine.

Hail to the Echoes

The eighth hole at Brae Burn Country Club, in West Newton, Massachusetts, stretched 185 yards across a rock-strewn ravine for the 1919 U.S. Open, a considerable carry with the equipment of the day. A wooden footbridge spanned the ravine for the convenience of those fortunate enough to have carried across.

According to Francis Ouimet, before teeing off in the first round, Willie Chisholm had played a few chip shots with Johnny Walker (Black Label). He further reported that the ball was not running well for Willie that day, and he had more than his share of bad breaks. "However," Ouimet commented, "when he made 5 on the par-3 sixth it seemed as though he had played himself back to his normal game. A steady 7 on the par-4 seventh more than confirmed this. There was much to look forward to at the eighth. It was only 185 yards long, and while the iron had to be played over a deep ravine, there had been some 2s and many 3s that day." Alas, but not by Willie.

After a fat tee shot, Chisholm's ball carried over the tiny stream threading through the bottom of the ravine, but it settled two or three inches beyond a large boulder. Rather than take an unplayable lie, Willie chose to play his way out and called for his niblick. He took one swipe at the ball, but his niblick met the rock first and bounced over the ball, giving off a few sparks and a sharp clang. This evidently was a new experience for Willie, so he tried again with the same result.

Afraid he would lose count, he asked Jim Barnes, his playing partner, to count along with him. With a successful shot, Barnes had taken to the footbridge and stood leaning against the rail looking down on Willie's misfortunes, ticking off the mounting strokes on his fingers.

Willie went back to his mining, but now he chose different tactics. Instead of attempting to blast his way through the rock, he tried to chip his ball away from it, then hack a path to the green. Stroke piled up on stroke until eventually Willie slashed his ball onto the green. With his heart pounding from the exertion, breath heaving through, and perspiration pouring off him like water through a decaying sieve, Willie dragged himself up the final inches of the incline; he had finally maneuvered his ball onto the green. His bad luck had not been left behind, though; Willie three-putted.

His head bowed and shoulders sagging, Willie asked Barnes, "How many was that?" "Eighteen," Jim answered with a smile. Chisholm's jaw dropped; he refused to believe he had taken so many strokes on a par-3 hole. "Oh, no, Jim," he howled, "that canna' be so. You must have counted the echoes." Chisholm had set a scoring record that would stand for 19 years.

Walter Hagen

Mike Brady led Walter Hagen by five strokes after fifty-four holes of the 1919 Open, but he shot 80 in the fourth round and slunk into the clubhouse to wait for Walter to finish. Playing loose golf himself through the first nine, Hagen reached the tenth tee knowing he had to par the second nine to earn a tie. A wild hook soared out of bounds on the eleventh, but two holes later Walter picked up a birdie. One more birdie and he would win. His last chance came on the eighteenth, a demanding par four with an out-of-bounds wall close behind a two-tiered green. Hagen played a useful drive that left him with a long-iron approach. Walter flourished on tight situations of this sort; no one handled them better. With the Open at stake he played a wonderful shot that flew onto the green and pulled up within eight feet of the cup. The gallery, packed tight around the green, waved their caps and cheered.

Walking ahead, Hagen sent word to bring Brady out to watch him finish. Then, with Brady watching, Hagen stroked his ball dead at the cup. The ball caught the lip of the hole and spun out. He had tied Brady, not beaten him; they would play off the following day.

On the second tee of the playoff Hagen told Brady, "Mike, if I were you I'd roll down my shirt sleeves." Suspicious, Brady asked, "Why?" "Well," Hagen answered, "the way they are now, everyone can see your forearms quivering."

Brady hooked his drive deep into the woods, double-bogeyed, and lost two strokes. Hagen held that two-stroke lead through sixteen holes. The seventeenth called for a right-to-left drive, but Hagen pushed a high soaring shot that plopped into a patch of soggy muck. With both men searching for Walter's ball, Mike found it embedded about four inches down in the soft turf.

Hagen's cause seemed lost now; in those days there was no embedded ball rule that would allow him to lift his ball without a penalty; he would surely waste his two strokes just to chop his ball loose. Walter Hagen, though, was a man of many resources. Since his ball had sunk so deep into the ground, Walter claimed a spectator must have stepped on it and that he could therefore lift it free. Since no official had seen anyone remotely near where Walter's ball lay, his claim was denied. Then Hagen claimed the right to identify his ball. Officials had no choice, they had to give him permission. Hagen bent over and plucked his ball free, wiped it clean of mud, and agreed that this indeed was his ball. Then he replaced it gingerly. It didn't sink into its former grave, and now he could play a shot. Even though he bogeyed and lost a stroke, he had saved the day. Hagen shot 77 and Brady 78.

Hagen made his first appearance in Britain with style and aplomb. Over for the 1920 British Open after having won two U.S. Opens, he put up at a posh London hotel and each day had himself driven to Deal, on the English Channel coast, in a chauffeured limousine.

On his first day of practice Walter stepped out of his car carrying a satchel holding his golf clothes and boldly attempted to change in the Royal Cinque Ports clubhouse, which as far as professional golfers were concerned was as sacrosanct as China's Forbidden City.

Turned away, Hagen was shown to the cramped changing quarters in the professional's shop. This, of course, didn't suit Walter at all. He turned his back, walked back to his car, parked just outside the clubhouse door, and changed in the back seat.

Hagen wasn't through getting his own back from the British establishment. After becoming the first native-born American to win the Open by shooting 300 at neighboring Sandwich in 1922 and beating both George Duncan and Jim Barnes by one stroke, Walter won £50. He gave it to his caddie.

A natty dresser, Hagen liked to wear white flannel slacks on the golf course. During a series of exhibition matches in Florida in 1922, he changed pants twice a day at a cost of $6 per day cleaning. One day he stumbled across a store selling white flannels for $5 a pair. He ordered a load and used the soiled pants to tip bellhops at hotels who supplied intelligence on ladies who attracted Hagen's attention. He not only saved a dollar a day on cleaning bills, but hundreds of dollars in tips as well.

Hagen was a master of gamesmanship. In an important match in New Orleans, he played on even terms through thirty-six holes with Jim Barnes, who by then had won the 1916 PGA, the 1921 U.S. Open, and the 1925 British Open. Barnes held the honor on the thirty-seventh, a 135-yard par three, but before choosing his club he glanced at Hagen and saw him holding a 6-iron. Thinking that might be a little too much club, Barnes pulled out his 7-iron and carried his ball over the green and down a steep slope.

Hagen grinned, dropped the 6-iron back into his bag and took out the 8-iron he'd planned to use all along. He won the match.

Walter wasn't exempt from the same treatment. Playing Gene Sarazen in the final of the 1923 PGA, at the Pelham Country Club, outside New York City, Hagen left his drive just a few feet short of Sarazen's on a par-5 hole. Stepping up to his ball, Sarazen quickly drew a mid-iron from his bag, the equivalent of a modern 2-iron. Hagen saw him and chose a mid-iron himself, but then left his shot twenty yards short of the green. Sarazen then pulled out his brassie and ripped a shot onto the green. Sarazen won on the thirty-eighth hole.

Confidence

Archie Compston, a burly English professional, had just polished off Hagen in a seventy-two-hole match at Moor Park, near London. Playing careless, almost frivolous golf, and dreadfully off his game, Hagen succumbed by the magnificent score of 18 and 17.

The match over, Hagen climbed into his rented Rolls-Royce—with chauffeur—for the drive to Sandwich and the 1928 British Open. Bob Harlow, his manager, huddled in a corner in the back seat, so disgusted he wouldn't even look at Walter. After a few miles of surly silence, Hagen spoke. "What's the matter with you?" he asked. Harlow barely turned his head, scowled, then turned away without speaking. Not taking Harlow's silence seriously, Hagen laughed, "You're not worried about that, are you?" referring to his abysmal defeat. "Hell, I can beat Compston any time he wants to play."

The next week Hagen shot 292 at Sandwich and beat Compston by three strokes as he won his third British Open.

Lost Trophy

Beginning in 1924, Hagen won four consecutive PGA Championships. Somewhere along the line, after winning either his second or third, he went out for a night on the town. When he felt he had had enough, he took a cab to his hotel, clutching the trophy to his bosom. In mid-ride he suddenly perked up, decided the night was still young, and told the cabbie to drop him at the next corner. Still fairly conscious, he gave the cabbie $5 to take the trophy to the original destination.

That was the last Hagen saw of the trophy. Everything was fine, though, so long as he kept winning. At the next two prize presentations, those familiar with his ways assumed his not bringing the trophy with him was Walter's way of telling the field he fully intended to win again.

Then came 1928. Leo Diegel beat Hagen, 2 and 1, in the semi-finals, at Five Farms, near Baltimore, and then went on to win the PGA. Alas, Walter finally had to confess he not only couldn't produce the trophy, he hadn't seen it in years and hadn't the faintest notion of where it might have wandered. The PGA had no choice; it had to buy a new trophy.

Time passed. Early in the 1930s an employee working in the Detroit warehouse of the L.A. Young Company came across a dust-covered crate in a dark corner and wondered what might lie hidden inside. He pried it open, and there, behold, sat the PGA Championship trophy.

Hagen evidently never knew or had forgotten that the trophy had been sent to his golf equipment manufacturing plant in Longwood, Florida.

When the business went bad and Hagen sold out to Young, the trophy, along with all the rest of the equipment, was shipped to the Young plant, where it sat building up layers of dust.

Of course once it was found, the original trophy was put back into circulation. Strangely, the PGA has since lost the replacement.

Throughout his match against Al Espinosa in the 1927 PGA semifinals, Hagen had conceded all Espinosa's putts within three feet of the hole. With Espinosa one hole up after thirty-five, Hagen overshot the thirty-sixth green while Al's shot settled twenty-five feet from the cup. Hagen chipped within a foot of the cup and Al conceded him the par. Espinosa's birdie attempt pulled up three feet short. Expecting to have the putt and the victory conceded, Espinosa looked toward Hagen, but Walter smiled, turned his back, and chatted with the gallery.

Forced to hole from a distance he hadn't putted from all day, Espinosa missed, losing the hole and allowing Hagen to draw even. One hole later Al three-putted, and Walter moved on to the final, where he met Joe Turnesa.

Once again Hagen conceded all putts inside three feet until the thirtieth hole. Unsettled, Turnesa missed a series of short putts on the final six holes, including one on the thirty-sixth that hung on the lip. Hagen won by one hole.

Hagen projected himself as the cool sophisticate, imperturbable, immune to the pressures that haunt mortal man, always in command of any situation, and supremely confident. Not quite. Walter won the 1928 British Open, at Sandwich, and the following year stood well ahead of the field after seventy-one holes. Only he evidently forgot. After driving down the middle of Muirfield's last hole and studying his approach, Hagen asked of the gallery, "What do I need to tie?" Somewhat taken aback, a Muirfield member told him, "As near as I can calculate, you need 10 to tie."

Next Hagen pulled a cigarette from his gold case, and with shaking hands tried to light it. He couldn't hold the match still. A member of the gallery helped him, and by the time Walter drew heavily on the cigarette and blew out a lungful of smoke, the tension had eased. He drilled a long iron to the heart of the green, made a cast-iron 4, and won by six strokes.

Hagen had supreme confidence in himself. Just before the singles matches of the 1929 Ryder Cup, at Moortown, near Leeds, Hagen, the captain of the American side, met with George Duncan, the captain of the British and Irish team.

Pairings for both the foursomes and singles are kept secret until they are announced, usually the evening before the first matches. Each captain names his players in the order they'll play without knowing how the other captain has lined up his players. No one knows, then, who he will play against until both line-ups are announced.

Going back to his hotel, Walter called a meeting of the American team, and while he smiled and rubbed his hands together he told them, "Duncan wanted to know if I could arrange for our captain to play their captain if he let me know what number their captain was playing. I said I thought it could be arranged. Well boys," Hagen went on, "there's a point for our side." The next day Duncan beat Hagen by 10 and 8.

Challenge Match

Hagen challenged the trick-shot artist Joe Kirkwood to an unusual one-hole match. The course began at the door of a clubhouse in Mexico and ended in the toilet bowl of their hotel room. They played stroke-for-stroke until they reached sight of their goal. Kirkwood eventually won. Explaining his unexpected victory over one of the game's greatest match players, he pointed out "Walter had trouble picking his approach off the tile floor in the bathroom."

Leo Diegel

Leo Diegel was a particularly high-strung golfer who reached his peak during the 1920s, a period dominated by Bobby Jones, Walter Hagen, and to a lesser degree Gene Sarazen. To control his nerves on the putting green, he invented the decidedly eccentric method of pointing each elbow parallel along the line of putt. While it made him look as if he'd grown wings, his method made anything other than a perfectly pendulum-like swing impossible.

The crowd was pulling for Leo to win the 1920 U.S. Open, played over the exceptionally fine Inverness Club course in Toledo. When he made the turn for home in 37 in the last round and climbed within two strokes of Harry Vardon and within one of Ted Ray, Chick Evans, who had won both the Open and Amateur in 1916, rushed out and took Diegel's bag, replacing his caddie.

Vardon fell apart when a storm lashed in from Lake Erie, and when Ted Ray dropped a couple of strokes, Diegel forged ahead. After sixty-seven holes he held a solid lead, three strokes ahead of Ray, the closest man to him. Par golf over the last five holes and he'd be an easy winner.

A topped drive on the fourteenth didn't seem too costly, but Diegel was on edge, ready to explode from the stress. As he prepared to play his second shot, a friend rushed up to him and told him Ray had bogeyed the seventeenth. That was the spark that caused Diegel's collapse. He slammed his club to the ground, and while his gallery gaped, wide-jawed, Leo screamed, "I don't care what Ray took. I'm playing my own game." Whereupon, he hooked his second shot into a bunker, double bogeyed the fourteenth, bogeyed the sixteenth and seventeenth, and tied Vardon for second place, one stroke behind Ray.

Open Door Policy

From the beginning, professional golfers had been treated as a lower caste, not acceptable in clubhouses or locker rooms; when an Open came to a club, professionals changed their shoes in their cars or under a convenient tree, and they took their lunch somewhere other than the club's dining room. Inverness broke that tradition in 1920, inviting the pros into their clubhouse, where they were free to use all the facilities.

Not long after the 1920 U.S. Open, a delegation of professionals led by Walter Hagen, a man of instinctive grace, presented the club a cathedral-chime clock well over six feet tall. Standing in the club's foyer ever since, its dark wood cabinet bears the rich patina of age, and its polished brass works gleam behind a beveled glass front. A brass plate fastened to the front is inscribed:

> God measures men by what they are,
> Not what in wealth possess.
> This vibrant message chimes afar,
> The voice of Inverness.

Roger Wethered

Jock Hutchison, a native of St. Andrews but a naturalized American citizen, shot 296 at St. Andrews in 1921 and forced a playoff for the British Open championship with the amateur Roger Wethered. Roger's sister, Joyce Wethered, had won the English Ladies championship by beating the invincible Cecil Leitch, until then the absolute monarch of British women's golf.

It was quite possible, though, that instead of playing himself into a tie, Roger might have won the championship outright except for an incident on the fourteenth hole of the third round.

After a decent drive, Wethered walked ahead and up an incline for a view of the green and to pick a landmark to indicate his line. He found what he wanted, and began walking backward toward his ball. Unfortunately, he miscalculated the distance and stepped on his ball. The accident cost him a penalty stroke. Hutchison destroyed him in the playoff, beating him by nine strokes. Jock shot 150, Roger 159.

Joyce Wethered

Bobby Jones believed Joyce Wethered may have been the best golfer he'd ever seen, man or woman. Miss Wethered was to win four British Ladies championships and five consecutive English Ladies championships, but in 1920 she was an unknown nineteen-year-old girl going against Cecil Leitch—the best of her time—in the English. Miss Leitch had won both the British and English championships in 1914, the last before the First World War, and the English again in 1919 (the British wasn't played because of a railway strike). Miss Wethered, on the other hand, had never before entered a tournament.

The two women met in the final, at Sheringham, on the Norfolk coast of England, where a railway line ran close by the seventeenth green. Predictably, Miss Leitch ran up a four-hole lead after the first nine and held it through the eighteenth, but Miss Wethered fought back in the afternoon, and pulled one hole ahead by the time they reached the seventeenth tee. Here the younger woman showed one reason why she was to become such a great golfer: she could concentrate totally on the job at hand. As she stood over a putt that would win the match, a train chugged past. The sound of the thundering locomotive caused the gallery to wince, but Miss Wethered calmly stroked her ball into the hole, and closed out the match, 2 and 1. Asked later why she hadn't waited until the train had passed, Joyce, surprised, asked, "What train?"

Five years later, a day or two after she had beaten Miss Leitch once again, this time in the final of the 1925 British Ladies championship, at Troon, Miss Wethered went salmon fishing in the Girvan, a few miles south. A novice at the craft, she was sent out alone with a gamekeeper. First, she caught a twelve-pound salmon, and then, almost as soon as her hook sank beneath the surface, another fish bit. This was a fighter. Trying to break free, he made rapid and violent rushes and tugged so hard at the line, Joyce's left arm, already weary from her exhausting march to the golf championship, felt, she said, "as if it would drop off."

After fighting the fish for fifteen minutes or so, Joyce finally called to the gamekeeper that she couldn't hold on any longer. For his part the gamekeeper kept urging her on. In a moment of inspiration, he called, "Imagine it's Miss Leitch you have at the end of the hook."

It worked. Startled and faintly amused that the gamekeeper knew anything about golf, Joyce held on for another five minutes and landed her prey.

Gene Sarazen

Gene Sarazen had been U.S. Open champion for a month when he played an exhibition match in Columbus, Ohio, on August 11, 1922. Back at his hotel after the match he was sitting in a restaurant waiting for his dinner when a passer-by spotted him through the window. The man stopped, then hurried to Sarazen's table and asked, "What are you doing here?" Startled at the man's apparent insolence, in an indignant tone Sarazen snapped, "What am I doing here? I'm having my dinner." The man said, "But you're supposed to play in the PGA tomorrow morning in Pittsburgh." Sarazen, in the more off-handed spirit of the period, had forgotten the PGA's dates.

Now he was even more startled. Desperate, he asked "What'll I do? How can I get there?" The man said, "There's a train for Pittsburgh at 7.15 tonight. You grab your bags and take a taxi to the station. I'll take care of the check." Sarazen raced to his room, piled his clothes in his bag, and sped to the station. He arrived as the train was pulling out, but he managed to leap onto one of the cars just in time.

His problems weren't over. Delayed along the way, the train pulled into Pittsburgh and hour and a half late, and Sarazen reached the Oakmont Country Club two hours behind schedule. Now the question was would the PGA allow him to play. It did, Sarazen said, "because I was the reigning Open champion. They knew I was a good draw."

Sarazen not only beat Tom Mahan in the first round he also won the championship, defeating Emmet French in the final, when French became the victim of his own accuracy. A brassie second to the twenty-seventh hole, an uphill par 5, hit the flagstick and caromed into a bunker. One hole down at the time, Sarazen won the ninth to pull even, and took the match, 4 and 3.

Sarazen had a fling as a comic actor in the movies. At the end of a 1923 exhibition tour, Sarazen and Jock Hutchison wound up in Hollywood. Sarazen by then had made a couple of two-reel shorts playing golf, but now movie people began telling him he had star potential and they could make him a bigger attraction than Rudolph Valentino, one of the film world's greatest heart-throbs. One agent told him, "You have perfect camera ears." (Another sneered, "He's got ears like Rin Tin Tin"—a German shepherd that starred in adventure films.)

Sarazen was given a number of screen tests, and one studio took him on location to the San Fernando Valley to watch a movie being filmed. He sat in a canvas chair while studio gofers brought him lunch on a tray, and young and luscious starlets told him how thrilled they were just to talk to him. Gene ate it up, asking himself, "Why should I kill myself hitting golf balls when I have better ears than Valentino?" Whenever he'd visit a studio, photographers would snap his picture showing the golf swing to the likes of Norma Talmadge, Jack Pickford, Larry Semon, Alan Hale, and Buster Keaton.

Finally he appeared in a movie with Keaton, the dead-pan comic. In the mode of the time the film was heavily slapstick, with lots of falls. Picking himself up and brushing the dust off his clothes after one fall, Sarazen asked the director, "Do you mean to say people laugh at things like this?" "Sure," the director answered. "What do you think makes Keaton a great star. It's not his profile."

After taking several more tests, Sarazen awoke one morning, and in a moment of clarity realized he was out of place in Hollywood. Before having breakfast he packed his bags and made arrangements to head back east and get on with playing golf for a living.

Writing of Sarazen after Gene had won the 1932 U.S. Open, Bob Jones commented, "When he is in the right mood, [Sarazen] is probably the best scorer in the game, possibly that the game has ever seen." Jones had written this after Sarazen had played the last twenty-eight holes of the Open in 100 strokes at the Fresh Meadow Country Club, a 6,815-yard course in Flushing, on Long Island—a course that no longer exists. Five strokes behind after thirty-six holes and unable to make up any ground through the early holes of the third, Sarazen began his rush with a birdie 2 at the ninth, a hole of just 143 yards. He played the second nine in par-par-par, par-birdie-birdie, birdie-par-par, shot 32, and posted 70 for the round.

The Open in those days closed with two rounds on the final day. He played the first nine of the last round par-bogey-birdie, birdie-par-birdie, par-par-birdie, scoring another 32, then came back in par-par-par, par-par-birdie, par-par-par, 34, and 66 for the round. Sarazen shot 286, equaling the record Chick Evans had set in 1916, the last Open before the United States entered the First World War, and won by three strokes over Bobby Cruickshank, who had lost a playoff to Bob Jones, in 1923.

In 1934, Sarazen stopped in Fiji on his way to join an exhibition tour around Australia. He drove to a golf course in Suva, the capital, and set about playing practice shots. A circle of Fijian caddies sat behind him ooohing and ahhing after each shot. Finally one of the young men said to Sarazen, "Mister, no one on this island hits a ball long like you." Smiling and half-joking, Sarazen said, "Don't you know who I am?" The boy shook his head no. "My name is Gene Sarazen," he said. The boy paused for a moment and replied, "We no hear of Mister Sarazen, but we hear of Mister Jones."

In the 1931 Ryder Cup singles, played at Skokie, in the Chicago suburbs, Gene and Fred Robson were even as they approached a par-3 hole. Up first, Robson played a nice shot about twenty-five feet from the hole, but Sarazen hit a wild hook. His ball banged against a stack of Coca-Cola boxes and bounced through the door of a refreshment stand. Gene found his ball nestling in a crack in the cement floor. Deciding first to pick up and concede the hole, Gene thought a little more and figured he might as well try to play through an open window that looked out onto the green. He had one problem: a refrigerator was in his way.

Sarazen persuaded the operator of the stand to help his caddie move the refrigerator. Sarazen nipped the ball cleanly off the cement, out through the window and onto the green. Not realizing what Sarazen had done, Robson three-putted rather carelessly, and then Sarazen holed his putt for a par 3. Walking off the green Robson said, "That was tough luck, Gene." Stunned, Sarazen said, "Fred, I made a 3." Now Robson was stunned.

You did?!" he said. "I thought you had played a hand mashie." The turnaround upset Robson so badly he lost the match by 7 and 6.

Of the world's four principal golf championships, only the PGA continued through the Second World War. As a representative of the Vinco Corporation of Detroit, a manufacturer of machine tools in which he held a small investment, Sarazen spent a lot of time in Washington during the war and became friendly with Robert Hannegan, the Postmaster General. The evening before he was to leave for Dayton and the 1945 PGA, Sarazen called Hannegan and invited him to dinner. Hannegan, however, had been invited to the British Embassy. Somehow he arranged for Sarazen to come along as his guest.

Following dinner, Sarazen became engaged in conversation with Lord Halifax, the British ambassador to the United States. Lord Halifax asked what Sarazen was doing in Washington. Gene answered, "I'm on my way to the PGA." Lord Halifax frowned for a moment, then said in a puzzled voice, "PGA. PGA. Well, that just goes to show you that no matter how closely you keep in touch with what's happening in Washington, the moment you turn your back the government has created a new agency."

Darwin a Player

As correspondent for *The Times,* London's leading newspaper at the time, Bernard Darwin was assigned to the first Walker Cup Match, at the National Golf Links of America, on eastern Long Island, in 1922. He wound up as a major participant.

Originated in the early 1920s by the United States Golf Association, the match was played between amateur golfers from the United States on the one side, and from Great Britain and Ireland on the other. The trophy—the Walker Cup itself—was donated by George Herbert Walker, the maternal grandfather of President George Bush. Walker served as president of the USGA in 1920.

Robert Harris had been appointed captain of the British side, but he burned his hand on the sea voyage over and couldn't carry out his assignment. Though never a player of championship quality, Darwin took over as captain and played in the match as well. Three down after the first three holes of his singles match, Darwin eventually defeated William C. Fownes, a former president of the USGA, who had won the 1910 U.S. Amateur championship.

Origin of Roping

Contrary to popular belief, roping golf courses came about not to give players more room but to discourage fans from scrambling for tees.

The wooden-peg tee we know today was patented in 1922 by Dr. William Lowell, a dentist from Maplewood, New Jersey, and a member of the Plainfield Country Club. Dr. Lowell originally made his tees with gutta percha, but because it was too brittle he eventually switched to white birch. Lowell painted the tees green at first, but he changed to red after a time, hence the name "Reddy tees."

Lowell eventually persuaded Walter Hagen and the trick-shot artist Joe Kirkwood to use them. Playing at the Shennecossett Club, in Groton, Connecticut, Hagen and Kirkwood strutted around the course with bright red tees stuck behind their ears. They'd leave them behind on each tee, and kids would scramble, grabbing them for souvenirs. The kids became so troublesome the club roped off both tees and greens to control the gallery.

In taking out his patent, Dr. Lowell had used his family lawyer rather than a patent attorney. Consequently, by 1926 more than 200 different brands were being marketed by competitors, and even though Dr. Lowell sued, his patent had been written so loosely he couldn't claim exclusive rights.

Laddie Lucas

Perhaps the best left-handed golfer ever developed in Britain, captain of a Walker Cup team, and a genuine war hero, Laddie Lucas was born in the clubhouse of Prince's Golf Club, set on the English Channel coast in Sandwich, southeast of London. The son of Percy Montagu Lucas, the club's first secretary, Laddie grew up roaming the course.

Prince's actually adjoins Royal St. George's; as a matter of fact, the fourteenth tee at St. George's stood so much closer to the old Prince's clubhouse than Prince's first tee, strangers often confused the two.

The 1922 British Open was played at Royal St. George's. Over from the United States came Long Jim Barnes and Jock Hutchison, naturalized Americans, and Walter Hagen. A few days before the Open, those three decided to try Prince's. Laddie persuaded his father to allow him to follow the great men around. Laddie had reached the age of six by then, and the players couldn't help noticing him trudging along through the full eighteen holes.

When the round ended Barnes asked the little boy if he played the game. When Laddie said proudly that he did indeed play, Barnes told him to be at the first tee at 5 o'clock that afternoon, and to bring his clubs with him.

Laddie was there well before the appointed hour, waiting for Barnes, who had won the U.S. Open the previous year. Jim brought two boxes of brand new golf balls imprinted with his name. Opening the boxes he told Laddie, "Tee these up and drive them off. You can keep every one that finishes on the fairway."

The wind from Pegwell Bay, the body of water bordering the Prince's shoreline, blew in powerfully from right to left as this little six-year-old boy took his stance. Accustomed to these conditions, Laddie played each shot perfectly, shading the right side and allowing the wind to drive the ball back toward the center. Astonished at the boy's golf sense, Barnes gaped as every one of the two dozen balls settled within the fairway's boundaries.

Laddie raced out and gathered his booty. He remembered later that these twenty-four balls lasted him two years, until he went away to school.

When war came to Britain in 1939, Laddie joined the RAF and became a fighter pilot. Zooming home from a mission over France, he was attacked by a German plane that swooped down, fired, and hit an ammunition drum in the right wing of his Spitfire, nearly ripping it off. Showing the same skill as a flyer he had shown as a golfer, Lucas somehow nursed the plane across the Channel and spotted Prince's below him. Knowing he couldn't reach his base, Lucas banked the plane as best he could and set it down just off the ninth fairway. As he climbed from the cockpit, he is said to have looked around at his plane mired in the rough and said, "I never could hit that fairway."

Warren Harding and Ring Lardner

The sportswriter Grantland Rice set up a game with President Warren G. Harding, a man who played golf with a passion. To round out the group, the President invited Under Secretary of State Henry Fletcher, and perhaps for comic relief, Rice brought along his close friend Ring Lardner, the humorist who also dabbled in golf writing. Introduced to Lardner, Harding said, "Rice is here to get a story. Why did you come?"

Reserved and occasionally taciturn, and a dubunker of heroes, Lardner at the same time had a sardonic and impish wit. Answering the President, he said, "I have a good reason; I want to be appointed ambassador to Greece."

Not lacking humor himself and anticipating a senseless response, the President asked, "Why?" Lardner answered, "Because my wife doesn't like Great Neck" (the town on Long Island where the Lardners lived). Suddenly wearing a grim expression, the President allowed, "That's a better reason than most people have."

They played at the Burning Tree Club, in Bethesda, Maryland. An impatient man, Harding had the habit of walking ahead after he had played a shot, not waiting for the others. Somehow taking the honor, Harding played his drive and strolled ahead without waiting for Lardner—who was up next—and stood down the fairway beside a dead apple tree. Lardner did not hit a good drive; the ball swerved off line in what Rice described as "a lusty ball with a slight slice." Shooting off to the right, Ring's ball crashed into a branch of the apple tree, the branch broke off, fell, and hit the President on the shoulder. Anticipating an apology, Harding waited as Lardner approached, but instead of telling the President how sorry he was to have put the nation's chief executive in mortal peril, with the aplomb of a French diplomat he said, "I did all

I could to make Coolidge President." Harding dropped his club and roared with laughter.

While Harding played an enthusiastic game, he didn't play a good one. At the end of nine holes, the Secret Service man keeping score reported that the President had shot 55 and Rice 44. Hearing the report, Lardner, as irreverent as ever, quipped, "Well, I'd rather be Rice than President."

Harding, by the way, presented the U.S. Open trophy to Jim Barnes in 1921, when the championship was played at the Columbia Country Club, in Chevy Chase, Maryland, like Burning Tree, just over the border from Washington, D.C. As he handed Barnes the trophy, the President told him, "I'd give anything to be in your shoes today."

Recognition at Last

Members of The Country Club, where well-connected Bostonians played, often take on new friends slowly. On a balmy Saturday morning one May during the 1920s, a man wearing white flannel knickerbockers, argyle socks, and a white shirt with a four-in-hand tie was spotted about to play a shot from a few yards ahead of the tee markers, strictly against the rules of golf and agitating to at least one Older Member.

The Older Member approached the man, a stranger to him, and reproached him for his obvious breach of the rules and of golf etiquette, and that simply wouldn't do at The Country Club. He then launched into a history of the club, emphasizing it had been formed in 1882, which made it the oldest country club extant in the United States, that it, along with four other old-line clubs, had founded the United States Golf Association, and that in 1913 it had been the site of the amazing victory by Francis Ouimet over the great English golfers Harry Vardon and Ted Ray.

Motioning for the miscreant to return to the tee markers, the Older Member pronounced: "If you can't abide by the rules of golf, the board of governors will accept your resignation immediately." Bewildered at first, and then worked up to a full head of steam, the offender stared at the Older Member while his anger grew. Fuming, he replied, "Look, you son-of-a-bitch, I've been a member of this club for six months and you're the first person who's spoken to me. Furthermore, I'm about to play my *second* shot."

Driving In

Until 1824, the title of captain of the Royal and Ancient Golf Club of St. Andrews went to the winner of the club's Autumn Medal, a tournament played over the Old Course each September. It seemed simple enough, but the system had its holes. As arbiter of all internal as well as external disputes, the club's best golfer wasn't necessarily the club's best statesman. A change seemed in order.

Beginning in 1824, then, the captain has been chosen by past captains.

In the morning of the last day of the Medal, while a tall grandfather clock inside the gray, stone clubhouse tolls eight, the hangman's hour, the new captain is led to the first tee by the old captain. There, before a sizable gathering of family, friends, club members, and those who like to see grown men writhe in misery, he plays a shot from the first tee of the Old Course while, with a shattering boom, a bombardier fires a cannon dredged up from a sunken wreck off the shoreline late in the eighteenth century.

Prince Leopold, a son of Queen Victoria, played in as captain in 1876 during the British Open championship. It was a strange day, with players from the Open and from the Autumn Medal teeing off in alternate groups, because the R and A remembered to reserve the course for the Medal but forgot about the Open.

Horace Hutchinson, who had won the second and third British Amateur championships, was so ill when he drove in as captain in 1908 he staggered across the street from what was then the Grand Hotel, played the required shot, then slouched back to his bed.

Caddies range down the fairway waiting to retrieve the ball as each new captain drives. The fortunate caddie who retrieves the ball returns it to the captain, who pays him a gold sovereign. Spectators can tell how the caddies regard the new captain's game by how far they station themselves downrange. When the Prince of Wales (Edward VIII) drove in in 1922, the caddies, who have a good idea of how far the incoming captain will hit the ball, were reported to have stood "disloyally close to the tee."

By tradition the incoming captain is offered a bracer before facing the ordeal. Evidently the Prince accepted not only his own but a few that might have been missed by those who had gone before him. He wove to the tee through a thick mist, and as Andrew Kirkaldy, the salty honorary professional of the day, bent to tee his ball, the Prince muttered, "This is a dreadful job."

Versions of what happened next vary, but it seems clear he hit a terrible shot. Bernard Darwin, the great British essayist, journalist and sometime golfer, claimed the Prince hit the ball "low and to the left." Others less inhibited said he actually heeled the ball. It evidently squirted behind him and rolled into the Valley of Sin, a deep depression on the left front of the 18th green. Seeing the ball come to rest, one spectator, who might have had as much to drink as the Prince, cried, "My God. If he holes it we'll have a new course record."

After the driving in, the new captain returns to the clubhouse where he entertains guests of his choice at a champagne breakfast. Later in the day he plays his round in the Medal. The Prince had a terrible day. Playing off a 15-stroke handicap, he hit his first drive off the toe and rolled it close to the out of bounds fence. A hole-by-hole description related, "Then he had a 7, then a 9 . . . He finished with a good 6. The Prince shot 119."

Cyril Walker

Even though he had won the 1924 U.S. Open, the most important tournament in the country, and beaten Bobby Jones in the process, Cyril Walker was a rather obscure golfer, hardly known outside the close group of men who played tournament golf. It wasn't surprising, then, that the man doing the introductions at the Los Angeles Open the following January couldn't remember what this unprepossessing man might have won. He had to say something, though, so holding his hand over the microphone so the spectators couldn't hear, he spoke directly to Walker and asked, "What state are you the champion of?" Offended, and not believing what he had just heard, Walker yelled loud enough to carry through the public address system, "What state? The whole damn forty-eight!"

Mac Smith at Prestwick

The 1925 championship turned out to be critical in the development of the British Open. It was played at Prestwick, on Scotland's west coast, thirty miles or so south of Glasgow. Macdonald Smith, a naturalized American citizen but a native of Carnoustie, Scotland, had his best chance to win a national championship. As smooth a swinger as ever lifted a club, Smith had been third to Arthur Havers in 1923 and third to Hagen in 1924.

A 69 in the second round had carried him to a two-stroke lead after thirty-six holes, and a 76 on the morning of the last day pushed him five strokes ahead. The passenger station for the main railroad line to Glasgow stands alongside the golf course. Train after train from Glasgow, Kilmarnock, Ayr, Troon, and Irvine emptied frenzied fans onto the Prestwick links throughout the morning, hungry to see a native son win.

Unfortunately for Smith, he was among the late starters. From 10,000 to 15,000 fans swarmed around Prestwick, most of them straining to watch Smith. They crowded around, barely giving him room to swing, raced along the fairways after each shot, and fought for vantage points, building up unbearable tension.

Smith cracked. He went out in 42, double-bogeyed the short eleventh, then finished in 82, falling like a stone into fourth place. Jim Barnes, another naturalized American, had played early and missed the crush of the crowds. He shot 74 and beat Smith by three strokes. Barnes shot 300, Smith 303.

The events of the day had a profound effect on the British Open. Even though Prestwick had originated the championship and had been the site of the first twelve, and then twelve more after St. Andrews and Muirfield joined the rota, the Royal and Ancient Golf Club of St. Andrews—which had taken over administration of the Open in 1919—determined that because it was such a very compact course it was no longer suitable for the Open. It never returned.

Along with every other important golf tournament, the British Open had been open to galleries without charge. Besides eliminating Prestwick, the R and A decided that by charging an admission fee it could control the crowd's size. In a third move, the R and A imposed a thirty-six-hole cut, reducing the field for the last day.

Head Start

Dale Bourn, a very fine English golfer who never quite reached Walker Cup status, reached the final round of the 1928 French Amateur, at Saint-Germain, in the outskirts of Paris. Since he had never before reached such lofty heights, he went out on the town the night before the match, but he forgot to leave a wake-up call for the next morning.

When he awakened, he saw the clock showing ten o'clock, his starting time. The match was to be played over thirty-six holes. Bursting with apologies, Bourn telephoned the club and asked to speak to Tony Vincent, his opponent. The rules evidently weren't so strictly enforced in those days, so when Vincent took the phone, Bourn said, "Terribly sorry, old chap, I'm a bit late, you know. Tell you what. I'll concede you the first nine holes and meet you on the tenth in an hour and a half." Vincent refused the offer, but he lived to regret it. When Bourn eventually arrived, he beat Vincent, 8 and 7.

Howard Hughes

Howard Hughes, who died an eccentric recluse, produced the great silent film *Hell's Angels*, a story of air combat during the First World War. The film featured a series of aerial episodes showing bombers, fighter planes, dirigibles, and balloons. During the filming one afternoon in 1928, the ground at the Lakeside Country Club in Los Angeles shook as perhaps 90 planes roared overhead simulating dogfights between Allied and German warplanes.

Everyone on the course stopped and stared at the spectacle. Someone cried, "It's a flying circus." Only Hughes knew exactly what was happening. He should have; the film was costing $1.4 million, but Hughes was preoccupied at the moment. He stood on the practice range belting balls. When someone approached to tell him he was missing a great spectacle, he didn't bother to look up. Instead he growled, "Don't bother me. I'm hitting 3-woods."

Hughes was a very good golfer, at one time holding a handicap of one stroke. He was determined to be good, too. George Von Elm, one of the great amateurs of the 1920s, said, "After a bad day, when he [Hughes] shot 76 or something like it, he'd stay out hitting balls for hours after dark with movie-like floodlights."

Hughes took lessons only from the best. He counted among his tutors Ralph Guldahl, who won two consecutive U.S. Opens, in 1937 and 1938, as well as the 1939 Masters Tournament. As Hughes was about to play in a club tournament, he contacted Guldahl by phone and spent most of the day discussing bunker shots, downhill putts, and how to play in the wind. Hughes won the tournament. When Guldahl opened his mail a few days later he found a check from Hughes for $10,000.

Jimmy McLarnin, who held the world welterweight championship in 1933, played against Hughes occasionally and told of a group of gorgeous Hollywood starlets known as H. H.'s Harem. McLarnin recalled a time when he and Hughes and a group of others were putting for money. "I won't say Hughes cheated to beat people, but there on Hughes's balcony above us appeared one of those girls, naked as a grape. We all six-putted."

Hughes once filed a multi-million-dollar lawsuit against Howard Hawks, another producer, contesting Hawks's rights to a British play. Before the suit reached the courtroom, Hughes telephoned Hawks at the Wilshire Country Club. "If you don't have a game today, let's play," Hughes suggested. Hawks swore and turned Hughes down flat. Hughes responded by proposing, "I'm willing to drop the suit if you'll give me some action." Hawks changed his mind, both men contacted their lawyers, and Hughes and Hawks had their game at Wilshire.

Recalling the match, Hughes said, "At first we played in complete silence, but before we finished we decided to work together on the play. I shot 71, beat him by a stroke, and saved a court battle."

A daring pilot as well as a golfer, Hughes was testing the FX-11 reconnaissance plane when it lost power flying over Los Angeles in the summer of 1946. He would have to bring the plane down quickly, but he could find only one open area within range—the Los Angeles Country Club. While Hughes managed a rough landing on the edge of the course, he fractured his skull, broke seven ribs, suffered a crushed chest, and sustained severe burns. Even though the plane was severely damaged, Hughes was most upset because his injuries kept him away from golf for more than a year—"A hell of a sorry note," he complained.

Hughes landed a plane on at least one other golf course. A member of the Bel-Air Country Club, off Sunset Boulevard, in Los Angeles, he had scheduled a round with Katharine Hepburn at one o'clock on a Friday afternoon in May of 1938, but at eleven o'clock he telephoned ahead and told the club professional that although he was calling from Santa Barbara, he'd definitely be at Bel-Air on time. After hanging up, the professional suddenly realized that Santa Barbara was a three-hour drive.

He needn't have worried. At five minutes to one Hughes circled the club in a single-engine plane, then set it down neatly on the eighth fairway.

Upset at what they considered a breach of sanity, offended members called the Los Angeles sheriff's office and had the plane grounded. Then they called an emergency meeting of the board of governors and assessed a fine against Hughes. He refused to pay; instead, he resigned from the club.

Harry Cooper

Harry Cooper was an exceptionally fine golfer, with a smooth swing and a quick gait that earned him the nickname Lighthorse, but he couldn't win a tournament of lasting significance. He nearly won the 1927 U.S. Open, but lost a playoff to Tommy Armour at the brutally difficult Oakmont Country Club, near Pittsburgh.

Everyone in those days labored under the shadow of Bob Jones, of course. In 1930, the year of Jones's Grand Slam, Cooper played remarkable golf at Interlachen, in Minnesota, perhaps the best of his career, certainly the best he had played in a U.S. Open. Statistically, from tee to green, he played better than Jones. Cooper hit an average of sixteen greens in every round, and he missed only one fairway in seventy-two holes. He putted terribly, though, claiming he couldn't get the ball into the hole with a rake.

Cooper had trouble with only one hole, the twelfth, a very long par five. He had gone for the green with a brassie for his second shot in each of the first three rounds, but every shot had caught a bunker guarding the front. Figuring he had a chance to win after eleven holes of the last round, Cooper decided to play safe and lay up short of the bunker with a 4-wood. Unhappily, Cooper miscalculated. Moving into the shot with all his fluid grace, Cooper met the ball squarely on the face of the club. Instead of falling short of the bunker, it carried over the green on the fly and settled into such a poor lie he had lost his chance to win. He shot 76 and finished in fourth place, six strokes behind Jones.

A Visitation

Playing at Royal St. George's, the American Don Moe fell four holes behind Bill Stout after the first eighteen holes of the 1930 Walker Cup singles. Stout had shot 68. After the luncheon break Stout started the afternoon round 3–3–3, two birdies and a par, and moved seven holes up with thirteen to play. Then Moe went to work. Birdie followed birdie as he relentlessly whittled away at Stout's lead. By the time they reached the eighteenth, Moe had pulled even. He wasn't through yet.

The eighteenth at Royal St. George's is a long and testing par 4 of well over 400 yards. After a good drive, Moe rifled a 3-iron that streaked directly at the flagstick and pulled up three feet from the hole. The putt dropped and Moe won by one hole. He had shot 67.

Changing his shoes in the locker room, Stout pondered the match. Without showing disappointment or irreverence he said, "That was not golf; that was a visitation from the Lord."

A Reticent Judge

During the 1930s a woman in Chicago filed a petition for divorce, claiming she had been reduced to golf widowhood by a husband suffering from terminal addiction to golf. The petitioner grumbled that despite his being a poor player with what she termed a "wretched drive," her husband spent most of his time playing golf. Her appeal fell on reluctant ears. In rejecting the petition, Judge Joseph J. Sabbath noted, "Golf is not yet grounds for divorce. Saying it is would set a dangerous precedent. I play golf myself."

Charles Blair Macdonald

A man of many talents, Charles Blair Macdonald, who drove everyone crazy until he won the 1895 National Amateur, expected nothing less than adulation, even from his family. He did indeed design a wonderful golf course on eastern Long Island, which, throwing modesty aside, he named the National Golf Links of America. Everyone agreed the National ranked as a masterpiece, one of the first modern courses in the United States.

While Peter Grace, Macdonald's nephew, agreed the National wasn't bad, he had reservations about the opening hole, a 320-yard par 4 with a slight bend to the right. Home from college and feeling pleased with himself, Grace confronted Macdonald and criticized the first, adding he could probably even drive the green.

A man with a short fuse and forbidding temper, Macdonald first turned purple, then snapped, "It's never been done." "Let's see, then," Grace smirked, and the pair of them stormed off to the tee. Grace teed up a ball, swung with all the controlled force he could command, and sent the ball screaming toward the green. It came down short, bounced a few times, then rolled on. Now in a force-nine rage, Macdonald turned away, and without saying a word stalked home, where he telephoned his lawyer and wrote Grace out of his will.

Close Call

At least one bunker of Australia's Royal Sydney Golf Club was built over a deposit of quicksand. On July 11, 1931, D. J. B. MacArthur, a club member, stepped into it when it was water-logged and began to sink. He tried to scramble out but he couldn't budge; the harder he fought the deeper he sank. His cries for help brought the rest of group rushing to him. Careful not to become stuck themselves, they hauled him out in the nick of time—by then he had sunk up to his armpits.

Leonard Crawley

Leonard Crawley, who spent his last days as the golf correspondent of London's *Daily Telegraph,* ranked among the finest games players ever developed in Britain. An exceptionally fine golfer, Crawley was named to the British Walker Cup team for the 1932 match, at The Country Club in Brookline.

An accident the last night aboard ship, though, almost cost him a chance to play; it would have had another team member not become drunk.

A box of matches exploded and burned his hand during a shipboard cocktail party. Even though the team went ashore two weeks before the start of the match, Dr. William Wheeler, a member of The Country Club, told Crawley he probably wouldn't be fit to play.

Crawley wasn't able to play any golf at all, but he kept himself fit by walking the course twice a day, or by playing tennis with some of the game's leading players who were preparing for the United States Doubles championship, also scheduled for the Boston suburbs.

Like every other visiting pair, Johnny de Forest and Tony Torrance were beaten badly in the foursomes, and de Forest left a note saying he would be out all night and might not be available for the following day's singles. Meantime, according to accounts, John Bookless, the Scottish champion, had been drunk since the ship left Liverpool. With no one else left, Torrance, the team captain, asked Crawley to play George Voigt in the singles.

After surveying his approach to the eighteenth hole, Crawley wanted to play his 4-iron, but his caddie insisted he needed but a 6-iron. The gleaming silver cup sat on a table directly behind the green. Trusting his own judgment, Crawley played the 4-iron and hit a gorgeous shot. His ball flew directly at the flagstick but, as his caddie had tried to warn him, he had played too much club. His ball carried the green, hit a paved road, then flew directly at the Walker Cup itself.

From the fairway Crawley heard a loud "clang!" Puzzled, he asked his caddie what had happened. "Well, boss," the caddie drawled, "I guess we've hit the cup." He had indeed hit the cup, and dented the lid. Nevertheless he beat Voigt by one hole.

The dent remained through 1938, when the British and Irish won their first match of the series. With the victory, the R and A kept the cup in St. Andrews, and when the Second World War began, stowed it safely away. When the war ended the club sent it off to be cleaned; whoever cleaned and polished the cup removed the dent as well, which upset Crawley deeply.

Speaking of the incident, Crawley said, "I must surely be the only man who has hit the cup, and I am therefore rather annoyed that the dent was removed."

Crawley was immensely popular with American golfers. The American team in the 1934 match, at St. Andrews, whiled away idle hours playing the card game hearts. The night before the foursomes, the Americans gathered around the table as usual when they heard a knock at the door. It was Crawley. Seeing the game in progress, he asked if he could join, and indeed he was welcomed. He sat next to Francis Ouimet, a mischievous man. Every time he dealt, Ouimet seemed to give himself the Queen of Spades, the most deadly card in the pack if you happen to hold it at the end of the hand. All through the night Ouimet managed to slip it to Crawley. Leonard took it in good grace and everyone enjoyed the evening.

As captain of the American side, Ouimet was responsible for arranging the order of play for his team, and so he placed himself third in the singles. By chance he drew Crawley as his opponent.

All Walker Cup matches, both foursomes and singles, were played over thirty-six holes then. Playing first-class golf through the first twenty-seven holes Ouimet stood eight holes up with nine to play.

Not one to give up the struggle as long as there were holes to be played, Crawley fought back. He drove the tenth green, a distance of 310 yards, and won that hole; took the eleventh when Ouimet missed the green, and pitched stone dead on the twelfth, winning his third consecutive hole.

Now Crawley stood five holes down with six to play. He barely missed a ten-foot birdie attempt on the thirteenth and they halved with par fours, leaving Ouimet dormie five.

The fourteenth is one of two par-5s on the Old Course. With two tremendous shots, Crawley reached the back of the green, but Ouimet needed three shots. Away, Crawley putted first and left his ball about three feet from the cup, in position to cut Ouimet's lead further. Ouimet followed with a putt that hung on the edge of the cup, directly in Crawley's line.

This was the era of the stymie. According to the rules, Crawley could neither have Ouimet's ball moved, nor could he knock it away and concede the next putt. He had to find a way either around or over Ouimet's ball. Crawley studied every possibility of holing his ball, found no conceivable way, then turned to Ouimet and with a resigned smile asked, "What is that, Francis, the bloody Queen of Spades again?"

Playing in a club match, Crawley glowered silently as his opponent repeatedly and blatantly cheated. Finally winning the match on the eighteenth hole, Crawley offered the obligatory handshake, and both men retreated to the clubhouse.

Once in the locker room, Crawley spun around, and with one quick blow to the jaw, knocked his opponent to the floor. While the man cowered expecting another blow, Crawley warned, "Don't ever do that again. Now get up and come and have a drink."

Rest for the Weary

Always spirited rivals, Walter Hagen and Gene Sarazen exchanged good-natured insults throughout the 1933 U.S. Open. When Hagen remarked after the first round that the fairways at the North Shore Country Club, near Chicago, looked a bit narrow, Sarazen riposted, "Well, they would look narrow to an old man of 40."

Then, at lunch between the last two rounds, they went at it again. Sarazen suggested that Walter should take a wheelchair to get around in the afternoon. Later in the day Hagen had his revenge. Storming home in a 66, he matched the Open record at the same time Sarazen was foundering. Sarazen had been playing well behind Hagen, and when Walter heard Gene was on his way to a 75 and 303, fully sixteen strokes behind Johnny Goodman, the winner, and eleven behind Hagen, Walter had an idea. He gave a clubhouse attendant $5 and sent him on a mission. When Sarazen huffed and puffed up the incline to the eighteenth green, the attendant was there waiting for him. In a loud voice he called, "Mr. Hagen, the old man, sent you this," and handed Gene a rocking chair.

Diegel at St. Andrews

Denny Shute and Craig Wood had each shot 292 in the 1933 British Open, at St. Andrews, setting up a playoff, when Leo Diegel stepped onto the eighteenth green two putts away from matching them. A nervous golfer who had trouble controlling his bizarre putting stroke under pressure, Diegel rolled his first putt within two feet of the cup, then stepped up again for the putt that would make tomorrow a three-man playoff. Drawing back his putter, Diegel missed the ball completely. He whiffed. Shaking from the embarrassment, he managed to hole out with his next stroke and drop into a tie for third place with Gene Sarazen and Sid Easterbook.

Ryder Cup

The Ryder Cup, a match originally between professional golfers from the United States on one side and from Great Britain and Ireland on the other, was named for Samuel Ryder, an English seed merchant, who donated the little gold trophy. Ryder had such a passion for golf that when he died he was buried with his mashie, his favorite club.

The American professionals arrived a month early for the 1929 match, the first in the series, and stayed at the Gleneagles Hotel. As a warm-up, Bob Harlow arranged an exhibition match with a group of first-class Scottish amateurs, which turned out financially successful.

Four years later the Americans arranged another match at Whitecraigs, on the south side of Glasgow. Drawn against Sam McKinlay, a newspaperman who was also a first-class golfer and had played on five international teams for Scotland by then, Hagen suggested they play for a shilling a hole. McKinlay accepted.

Off to a quick start, Hagen played the first two holes 2–3 and won them both, but McKinlay fought back, and stood one hole up with one to play after the seventeenth. The home hole at Whitecraigs was driveable in those days, with the green set beside a stone boundary wall. Reaching back for all the power he could muster, Hagen lashed into his ball with a slightly open clubface. It curled off to the right, cleared the course boundary, and was last seen bounding down the road toward Giffnock, the next town. McKinlay collected.

The home sides won the first few Ryder Cups, so when the 1933 match was played at Southport and Ainsdale, in the resort town of Southport, on England's west coast, Walter Hagen tried to upset the British and Irish side. As captain, he was to exchange line-up cards with J. H. Taylor, the British captain. Twice Taylor arrived at the appointed spot at the appointed time, and twice Hagen stood him up. Angry now, Taylor proclaimed that if Hagen didn't show up a third time, he would cancel the match. Hagen did show up, but the strategy didn't work. The British and Irish won when Denny Shute missed a four-foot putt.

The following week, Shute won the British Open, becoming the last American winner until after the Second World War.

Gloria Minoprio

Until Gloria Minoprio came along, women played golf in traditional dress—skirts and blouses, with sweaters when cool weather called for them. Then, in 1933, Miss Minoprio showed up for the English Ladies championship, played at Royal North Devon Golf Club, known widely as Westward Ho!

Miss Minoprio, a trained conjurer, stepped onto the tee wearing well-cut, close-fitting navy blue trousers with stirrups, a navy turtleneck sweater, and a navy woolen cap. She wore dead white make-up. Her outfit enraged the Ladies Golf Union, which oversees major women's amateur competitions in Britain. The LGU claimed Miss Minoprio's outfit undermined the "dignity and deportment of the game."

A solitary figure who spoke to no one and played as if she lived her life in a trance, Miss Minoprio had little use for a caddie. He carried only a small bag of balls and two clubs—a 2-iron and a spare 2-iron in case she broke one. She drove with it, putted with it, played her approaches with it, and played it reasonably well from sand bunkers.

With a sound, rhythmic swing, she hit the ball nicely if it sat on a peg tee or on a good lie, played solid run-up shots from just off the green, and she could putt well. She had trouble, though, from tight lies in the fairway or rough, or when she needed a high pitch.

Miss Minoprio lost her first-round match by 5 and 4.

The following year she won her first match by 2 and 1, then lost to Mary Johnson, an eventual finalist, by 7 and 6 in the second. She played in six

English Ladies championships, never winning more than one match. Her record gave Henry Longhurst, one of Britain's leading writers on golf, the opening to use the headline, *Sic Transit Gloria Monday* until she won a match late in her career. Then he changed it to *Sic Transit Gloria Tuesday*.

Miss Minoprio ended her career by losing in the first round of the 1939 British Women's Amateur.

At about this time a woman teed off in another Ladies Golf Union tournament with the sleeves of her blouse rolled up to her elbows. As she passed, she heard a spectator sneer, "She looks as though she had just done a day's washing."

Without changing her stride, she snapped back, "And it would not be for the first time."

Bobby Cruickshank

A man of indomitable spirit, Bobby Cruickshank had seen his brother blown apart by a German shell during the First World War, was captured and escaped from a prisoner-of-war camp and worked his way through enemy lines to rejoin the fighting. With these grim memories, golf tournaments held little terror for him. He played Bobby Jones stroke-for-stroke through the 1923 U.S. Open but lost on the last hole of a playoff, and although he hadn't won he had been a serious challenger in several others.

In 1934 it looked as if he might finally win. He led by two strokes as he approached the eleventh hole of the Merion Golf Club, a great short par 4 of little more than 360 yards with a stream called Baffling Brook flowing across the front and around the right side.

Cruickshank's drive settled in a divot hole; this would be no easy approach. As he moved into the ball, he tried to hit down sharply to be sure he dug it out, but he dug a little too deeply. It was obvious from the start that the ball would not carry to the green. Instead it plummeted into the creek bed. Cruickshank's shoulders sagged as he saw the ball fall, but suddenly it shot upward again after bouncing off the rocky bottom and popped onto the green.

Elated, Cruickshank flung his club heavenward, tipped his white linen cap, and cried, "Thank you, Lord." Just then his ears rang and he slumped to the ground dazed. The head of his iron had come down squarely on the top of his skull. Cruickshank was never the same. He finished the third round with 77, shot 76 in the afternoon, and tied Wiffy Cox for third place, two strokes behind Olin Dutra, one behind Gene Sarazen.

Henry Cotton

In the years between Hagen's first victory, in 1922, and Denny Shute's triumph in 1933, only one Briton had won the British Open championship—Arthur Havers beat Hagen by one stroke at Troon in 1923. Then along came Henry Cotton at Royal St. George's in 1934.

The championship was still being played over two days, the first thirty-six holes on Thursday and the second thirty-six on Friday. Cotton shot 66 in the qualifying round, 67 in the first round, and then 65 in the second. In shooting his 65 he played such exquisite irons that of his eleven birdies he holed not one putt from outside nine feet.

Playing under difficult conditions of wind and rain the next day, Henry shot 72 in the morning and opened a lead of ten strokes with eighteen holes remaining. He stumbled badly on the first nine of the second round, shot 40, righted his listing ship somewhat on the homeward nine and finished with 79. With 283, he beat Sid Brews, of South Africa, by five strokes. A Briton had finally won the championship once again. No American would win again until Sam Snead, in 1946.

Along with a number of other former champions, Harry Vardon had come to Sandwich to watch, but he was taken ill the last day and stayed in his hotel room. When the day ended, Cotton took the old trophy to Vardon's hotel and showed it to the man who had held it six times. Cotton reported later, "There were tears glistening in both our eyes."

The Double-Eagle

At about 5.30 on a chilly April afternoon in 1935, Gene Sarazen hit what is without question the most famous shot ever played, and quite possibly the most decisive as well. Played with a 4-wood, the shot carried about 230 yards, rolled a little, and dived into the cup of the fifteenth hole of the Augusta National Golf Club. This was the double-eagle, the shot that tied Sarazen with Craig Wood and set up Gene's victory in the playoff for the Masters Tournament the following day.

Because he had been scheduled to leave the country for an extensive exhibition tour through Australia with Joe Kirkwood in 1934, Sarazen had declined his invitation to the first Masters, but he made certain he would play in 1935. Installed as co-favorite with Wood following a set of sensational practice rounds, Gene trailed Craig by two strokes as he approached the fifteenth, a par 5 of 485 yards, reachable with two solid shots.

Paired with Walter Hagen, Sarazen drilled an exceptionally long drive over hard ground that might have put him in range of the green and allowed him to pick up at least one stroke on Wood. As Sarazen and Hagen walked toward their balls they heard an ear-splitting roar from the home green. Soon word reached them that Wood had birdied the final hole and posted a score of 282. Now Sarazen stood three strokes behind.

Hearing that Wood had birdied, Hagen said, "Well, that's that," then played a safe pitch short of the pond. Sarazen, though, wasn't ready to give up. Telling himself, "Shots might go in from anywhere," he spoke to Stovepipe, his caddie, named for the tall silk hat he wore. "What do I need to win?" he asked.

Somewhat surprised that Sarazen could even think of winning when he stood three strokes behind with four holes to play, Stovepipe questioned Sarazen, "What do you mean, Boss? To beat Craig Wood?"

Sarazen nodded and Hagen began to giggle.

"Oooh, you need four 3s, Mister Gene—3, 3, 3, 3."

Reaching his ball, Sarazen saw that it lay about 235 yards from the green, over the crest of a slight rise and beyond a wide pond. It could be reached with a good shot, but Gene's ball sat in a slight depression. Sarazen and Stovepipe agreed a 3-wood wouldn't get the ball up from the close lie, and so Sarazen drew out his 4-wood.

Out of contention, growing impatient, and somewhat incredulous that Sarazen thought he had a hope of catching Wood, Hagen called, "Hurry up, Gene; I've got a date tonight." Sarazen ignored him; Walter had been reminiscing about old times all day.

Before he played the shot, Gene reached into his pocket and drew out an elaborate ring a friend had given him as a lucky charm the previous evening, claiming Benito Juarez had worn it the day he was elected president of Mexico. He called Stovepipe over and rubbed the ring on his head for good luck.

About twenty people clustered around the green as Sarazen stood up to the ball. He toed in the clubhead to decrease the loft and give him extra distance, then rode into the shot. As soon as he made contact he knew his ball would clear the pond. As the ball soared into the clear blue sky, Gene raced to the crest of the rise to watch. The ball hit the ground just short of the green and ran toward the pin.

Bobby Jones had finished his round by then and was standing on a mound set between the fifteenth and seventeenth fairways. As the ball darted toward the flagstick, set in the right rear of the green about fifteen feet from the edge, Jones said to himself, "My God, he's going to have a chance at a 3."

The ball rolled toward the cup as Sarazen stood on the hilltop straining to see how close it would finish. Just then the small gallery roared and began to jump up and down. Sarazen knew then the ball had tumbled into the cup. He had scored 2 on a par-5 hole, and with one shot he had caught Wood. The following day he won the Masters in a thirty-six-hole playoff, shooting 144 against Wood's 149.

If not the greatest shot ever played, this was certainly the most famous.

Dizzy Dean

Looking over the Oakmont Country Club in Pittsburgh just before the 1935 U.S. Open, Dizzy Dean, the great pitcher, said, "If I didn't have to throw that tomato for a living, I'd take this game up in a serious way and win all the championships. And this is one course I'd tear wide open."

Prediction

Playing Oakmont's thirteenth hole in the final round in 1935, Sam Parks not only didn't know how he stood in relation to the leaders, he had been making definite efforts not to know. Coming into the championship Parks had hoped for nothing more grand than to place among the top ten, but after three rounds he had tied for first place with Jimmy Thomson, a very seasoned player who was among the longest drivers the game had seen. Sam's nerves jangled and the tension ate away at him.

The pin had been cut toward the back, where the green rose before tumbling down into a sand bunker, one of the nearly three hundred that made Oakmont the most penal course that had ever served at the Open site. Hit too hard, Parks's ball had run into the bunker behind the green and settled on a worm cast. Sam laid back his very heavy sand wedge and tried to blast his ball onto the green. Hit a little thin, the ball barely cleared the sand, hit the grass at the top of the bunker, which took some of the steam off it, then slowly rolled downhill and stopped about eight feet past the cup.

Parks holed it for the par three, then began his walk toward the fourteenth tee. Along the way he passed Grantland Rice, the most widely read sportswriter in the country. As he walked by, Parks heard Rice say to Fred Brand, Sr., a stalwart Oakmont member, "Say, this young fellow looks like he might just win this thing," and then followed him the rest of the way.

Rice was right. Even though Parks shot 76 that day, with 299 he had the only score under 300 for the seventy-two holes. Thomson placed second with 301.

Parks became the symbol for the upset champion.

The Trials of Alf Padgham

An early starter on the last day of the 1936 British Open, Alf Padgham left his clubs in the Hoylake golf shop overnight. When he arrived the following morning, he found the door locked and no one in sight; the young man assigned to open the doors had overslept. To avoid being disqualified, Padgham had no option—he broke down the door.

Later, in the morning round, he cut his tee shot into heavy rough beside two elderly ladies enjoying the warm sunshine beside the fifteenth hole, a testy par 4 of 460 yards. One of the women picked up Padgham's ball and set it atop a nice tuft of grass. She was seen by Leonard Crawley, a first-rate golfer who had played on Britain's Walker Cup team. Approaching her, Crawley told her in no uncertain terms, "You must not do that sort of thing." Unperturbed, the woman smiled meekly and told Crawley, "Oh, but Mr. Padgham's ball was in such a horrid place."

From this heaven-sent lie, Padgham rifled a 3-wood onto the green and holed a vast putt for a birdie 3. He shot 71 in the morning, another 71 in the afternoon, and with a 72-hole score of 287 won the championship by one stroke over Jimmy Adams, the Hoylake professional.

Two years later Padgham had other adventures at Sandwich during the 1938 British Open, which was played through a raging gale. The wind ripped the canvas covering of the exhibition tent to tatters and flung merchandise willy-nilly over the landscape. With the wind at his back as he played the eleventh hole, a par 4 of 380 yards, Padgham reached the green with his drive and holed the putt for an eagle 2. Reaching the fourteenth, a 508-yard par 5 that played directly opposite and into the wind, Alf hit four drivers and still hadn't reached the green.

Glenna Collett Vare

At a time when Jones and Hagen ruled men's golf, Glenna Collett Vare was devastating women's competition in the United States. She won six U.S. Women's Amateur championships during the 1920s and 1930s, but she never could win the British. She made her last serious attempt in 1936, a year after winning her last American championship.

The British was played at Southport and Ainsdale, on England's west coast, but Mrs. Vare, in her middle thirties and past her best years, went out in the first round, losing to Charlotte Glutting, one of the bright lights of American women's golf. Although she lost, Glenna showed how the game is played.

Both women had missed the eighth green, and Mrs. Vare played too strong a recovery, knocking her ball onto a steep bank. Drawing a club from her bag, she addressed her ball, then stepped away and changed clubs. When she had walked away, her ball rolled down the slope and stopped several yards closer to the hole. Glenna's little chip shot missed the cup, and she picked up her ball, conceding the hole to Miss Glutting.

Hidden in the gallery, the referee hadn't seen Glenna's ball move, and he thought the hole had been halved. No, Glenna insisted, she had incurred a penalty stroke because the ball had moved after she had addressed it.

Miss Glutting refused to accept the conceded hole, but Glenna had her way and eventually lost the match.

As a player of international standing, Mrs. Vare didn't enter club events until late in the 1930s, when she decided to play for the first and only time in the Philadelphia Country Club's ladies club championship. That she would win seemed certain, and perhaps in honor of having a woman who had won six national championships as its winner, the club had ordered a new trophy from the famous Philadelphia jewelry store Bailey, Banks and Biddle. The store promised the trophy would be delivered by the time the final match ended.

Sure enough, Mrs. Vare made it to the final match, and by the time it reached the seventh she stood several holes ahead. Still the trophy hadn't arrived. A call to the store brought the assurance that the trophy was even then on its way.

Half an hour later, still no trophy.

By the time the match reached the fourteenth hole, Mrs. Vare had so many holes in hand it was obvious the match would end very quickly, and still the club had no trophy to present.

Some times men feel the touch of genius. The chairman of the golf committee told a member of the board of governors he would pop home, grab a trophy one of his dogs had won, polish it brightly, present it to Mrs. Vare, and explain later. It seemed a good idea at the time.

Sure enough the chairman sped back to the club and placed a gleaming silver bowl on a table near the eighteenth green.

Mrs. Vare won, of course, and after a brief ceremony took possession of the trophy. Glenna accepted it with her usual grace, took a look at it, then put it back on the table and walked away in a huff. Bewildered, the chairman of the golf committee only then took the time to read the inscription: "Best Bitch in Show."

Nelson and Guldahl

Byron Nelson was always capable of wild bursts of scoring. Trailing Ralph Guldahl on the final day of the 1937 Masters, Nelson birdied the difficult twelfth hole, a par 3, and eagled the thirteenth, a par 5, just after Guldahl had played those two holes in 5 and 6. Nelson picked up six strokes on those two holes and won by two strokes.

Hagen at Home

In addition to his five PGA Championships, Hagen had won four British Opens by the end of 1929. When the United States won the Ryder Cup at Southport and Ainsdale, in England, in 1937, Hagen, as captain, spoke to the British gallery, saying, "I'm proud and happy to be the captain of the first American team to win on home soil."

Quickly someone reminded him he was in England, not back in the United States. Ever equal to the challenge, Hagen smiled and said, "You'll forgive me, I'm sure, for feeling so much at home over here."

A huge gallery crowded around the fifteenth green during the match between Gene Sarazen and Percy Alliss. Misjudging his approach, Sarazen carried his ball over the green and into the lap of a women spectator. Without a thought she tossed it onto the green so close to the hole that Sarazen scored an easy birdie.

Disappointed that he wasn't named the 1937 captain, Sarazen swore he would never play in another Ryder Cup. In a bitter statement he lashed out at George Jacobus, the president of the PGA, who had chosen Hagen for the sixth consecutive match, saying, "The failure of George Jacobus to name me captain of the team is an insult and a wound that will never heal. I think everyone will agree my record stacks up favorably with that of every other pro, past or present. I won every worthwhile championship, and I guess I did my share on every Ryder Cup team we ever had. But, no. Jacobus didn't see fit to let me achieve my last and profoundest ambition."

Sarazen never played in another match, but not because he turned down the honor. The next Ryder Cup was scheduled for the United States late in 1939, but it was canceled because war broke out in Europe on September 1. By the time the match resumed, in 1947, Sarazen was no longer competitive enough to warrant an invitation. He was never appointed captain.

Ralph Guldahl

Ralph Guldahl nearly won the 1933 U.S. Open, but he bogeyed the last hole and lost to Johnny Goodman, an amateur, by one stroke. He placed eighth the following year, but he fell to fortieth place at Oakmont in 1935 and played so poorly in other tournaments that he announced he was through with golf and went home to Dallas. His putting touch, always unreliable, had deserted him entirely. For a time he worked at odd jobs like selling cars, but because his young son, Buddy, had developed a lingering cold, Guldahl moved his family to the California desert. While he was there he played in a few tournaments, winning practically no money. Then he ran across Robert Woolsey, who was in the movie business, and the actor Rex Bell, an enthusiastic golfer. After they had played a few rounds together, Guldahl had impressed Bell and Woolsey so much they staked him to $100 (when $100 meant something) and sent him to the True Temper Open in Detroit in the spring of 1936. He won $240 there, placed eighth in the Open, worth $137, won $360 in another tournament, then won the Western Open. At the end of the year he had averaged 71.65 strokes per round, the best on the tour. The following year he won the Open at Oakland Hills, near Detroit. He was the best player in the game, all because of a $100 stake by two men who had faith in him.

By 1938, only Willie Anderson, in 1903 and 1904, Johnny McDermott, in 1911 and 1912, and Bobby Jones, in 1929 and 1930, had won the Open in consecutive years. After winning the 1937 championship, Ralph came back and won the following year at Cherry Hills, in Denver.

With 69 in the last round, Guldahl had the championship well in hand as he climbed the hill to the green of the seventy-second hole, and he knew a battery of photographers waited to shoot him holing the final putt. A tall, powerful man with dark, wavy hair, Guldahl wanted to look his best for the pictures. Just before stepping onto the green he stopped, pulled a comb from his pocket, and combed his hair.

Unhappy Record

Henry Picard shot two consecutive rounds of 70 and held the thirty-six-hole lead in the 1938 U.S. Open, but he drew not nearly so much attention as Ray Ainsley, an obscure golfer from Ojai, California. Ainsley had just surpassed Willie Chisholm as the man who posted the highest score for one hole ever recorded in the championship. Flailing like a headsman gone mad, Ainsley took nineteen swipes at his ball before it sank inside the hole.

He committed his offense on the sixteenth hole, a par 4 of 397 yards with a stream purling along its right flank. Ainsley's drive drifted off to the right and sank below the waves. He could have lifted his ball at a penalty of one stroke, then gone on his way, but Ainsley would have none of it; no easy way out for him. Instead, he waded in and began the longest sustained punishment one ball has ever endured in so important a competition.

As the strokes mounted, the official scorer who walked with each pairing in those days asked for help from Bud McKinney, Ainsley's playing partner. Covered from head to foot with sand, and squirming inside his sopping wet clothes, Ainsley continued whacking away. With each blow the swift current carried the ball farther downstream, Ainsley wallowing in its wake. With each blow rocks, sand, and weeds piled up on the stream banks, but still the ball escaped the full fury of Ainsley's wrath.

But no mere golf ball is equipped to stand up to an assault so determined as Ainsley's. Finally it sought shelter in a peaceful eddy, but Ainsley found it, caught it squarely with a final blow and sent it flying behind a tree on the far side of the green. Still Ainsley fought on. Eventually he nursed it onto the green where, with a final soft pat, he nudged it into the hole. Nineteen strokes.

Ainsley had actually set two records that day. Not only did he set the one-hole record, but so far as anyone knows, he still holds the record for the greatest difference between rounds. He had shot 76 in the first round, 96 in the second. As Ainsley sat moaning over his humiliation, Morton Bogue, the chairman of the USGA's rules of golf committee, asked why Ray hadn't accepted the penalty of one stroke and lifted his ball from the churning waters.

Ainsley responded, "I thought I had to play the ball as it lay at all times."

Movin' On

Frank Strafaci grew up in Brooklyn, New York, and became quite a good golfer, good enough to win the 1935 Amateur Public Links championship shortly after he had turned nineteen. He played so well he qualified for the 1937 National Open and placed ninth, ahead of Byron Nelson, who had won the Masters earlier in the year, and Gene Sarazen and Sam Parks, who had won, respectively, the Masters and the Open the year Frank won the Public Links.

As a consequence of his placing so high in the Open, Strafaci was invited to the 1938 Masters, but he played so badly he had fallen well out of contention after the first three rounds. Adding to his problems, rain had caused Masters officials to postpone one round, and when a second had to be held over, Frank realized the fourth round of the Masters would conflict with qualifying for the North and South Amateur, in Pinehurst, North Carolina, a few hundred miles north of Augusta.

It was one of the Augusta National's understood policies that those invited to the Masters were expected to play through to the end, but Strafaci found this a little confining. A lifelong amateur, the North and South meant more to him than competing against professionals in the Masters, so he packed his bags, withdrew from the Masters, and arrived at Pinehurst in time to win the North and South.

Lefty Stackhouse

Wilbur Artist (Lefty) Stackhouse, a small, crusty, rawboned Texan with whipcord muscles and leathery skin who played the PGA Tour in the 1930s and 1940s, lived with an insatiable thirst and a raging temper. Fortunately for the rest of mankind, he aimed most of his eccentricities against himself. Failing to carry a water hazard, Lefty dived head first into the rocky stream bed. Another time he missed a gimme putt on the final green, stood up, roared, then ran straight ahead to a tree and butted his head two or three times.

He was equally liberal with punishment to other parts of his body. After duck-hooking a drive, Stackhouse walked to a rose bush and whipped his right hand back and forth through the thorns, tearing long gashes and shouting, "That will teach you to roll on me." Then, his hand dripping blood, he held up his left hand and cried, "Don't think you're going to get away with it, either," then whipped it through the bush as well.

Leading a tournament late in the second round, he sent a shot clattering through a stand of palm trees. It didn't come down, but Lefty didn't give up easily. Jumping up and down, he made three or four swipes at it, but couldn't reach it. Finally giving up, Stackhouse tossed his club to his caddie and told him he'd see him at the clubhouse, and without a further word began walking in a straight line. Approaching a water hazard, Lefty held a steady course, walked in one side and out the other, sopping wet.

Stackhouse was known to have a drink. He always claimed he had two idols—Walter Hagen and Tommy Armour. He once said, "They always had a bottle of Scotch sitting in front of them, and I wanted to be like them. If I had a good round, I would celebrate."

After once celebrating with Ky Laffoon, Lefty woke up nine hours later, lying on his bed but still fully dressed, shoes and all. He felt terrible, so he called for room service and said, "Send me two beers before I die."

Shelley Mayfield, who grew up to play the pro tour during the 1950s, caddied for Stackhouse in Seguin, Texas. Lefty stood four under par after seven holes, but even though he had a chronic hook, he pushed his next drive into a lake. As the ripples spread, Stackhouse snatched his bag from Mayfield's shoulder and threw it into the lake, then took off his shoes and socks, rolled up his pants legs, and walked barefoot back to the clubhouse through a 250-yard field of bull nettles.

Trying to stimulate attendance, promoters of a tournament in Knoxville, Tennessee, persuaded Sergeant Alvin York, the legendary World War I hero, to come along simply as an attraction. A quiet, unworldly man from the backwoods of Tennessee, York had only a vague idea of golf. He was standing behind the eighteenth green when Stackhouse staggered up the fairway. Lefty had begun celebrating earlier in the day and had a serious list to starboard as he stumbled up the grade in the general direction of his ball, which lay in a bunker. He reeled toward the edge, teetered on the brink, then crumbled into a heap and fell flat on his back in the sand, his arms spread wide, and his eyes, wide as an Indian fakir's, staring at the sun.

After a short rest Lefty pulled himself together, played a successful explosion shot, then crawled onto the green. Standing over his ball as he prepared to putt, Lefty fell flat once again. He struggled to his feet, then fell again. Sergeant York watched in wonder, then turned to his companion and said quietly, "I didn't know golf was such a violent game."

How About a Game?

Hotel accommodations in St. Andrews were not only scarce in the years leading up to the Second World War, they were expensive as well, often too expensive for the professionals who had come for the British Open. Often they stayed in neighboring villages while the more affluent amateur golfers, for whom the championship was a social occasion, frequented the costly hotels either bordering the Old Course or close by.

Not only was housing hard to come by, but so were starting times on the Old and New Courses, where the field would have to qualify for the Open proper. Jimmy Alexander, the one-armed custodian of the first tee, passed them out as if each were a gold sovereign—no matter that a player might find a practice round useful.

As the 1939 championship drew near, a player who had been unable to book a time for practice approached R. H. Foster, a well-known scratch man and Open entrant, while he sat at dinner in Elie, about twenty miles from St. Andrews.

"Excuse me, Mr. Foster," he said, " but I understand from Jimmy that there are only three of you teeing off on the New tomorrow morning. By any chance could I join up with you? I haven't played that course before."

"Of course you can," Foster said.

Then the player introduced himself. His name was Richard Burton. The following week he shot 298, beat the American Johnny Bulla by two strokes, and won the British Open.

On September 1, 1939, Germany invaded Poland, igniting the Second World War. The Royal and Ancient Golf Club of St. Andrews canceled the British Open for the duration.

Richard Burton reigned as British Open champion longer than any other man. When it resumed, in 1946, he had been waiting seven years to defend, five of them in the Royal Air Force. The 1946 British Open was played at St. Andrews once again, and when his turn came to tee off, Burton demonstrated how the tension can build up after so long a wait. His opening drive flew over the fence to the right and into a refreshment stand. Burton shot 74 in the opening round and followed with three 76s. With 302, he placed twelfth.

Because the war deprived him of the opportunities that could have earned him a great deal of money, Burton was considered an unlucky champion. After his discharge from the RAF, Burton was asked if he indeed considered himself unlucky. "Unlucky?" he answered. "I came through the war, didn't I? That's a lot better than many did who were at St. Andrews the day I won."

A Failed Quest

Henry Gonder, a club professional, spent eighteen hours and twenty-five minutes one day in 1940 deliberately trying to make a hole-in-one on a 169-yard hole. After 1,817 shots, he gave up.

Ky Laffoon

Perhaps part American Indian, and then again perhaps not, Ky Laffoon could be called flaky. A man with a Krakatoan temper, Laffoon, like Lefty Stackhouse, was among the tour's characters. Although he was an outstanding player, he never won anything of consequence, but his antics made him immortal. At one time he swore he'd never again score higher than 72, shot 67–69–65 in the first three rounds of a tournament, and then facing a short putt on the eighteenth green of the fourth round, picked up his ball and walked off. The putt would have given him 73.

Missing a five-footer in a Sacramento Open, Laffoon whacked his mallet-headed Schenectady putter against his foot with such force he snapped the putter's shaft and broke one of his toes as well. Limping and using the broken putter, whose remaining shaft was barely long enough to grip, Laffoon bent over to hole his tap-in, didn't bend low enough, missed the ball, then stuttered, "Wh . . what the h . . . h . . . hell is h . . . h . . . happening here?"

Since he liked to chew tobacco while he drove his yellow Cadillac, he wound up with a two-toned car—the basic yellow streaked with brown tobacco juice.

He often had disagreements with his putter and occasionally felt a little punishment might be due. With Ben Hogan snoozing in the back seat as they drove from one tournament to another, Laffoon once tied his putter to the door handle and dragged it along the road ("The dirty son of a bitch deserved it," he insisted), trailing sparks, which flew into the back seat, frightening Hogan awake.

Laffoon played at a time when a player had to finish among the top 10 or 15 places to win money. He needed a par 4 on the closing hole of one tournament to win enough money to get to another. Ky misjudged his pitch to the closing hole. His ball hit the back of the green, bounced over, rolled down a bank, and sank into a pond.

Laffoon's face turned red. He gripped his wedge, then flung it toward the same pond, screaming, "Drown, you son of a bitch."

Because of his tantrums, his wife eventually warned she'd leave him if he ever again made a fool of himself on the golf course. Not long after the ultimatum, Laffoon mis-hit a ball into a forsythia bush. One swing failed to bring the ball to earth. Another failed, and then still another swipe had no noticeable effect on the ball's position. Blood rising to full boil, teeth grinding, and a low growl rising from within, Laffoon wrapped his massive hands around the forsythia and ripped it from the ground, roots and all. Watching in growing dismay, his wife spun away and began a rapid walk toward the parking lot.

Sanity breaking through his crimson rage, Ky suddenly awoke to his sad plight and raced after his wife clutching the bush and pleading, "I wasn't mad about missing the ball, Honey, it's just that I hate forsythia."

Gravely ill, Laffoon eventually took his own life with a shotgun.

Jimmy Demaret

During the Depression years, many professional golfers who held club jobs played tournaments as a sideline. When Jimmy Demaret won the 1940 Masters, he had to hurry back to the Brae Burn Country Club, in Houston, where he held the pro job. Asked why, Demaret said, "I was scared somebody might steal my job."

The following year, Demaret, as the previous year's winner, was paired with Bobby Jones, the tournament's host, in the first round. After driving on the thirteenth hole, a par 5 of 485 yards then, Demaret first drew out a wooden club, then played an iron instead. Seeing Jimmy lay up short of a creek that knifes across the fairway only a few yards short of the green, Jones asked if he had thought of going for the green with his second. Demaret answered, "Bob, I couldn't have got there from where I was with a Greyhound bus."

Demaret was standing at a bar when he was told Ben Hogan had said that if Jimmy had practiced as hard as Hogan practiced, he'd win every tournament he entered. Obviously pleased, Demaret smiled, raised his glass and said, "I'll drink to that."

Whose Honor?

Dr. Walter C. Ratto, of Rio De Janiero, accompanied by Mario Gonzalez, of São Paulo, sailed for the United States and the 1941 Open, at the Colonial Country Club, in Fort Worth, Texas. Unfortunately, because they sailed too late to play in the qualifying rounds, it seemed likely for a time they would realize nothing from the venture except a long sea voyage. But they didn't give up. After a series of telegrams between USGA headquarters, in New York, and the offices of the Brazilian Golf Federation, in Rio, with the Brazilians pleading for special exemptions in the name of good relations with our neighbors to the south, the USGA allowed them to play.

Gonzalez was a first-class player. Dr. Ratto wasn't. Dr. Ratto had both a bad and a puzzling beginning. In the first round he was paired with an American professional. Obviously nervous as he stepped onto the first tee, he took his position at the ball, waggled his club a few times, then flung himself into his swing. It was not a thing of beauty. Catching the ball on the wrong lurch, he hit it off the toe of his driver. It squirted dead right, slammed solidly into the trunk of a tree, rebounded back toward Ratto, streaked past him, then rolled dead behind the tee. Now his ball lay farther from the hole than when he teed it up.

This prompted an interesting question. The American professional looked Dr. Ratto squarely in the eye and asked, "Well, now, Doctor, is it my honor or are you still away?" Dr. Ratto shot 90 that day and 100 the next. He didn't survive the thirty-six-hole cut. Nor did Gonzalez.

Royalty No Barrier

Both the Prince of Wales and the Duke of Kent were students of Archie Compston, a crusty, brash, outspoken, but generous professional based at Coombe Hill, a shortish, though narrow and hilly course near London. As the duke and Compston walked off the first tee together at the beginning of a round, Compston shouted at him, "Don't carry your drive over your shoulder. It's not a flipping gun, man."

Compston and the prince often played together at a series of courses near London—Coombe Hill, Sunningdale, and Royal Berkshire. After playing Royal Berkshire with Rex Hartley, a member of two Walker Cup teams, and Bob Foster, a businessman, the prince led the group from the bar into the dining room. While Foster settled the bar bill, Commander

Sarrell, the club secretary, approached him and said, "Will you tell Compston professionals are not allowed in the clubhouse."

Foster replied, "You had better tell that to the prince. Compston is his guest."

Sarrell did just that.

The following week the club's name reverted to simply Berkshire. The Royal had been removed.

Regular attendance at Coombe Hill by the Prince of Wales prompted two American oilmen to buy the club. They rebuilt the clubhouse into a modern, functional building with roomy locker rooms with showers and all the other conveniences sadly missing from most British clubhouses of the time. They also built a private locker room, shower, and toilet for the prince. They had wasted their money; the prince never used it. The two Americans died early in the Second World War. While it was coming in to land in Lisbon, their flying boat crashed. The singer Grace Moore died with them.

Pine Valley

Tucked into the sandhills of southern New Jersey, within half an hour's drive of Philadelphia, Pine Valley Golf Club is not only one of the most exclusive of all golf clubs, but it ranks among the best five or six in the world. The club was the creation of George Crump, a wealthy amateur golfer who owned the Colonnades Hotel, in Philadelphia.

Along with a group of friends, Crump, an ardent and accomplished golfer, had grown so weary of the train ride to Atlantic City for a winter weekend of golf he determined to build his own course on 184 acres of useless scrubland. The land was bought in 1912; Crump went to work the following year.

Not only did his ideas show a touch of originality and a genuine sense of sound architecture, he brought the accomplished golf course designer H. S. Colt over from England to help. Colt had been responsible for the great courses at Sunningdale, Rye, and St. George's Hill.

During construction of Pine Valley, Crump literally became a recluse. He pitched a tent in the middle of this wasteland and never seemed to leave. He eventually built a small cottage on the banks of the stream that separates the fifth tee from the green.

Crump's commitment to Pine Valley never weakened. During one three-year period he spent more than $250,000 of his own money on its construction, without any hope of getting it back. Working from the plans that he and Colt had drawn up, Crump personally directed the day-by-day work. The work dragged on year by year, and then Crump died in 1918 at the surprisingly young age of 46 without ever playing the course. The twelfth, thirteenth, fourteenth, and fifteenth, a magnificent swing of demanding holes, were still to be built. Dedicated to finishing the project, Crump's partners in the enterprise brought in Hugh Wilson, who had laid out the great course at Merion. Wilson and his brother Alan finished the course.

There is a basic principle to playing a match at Pine Valley: never pick up, never give up. With twelve holes to play during the 1936 Walker Cup Match, the American pair of Harry Givan and George Voigt stood seven holes up in their foursomes match against Ireland's Cecil Ewing and England's George Hill. Ewing and Hill rallied to halve the match. Instinct leads us to suppose Ewing and Hill littered the landscape with birdies. They didn't. During this gallant comeback they shot a scintillating two over par. Over those same twelve holes, the Americans struggled home nine over par.

In the late 1930s, Pine Valley invited a dozen of the game's leading professionals to team with club members in an annual seventy-two-hole invitation tournament. Among those who accepted the invitations were Byron Nelson, Sam Snead, Gene Sarazen, Craig Wood, Vic Ghezzi, Leo Diegel, Horton Smith, Ed Dudley, Jimmy Thomson, and Jimmy Hines.

In more than 450 rounds, Pine Valley's par of 70 was broken just twice: Craig Wood shot 69 in 1938 and Ed Dudley shot 68 in 1939. In 1937, when the course measured only a bit more than 6000 yards, Snead and Thomson, the longest drivers in the game, tied for first place at 302, an average of more than seventy-five strokes per round and twenty-two over par.

Each year the club honors the memory of its founder with the Crump Cup, a match-play competition that attracts a field loaded with Walker Cup players. Guests may own handicaps as high as seven strokes, but club members must play to a three. Each year someone figures a ringer score of the *worst* scores on each hole. In 1983 the Crump Cup field put up a worst-score-per-hole total of 79–79—158 for eighteen holes against a par of 35–35—70.

Neither the second nor the third hole offers any water or out of bounds. Someone shot 12 on the second, a straightaway 360-yard par 4, and another posted 9 on the third, a par 3 of 180 yards. Playing into the full force of a strong wind, one man reached the front of the fifteenth green, a par-5 hole that plays 603 yards uphill, with his fourth stroke, and then made 12. He chipped six times. His first putt ran out of momentum before it reached the upper plateau of this slanted green, turned around, and rolled back off. His sixth chip stopped close enough to the cup for him to hole a putt.

Everything seemed to work perfectly for Bob Lewis in the qualifying round for the 1981 Crump Cup. A member of two American Walker Cup teams and runner-up in the 1980 Amateur, Lewis hit seventeen greens;

on the ninth, the only green he missed, his ball sat on the green's collar just a tantalizing foot off.

Out in 31, he came back in 33 and set the course record at 64. The next day he lost his first-round match. Two years later, when he returned, Pine Valley was waiting for him. He shot 82. It was waiting for everyone that day, though. Only five men beat him.

John Arthur Brown, a Philadelphia lawyer, ruled as Pine Valley's president from 1925 until his death, in 1977, at the age of ninety-two, still a virile and physically strong man. Brown ruled the club as Russia was ruled by Ivan the Terrible. He decided everything, and every member had better behave on the premises.

The club offers accommodations to its members in what it calls the dormitory, a building just a few yards from the clubhouse. One member brought a group of guests for a few days of golf. They partied so loud and so long that word reached Brown. At breakfast the following morning, Brown approached the member and told him, "I've decided to accept your resignation."

Another member brought a group of guests who struggled so badly through the first four holes they held up several other four-balls behind them. Again word reached Brown. As the offending players stepped off the fourth green, which sits alongside the clubhouse, Brown flagged their caddies and told them to take their clubs to their cars, saying, "These gentlemen are through for the day."

Brown cared nothing about how powerful a man might be. Ben Fairless, the chairman of U.S. Steel, decided the club needed a better golf shop than the crowded little room behind the club's reception desk, and that he was the man who would build it. He knew just where it should sit—on the hill behind the eighteenth green. It would feature a patio beside it where members could lounge while their comrades struggled with the closing hole.

Saying nothing to Brown, Fairless summoned Eb Steiniger, the green superintendent, and ordered him to drive stakes into the ground outlining the foundation for the new shop. When Brown saw them he sent for Steiniger, telling him to come immediately. A man with a powerful voice, Brown barked, "What the hell's going on here?" Told that Fairless had ordered him to plant the stakes, Brown stormed into the clubhouse, confronted Fairless and told him, "We don't need a new golf shop at Pine Valley, and we certainly don't need you building it. And furthermore, I'm not even sure we need you in this club."

A very important investment banker from New York who also held a high position in Washington violated Brown's idea of proper conduct so badly

Brown expelled him, sending off a terse note. When the mail arrived, the member telephoned Brown and cried, "You can't do this to me today."

Brown roared back, "You're right; I did it yesterday."

J. Wood Platt, a Pine Valley member, ranked among the best golfers Philadelphia ever produced. He started one round by playing a nice drive on the first hole, a par 4 of 427 yards that turns sharply right in the drive zone, and following with a 4-iron close enough to hole a birdie putt. The second hole is a straightaway par 4 of just 367 yards, but its green, which rolls and tumbles like the waves of a troubled sea, sits high above the fairway atop a sharp rise pitted with bunkers. After a solid drive, Platt holed his 7-iron second for an eagle 2.

Pine Valley's third, a par 3 of 185 yards, plays across sandy scrubland from a high tee to a scalloped green tilting decidedly from right to left. Miss the green and you might need two more shots to recover. Platt solved that problem by holing his tee shot.

On to the fourth, a difficult par 4 of 461 yards where the drive must clear a hill rising high above the level of the tee. From there the ground falls away before leveling off for its run to the big green. Platt reached the green with a 4-wood second, then holed a thirty-foot putt for another birdie. Now he stood six under par for four holes of one of the toughest golf courses in the world.

The path from the fourth green to the fifth tee winds past the clubhouse. To bolster his courage before facing the fifth, a horrific par 3 of 226 yards across a river to a small green set in a pocket of trees where even a slightly mis-hit shot could result in a lost ball, Platt ducked into the bar for a quick drink. One drink led to another. Platt didn't leave the bar the rest of the day.

He Glows in the Dark

Eben Byers, a wealthy industrialist from Pittsburgh, won the U.S. Amateur championship of 1906 after losing to Louis James in 1902 and to Walter Travis in 1903. A small man, not quite five feet-five inches tall, he was twenty-six the year he won.

The son of the founder of A.M. Byers Company, one of the nation's largest steel manufacturers, Byers came to be chairman of the board as well as a director of Westinghouse Electric and Manufacturing. He was a socialite, a horse-racing enthusiast, and according to some sources an ardent ladies' man.

A 1901 graduate of Yale, he hurt his arm when he tumbled from the top berth of his Pullman compartment on his way home from the 1927 Yale-Harvard football game, where he'd evidently overdone his post-game celebration. He was forty-seven at this time.

Not only had Byers bruised his arm, over the next few weeks he complained as well of muscular aches and a run-down feeling that undermined his performances both athletic and sexual. A Pittsburgh physiotherapist recommended an over-the-counter patent medicine called Radithor.

Marketed by a con-man named William J. A. Bailey, Radithor promised to cure 150 maladies, but it was sold mainly as an aphrodisiac. Looking like nothing more than a bottle of water, the medicine promised to improve blood supply to the pelvic region, which in turn produced "tonic effects upon the nervous system and generally results in great improvement to the sex organs."

The description came from a pamphlet "Radithor, The New Weapon of Medical Science" circulated to doctors in the middle 1920s. Bailey claimed Radithor was the result of years of laboratory research. Essentially, though, it was nothing more than distilled water laced with radium, the magic ingredient.

Selling for one dollar a bottle, it became popular with the well-to-do, who could afford it. Sold as a health drink, it was based on the theory that taken in mild doses, radiation provided a metabolic kick to the body's endocrine system and infused depleted organs with energy.

For two years after being advised to take Radithor, Byers gulped down two bottles every day, nearly 1500 half-ounce doses. They produced astonishing results. His health evidently improved so remarkably he sent cases of it to his business partners and girlfriends. He even had it fed to his racehorses.

By October of 1930, though, he had quit, complaining to his doctor that he had lost that "toned-up feeling." He began to lose weight. He had headaches. Then his teeth fell out.

By then the Federal Trade Commission had begun hearings on other radium-water remedies and asked Byers to testify. He was too sick. Instead, the FTC sent a lawyer to take a statement at his mansion on Long Island. Seeing Byers, the lawyer was aghast.

He reported Byers's whole upper jaw, except for two front teeth, and most of his lower jaw had been removed. The lawyer went on, "All his remaining bone tissue was disintegrating, and holes were actually forming in his skull."

He died on March 31, 1932, in a New York City hospital from radium poisoning. He was fifty-one. He had lived four and a half years after experimenting with Radithor.

Using complex computer techniques, scientists calculated the level of Byers's exposure. Examining the study, Dr. Roger M. Macklis, of Boston, said, "The mystery is how did Byers survive so long, feeling so good with such a lethal burden in his body. He took enough radium to kill four people."

Respect the Sabbath

Dr. Everett Herrick, a God-fearing Methodist, served as the first president of the Maidstone Club, an upper-crust club on eastern Long Island,

New York, that, in concert with the National Golf Links of America and the Shinnecock Hills Golf Club, anchors the social set that summers in the Hamptons—Easthampton, Southampton, Bridgehampton, and every other Hampton extant.

A wealthy landowner, Dr. Herrick allowed some of his property to be used for the infant Maidstone Club rent-free with one quite strong provision: No golf could be played on Sundays. The club accepted the terms, but when Dr. Herrick agreed to sell the land to the members, they lifted the ban on Sunday golf. As a compromise, they ruled too that games couldn't begin until after 12:30, a time when members' weekly obligation to the church could be fulfilled.

Herrick was also a health-food nut and a teetotaler. When he died, in 1914, he bequeathed the club $7500, again with a serious provision. If intoxicating beverages were ever sold on club grounds, the money would revert to the East Hampton Free Library. Everything went well through the period of Prohibition, but when the Eighteenth Amendment was repealed, in 1933, an anonymous member sent the club a check for $7500, which meant that the Maidstone bar was soon as well stocked as the public library.

Enough Is Enough

V. L. Whitney, who represented an oil company in India, held membership in the Royal Calcutta Golf Club. An enthusiastic golfer, although not a good one, he found himself grouped in a four-ball one day, and started off 3–4–4, three consecutive birdies at the time.

After his third birdie, Whitney turned to his partner and said, "I'm not going to spoil this. Besides, I've already done more than you could possibly expect of me for a full round." Then he turned and walked back in.

Unscreened Entry

Watching the 1933 Western Open, at the Olympia Fields Country Club, on the outskirts of Chicago, John Shanahan, a spectator, saw a familiar face, a man he had played with at the public Cog Hill course a few weeks earlier. Later in the day Shanahan asked a tournament official if the Western Golf Association screened entries. Assured the association did indeed screen its entrants, the official asked why.

"Because Vince Gebhardi is playing," Shanahan answered.

"So who is Vince Gebhardi?"

"He's Machine Gun Jack McGurn," Shanahan said. An alleged hit-man with the Capone gang, McGurn was listed as Public Enemy No. 5. An accomplished golfer, he had entered the Western under his own name, claiming to be a professional from the Evergreen Golf Club, a daily fee course in suburban Chicago.

In the early 1930s, before the birth of the Masters, the Western Open ranked as the third most important tournament in American golf, behind the Open and the PGA. Most of the great players were there—Macdonald Smith, who eventually won the tournament, Jock Hutchison, and Tommy Armour were out on the course, and a gallery of several hundred spectators clustered around the eighteenth green. Suddenly a murmur spread through the crowd as the fans noticed a phalanx of uniformed policemen filter out of the clubhouse and lounge behind trees, trying to remain inconspicuous. Alerted that McGurn (Gebhardi) was playing in the Western, a municipal judge had issued a warrant for his arrest under a recently enacted "criminal reputation law." Two detective sergeants, John Griffith and John Warren, from the county highway force, were dispatched to take McGurn into custody, but realizing that arresting a man with a reputation for such violence might not be so easy, Griffith and Warren asked for help from Lt. McGillen, of the Homewood station of the highway patrol. McGillen brought the five uniformed policemen with him.

McGurn was playing great golf. He had opened with 83 in the first round, but now he stood one under par after six holes of the second. Dressed in light gray flannels and a white shirt, he strode along with confidence while his wife walked alongside. Louise Rolfe, called the "blonde alibi" because she saved him from prosecution in a killing known as the Clarke Street massacre, was a golfer herself; she had played in a woman's tournament at Olympia Fields only a few weeks earlier. Now she walked with Jack, wearing a tight white dress, white hat, and anklets, flashing a three-carat diamond ring on one hand and a glittering wedding band on the other. McGurn had given her the wedding ring to escape a federal sentence for violation of the Mann Act.

Seeing McGurn and Louise, walking together so openly, McGillen said, "The heat must be off or McGurn wouldn't bring her out here. If the Tougheys (a rival gang) were after him, they could shoot both of them like sparrows."

As the McGurns, both deeply tanned, approached the seventh green, McGillen stepped up and read the warrant. McGurn listened while the warrant was read, then asked politely if he could be allowed to finish his round. McGillen, just as civily, agreed. After all, there was no need to upset the golf tournament.

With McGillen striding alongside him all the way, McGurn double-bogeyed the seventh hole. Meantime, Louise, fuming, could hold herself in check no longer. "Whose brilliant idea was this?" she screeched at McGillen. He simply shrugged his shoulders. Then the situation broke down completely. As McGurn played the eighth hole, a photographer darted from the gallery and snapped three quick pictures. McGurn froze and tightened his lips while his club shook in his hands. He shot 11, seven over par on one hole. The hole over, McGurn rushed up to the terrified photographer, grabbed him by his shirt front, shook him, and snarled, "You've busted up my game." Meanwhile, hand on hip Louise posed prettily.

Up ahead, one player who'd had enough of his own sour game, quit in mid-round, and McGurn and his partner, a young man from a nearby town, were joined by Arthur Tilley, an Olympia Fields member who had been golf champion of the Chicago Bar Association. Having an attorney with him seemed to calm McGurn; he played the second nine in 41 after going out in 45, shot 86 for the day, and missed the cut anyway.

The golf formalities over, McGillen tapped McGurn on the shoulder, the uniformed men closed in, and they walked away. McGurn was allowed to drive his own car to the police station. McGurn and Louise sat in front, McGillen and Sergeant Griffith rode behind in the rumble seat.

Al Capone was known to play golf, too, although not as well as his henchman, nor with such favorable results. And he always carried a pistol in his golf bag. While he was playing the Burnham Woods course, near Chicago, in 1928, his caddie dropped his bag on the ground. His pistol went off and shot Capone in the foot.

Questionable Tip

George T. Dunlap, a likable American Walker Cup player, reached the semi-final of the 1933 British Amateur, where he lost by 4 and 3 to the Hon. Michael Scott, at Hoylake, in England. In losing the match, Dunlap had become an enthusiastic fan of Scott's, and indeed stayed on and cheered him on to a 4 and 3 victory over T. A. Bourn the next day.

The match over, Dunlap bought a dilapidated second-hand car, and with his wife, dressed more appropriately for a Rolls-Royce, rattled away to St. Andrews. Arriving after only two breakdowns, Dunlap shot 306 in the British Open, tying for thirty-fifth place. The championship over, Dunlap gave the car to his caddie.

English Ingenuity

Henry McLemore, an American sports reporter, was assigned by his newspaper to cover the 1934 Walker Cup Match and then the British Amateur, at the Prestwick Golf Club. After the United States won the Walker Cup at St. Andrews, McLemore arrived at Prestwick, entered through the gate marked "Greengrocer and Press," and approached the club secretary. After introducing himself, he asked the secretary for a press badge, or something that would allow him access to the clubhouse and locker room so he could talk to the players.

Listening to this unexpected visitor, the secretary, adjusted his tie, cleared his throat, refolded his handkerchief, and generally fumbled around until finally saying he was sorry, but all available press credentials had been issued to the British corespondents. "It's distressing, old boy," the secretary said, but don't be alarmed. For three pounds you can join the club for a week and jolly well see everything that is going on."

Unfortunately, however, McLemore's membership did not allow him access to the clubhouse; no working-man could enter that sacred place.

When Lawson Little played the Scotsman James Wallace, a bricklayer from the nearby town of Troon, in the final match, Prestwick members gathered in the clubhouse stood and raised their glasses in a toast: "Here's to Lawson Little. May an artisan never win our championship."

Little did indeed win.

Joe Ezar

Joe Ezar was an eccentric character better known as a trick shot artist than a competitive golfer. During August of 1936 he traveled to Sestrieres, in the Italian Alps, for the Italian Open. He had just won the German Open, and from his prize winnings bought himself a camel's hair topcoat, its edges bound in leather. He wore it everywhere, even when he played, slinging it across his shoulders and handing it to his caddie as he played his shot.

On the evening before the third round, Ezar went through his trick shot routine for the gallery, hitting hooks, slices, high balls, low balls, two balls at once, standing three balls one on top of the other and hitting the center ball, hitting a ball and catching it, and an assortment of other weird and wonderful displays of extraordinary coordination. It was a first-rate performance, so good it impressed the other players. It also impressed the club president, a non-golfer. Handing Ezar his fee, he said, "It's a wonder with your skill you don't break the course record."

The hair on the back of his neck rising, Ezar said, "What is the record?"

Henry Cotton had just set it by shooting 67s in the first two rounds.

Ezar then said, "How much would you give me if I do break the record."

Smiling, the club president said, "One thousand lira for 66."

"How much for a 65?"

"Two thousand lira."

"And for 64?"

Laughing now, the president said, "Four thousand lira for 64."

"Right," Joe said, "I'll do 64."

The deal made, Joe asked the president for a slip of paper. On it he wrote down the scores he would make hole-by-hole to shoot his 64.

With the club president stalking him the next day, Ezar started out right on schedule, matching the scores he had predicted on each hole through the eighth. He ran into trouble on the ninth, a par 3. Lying two at least fifty yards from the hole, he told the president he must hole this shot to maintain the schedule. He pitched into the cup.

He had gone out in 32, then came back in 32, scoring each hole just as he had announced while the club president, his playing partner, and several other spectators watched.

Cotton won the tournament and Ezar finished second.

The *Queen Mary*, the largest ship afloat at the time, sailed for the United States for the first time in May of 1936. Entering his cabin on the maiden voyage, the British sportswriter Trevor Wignall saw a golf bag lying on his bed with the name Joe Ezar emblazoned on its side. Assuming it had been left in his cabin by mistake, Wignall thought no more about it and wandered on deck while the ship slipped its moorings and tugs started the *Queen* on her way.

The ship finally free of the tugs and heading out to sea, Wignall returned to his cabin, opened the door of his wardrobe, and looked squarely into the face of smiling Joe Ezar. Flat broke, he had stowed away.

After Wignall convinced Joe he should give himself up and get off the ship, Ezar twice approached the purser, who was extremely busy, and both times he was told to wait until the ship left Cherbourg, the first port of call. By then it was too late to get off. Ezar eventually was assigned a tourist-class cabin and helped pay his passage by giving exhibitions.

Golf and Hitler

While the 1936 Olympic Games were under way in Berlin, golf enthusiasts organized an international foursomes competition in the spa town of Baden-Baden. The trophy, a large brass salver inlaid with yellow stones, was to be presented by Adolf Hitler only if the German team won.

With one round left, the Germans held so comfortable a lead a message was sent to Hitler telling him he should start out for the course. Immediately after lunch, though, the English pairing of Tony Thirsk and Arnold Bently began burning up the course, set a course record, and passed the Germans.

Nazi officials went into panic. Hitler was on the way, Germany had lost, and they had no means to cutting him off. Then Hitler arrived, confident the Germans had won. Told he'd have to present the trophy to the English instead, Hitler flew into a rage, ordered his driver to turn the Mercedes around, and raced back to Berlin. The English accepted the trophy from a group of embarrassed officials.

Prodigal's Return

After losing to Johnny Fischer on the thirty-seventh hole of the 1936 U.S. Amateur championship, Scotsman Jock McLean stood on a pier in New York waiting to board ship for his return to Britain. A well-wisher stepped up and asked McLean if he would autograph a dollar bill as a memento. He signed his name and the date, October 13, 1936.

McLean gave up amateur golf sometime later and took the job as professional at Gleneagles, a posh hotel and golf course complex in Perthshire, Scotland. During the summer of 1957 an American tourist exchanged a number of dollar bills into pounds at a bank close to Gleneagles. Someone at the bank noticed a signature on one of the bills and called McLean. He identified the bill as the one he had signed twenty-one years earlier and bought it from the bank. It remained among his mementos.

Tommy Armour

A veteran of the First World War, Tommy Armour, a Scot, lost the sight of an eye during combat. His tank hit by enemy artillery, he climbed out and strangled a German soldier with his bare hands.

The war over, he became an outstanding player, winning the 1927 U.S. Open, the 1930 PGA Championship, and the 1931 British Open. He became even more famous though as an instructor. He was also known to be impatient with incompetence.

While he held the job as golf professional at the Medinah Country Club, near Chicago, Armour occasionally amused himself during lessons by shooting at chipmunks with a.22 caliber rifle. Giving a lesson to a particularly uncoordinated member one day, he seemed to spend more time shooting than teaching. Finally the exasperated member snapped, "When are you going to stop that and take care of me!?" The rifle still at his shoulder, Armour turned slowly toward the member and drawled, "Don't tempt me."

Armour scored twenty-five holes-in-one during his long and glorious career. While scoring even one ace would impress most golfers, it meant nothing to Armour. Taking the philosophical approach, he said, "I never attached any importance to holes-in-one. I never felt the shot was anything but pure luck."

Willie Turnesa

After the morning round of their semi-final match in the 1938 U.S. Amateur, Willie Turnesa led Ed Kingsley by six holes. Kingsley struck back in the afternoon, won two holes, but still trailed by four as they approached the twelfth, a par 5 of more than 600 yards.

Both men drove well, but while Kingsley drilled a long and accurate 3-wood into the center of the fairway, Turnesa pushed his second shot into a grassy lie in the thick and tangled rough. Dampened by the high grass, Willie's third shot fell twenty yards short of the green while Kingsley's third bit and held the green. Turnesa reached the green with his fourth and faced an uphill twenty-foot putt for his par.

Turnesa had a little red-haired caddie who was about fourteen years old. When Turnesa's ball drifted into trouble, Willie saw his face flush a vivid crimson. Then tears welled in his eye and began streaming down his cheeks. Concerned, Turnesa frowned and asked, "Son, what's the matter?" His chin trembling, the boy said, "I hope we don't lose."

Describing the scene, Turnesa said, "I almost started crying myself. I told him, 'Don't you worry, Red. Somehow I'll win.'"

Willie holed his putt, Kingsley three-putted, and instead of losing the hole, Willie won it and then led by five holes.

"See, Red," he said, "I told you so."

The caddie answered, "Okay; you convinced me." Then he cried again.

Turnesa won the match, 4 and 3, and won the championship by defeating Pat Abbott, a movie bit-player, 8 and 7.

Turnesa won his nickname Willie the Wedge during his match against Abbott. Off-color approaches left him in thirteen greenside bunkers over the twenty-nine holes of the match, and yet he lost only two of those holes. Over the first eighteen holes, Willie played into eight bunkers and had eight one-putt greens, losing only the second hole. In the afternoon he played into five more bunkers, had seven one-putt greens, and lost just the third hole.

Bing Crosby

At the peak of his career as an entertainer, during the 1930s and 1940s, Bing Crosby ranked among the best golfers in Hollywood, with a handicap said to range from scratch to two strokes. He was so good and so dedicated that on a number of occasions he entered the U.S. Amateur during the era of Bud Ward, Dick Chapman, Johnny Goodman, Charley Yates, and Willie Turnesa.

The 1940 Amateur was to be played at Winged Foot. Crosby was among sixteen players competing for four places in the thirty-six-hole sectional qualifying rounds at the Bel-Air Country Club, in Los Angeles. Bing shot 152, the same score as Randolph Scott. Theirs were the sixth-best scores in the field.

Things then began happening later in the day. Stan Moss, who had qualified with 148, withdrew, leaving an opening. Cape Norcross, first alternate, with 151, announced he wouldn't play; and then both Scott and Jack Nounnan withdrew as well. Crosby would go to Winged Foot, along with other Bel-Air qualifiers Pat Abbott, Bruce McCormick, a first-class player who had been a factor in other Amateurs, and Bob Goldwater, the brother of Barry Goldwater, the U.S. Senator from Arizona.

Only 150 players qualified in those days, and even those who had survived sectional rounds had to play another thirty-six holes at the site to determine the sixty-four who would advance to match play. Crosby was paired with Pat Mucci, a New Yorker, and Billy Bob Coffey, from Fort Worth, Texas. They played immediately ahead of Jess Sweetser, one of the best amateur golfers ever, Ray Billows, loser to Bud Ward in the 1939 final, and Bob Clark, the current public-links champion.

News that Crosby was at Winged Foot created a sensation. His fans, mostly women, swarmed all over the course, straining to catch sight of him. The crowd grew so large and so unruly the club called in New York State troopers to protect him and his partners.

Crosby shot 83 in the first round. The next day even larger and more unmanageable crowds turned out. Trying to help the golfers move through the gallery, marshals grabbed the long bamboo poles normally used to sweep early morning dew from greens to create a box around them.

Crosby played better, but late in the day it became obvious he wouldn't qualify. On the last hole, a 415-yard par 4 then, Crosby played a good drive, but as he walked toward his ball the crowd broke through the cordon and swarmed around him. It took the troopers fifteen minutes to clear the fairway so they could finish. Crosby made 7 on the eighteenth and shot 77 for the round. He missed qualifying by five strokes.

Bing never entered another Amateur, but he cherished his player identification badge, a bronze medallion given to everyone who reaches the championship proper. After he died, his widow, Kathy, had it made into a necklace.

In 1981, forty years after Bing had failed to reach match play, his son Nathaniel won the U.S. Amateur. He wore Bing's medallion around his neck. In times of stress he reached for it and rubbed it with his fingers, much as Aladdin with his magic lamp. It seemed to work.

The Crosby

Heavy rains plagued the first Bing Crosby Pro-Amateur tournament, in 1937, at Rancho Santa Fe, in San Diego. The rains rushed along a river running through the course with such power, the flood washed out a bridge leading to the green of a par-4 hole. To solve the problem of how to reach the green, policemen and firemen wearing hip boots carried players across the river on their backs.

Sam Snead had a reputation as an unsophisticated hillbilly. Some of it was deserved. After he won the 1937 Crosby, Bing handed him a check for $750. Sam wouldn't take it. Instead he said, "If you don't mind, Mr. Crosby, I'd rather have cash."

Johnny Weismuller, the movie Tarzan, was a regular participant. After hitting a ball into a cypress tree, Weismuller climbed into the branches, hit his ball out, and then, hanging by one arm and swinging back and forth, he pounded his chest with his other hand and screamed the Tarzan yell. Playing through a drenching rain during the 1952 tournament, the man who had won three Olympic gold medals as a swimmer cracked, "I've never been so wet in my life."

Snow covered Pebble Beach during the 1962 Crosby. Stunned by the scene when he awoke and saw the golf course covered in white, Jimmy Demaret cried, "I know I was drunk last night, but how the hell did I get to Squaw Valley"?

Later in the day Harvie Ward, buffered against the cold by an overcoat, tossed it aside as he strode onto the second green. As the coat landed on the back fringe, a pint bottle of whiskey slipped out of his pocket and began sliding down the icy green. While the gallery roared, Ward picked up the bottle and marked its position on the green.

Mysterious Montague

John Montague was a mysterious man. An accomplished golfer, he gambled heavily, sometimes in bizarre matches occasionally involving unusual instruments. Sitting around after he had lost a $5 Nassau to Montague at the Lakeside Country Club, Crosby complained, "You don't give me enough strokes."

Laughing, Montague teased, "I can beat you with a shovel, a bat, and a rake."

Crosby accepted the challenge and they marched to the first tee. Taking his bat, Montague boomed a shot down the fairway, then hit another, but the ball swerved off to the right and dug into a bunker. Crosby, meantime, played two conventional shots to about thirty feet from the hole. Montague shoveled his ball onto the green, and after Bing rolled his first putt three or four feet past the hole, he knocked it away, saying, "That's good, pal." Using his rake as a pool cue, Montague rolled his ball smack into the hole. Crosby gave up and walked back to the clubhouse.

Clutch Shot

Jim Ferrier, an Australian who later won the PGA Championship, was locked in a battle with Hector Thomson, a quiet Scotsman, for the 1936 British Amateur over the Old Course at St. Andrews.

No grandstands had been erected, but they weren't needed in those days. Spectators bunched along the railing that separates the last fairway from the road running alongside it, R and A members crowded inside, others leaned out the windows of the buildings lining the road and from the old Grand Hotel, directly behind the eighteenth green. Some more daring fans climbed to the rooftops and clung to chimney pots, the better to see the action.

Thomson led by one hole after the seventeenth. Both men split the eighteenth fairway with their drives, and then Ferrier lofted a soft pitch that settled close enough to the hole for an assured birdie. Unless Thomson birdied as well, he would be taken to extra holes.

Thomson responded with a glorious shot. It soared toward the green, hung for a moment over the front edge, bounced a few times, and looked as if it might drop into the hole. It skirted the lip and stopped, hanging over the edge. Ferrier stepped up to Thomson and shook hands, conceding the match.

Short and Succinct

A member of the United States Walker Cup team, Charley Yates, from Atlanta, fought his way through the 1938 British Amateur and beat the Irishman Cecil Ewing by 3 and 2 in the final match, at Troon. The championship won, Yates asked a Scottish friend to send a telegram to his boss in Atlanta announcing his victory. A modest and succinct man, Yates sent two words. The message read: "Fortune smiled."

The Stymie

Outlawed following a joint conference on the rules between the USGA and the R and A, the stymie had been a part of golf since the game's beginnings. Essentially it meant that in match play, a ball lying between the player and the hole couldn't be moved and could act as a barrier. One stymie during a match was rare enough, but Dick Chapman broke all rules of probability in the semi-final round of the 1938 North and South Amateur. He had four within two holes.

Two holes up with four to play, Chapman played a nice tee shot to the fifteenth, a strong par-3 hole, but his opponent's shot settled inside Dick's in a direct line to the hole: stymie no. 1. Chapman lofted a pitch over his opponent's ball, but his opponent's first putt also pulled up in a direct line to the hole: stymie no. 2. Chapman lost the hole: one up with three to play, and on to the sixteenth, a short par 5.

Chapman reached the front of the green with his second shot, but once again his opponent's ball rolled dead ahead of Chapman's and again in a direct line to the hole. Another pitch over the offending ball, and then his opponent's first putt blocked him once again. Again Chapman lost the hole, and when his opponent won the eighteenth he won the match.

Chapman evidently was stymie prone. Playing C. D. Lawrie in the fifth round of the 1948 British Amateur he was stymied on the second, fourth, eighth, fifteenth, and nineteenth holes. Dick negotiated only one of those; he chipped over Lawrie's ball on the eighth, where it sat about eight inches from the cup. Lawrie won, of course.

Part Five

Byron Nelson to Arnold Palmer

War carries its price, and the Second World War had effects so widespread they can only be guessed at. It interrupted, certainly, and most likely changed the careers of three men who stand at the very top of any list of great golfers of the ages—Byron Nelson, Sam Snead, and Ben Hogan.

War began sweeping through Europe in 1939. By then Nelson had won the 1937 Masters and the 1939 Open. Snead had already missed two cast-iron chances to win the Open, and before he entered the navy he would win the 1942 PGA. Hogan, the late bloomer of this modern triumvirate, was just coming into his own. He won his first tournament in 1940, and quickly won two more, three within two weeks. In 1942 he had lost a storied playoff to Byron Nelson in the Masters, and in the wartime substitute for the U.S. Open, he played the last three rounds at the Ridgemoor Country Club, in Chicago, in 62–69–68, shot 271, and won the Hale America National Open by three strokes over Jimmy Demaret.

The United States entry into the war caused all but the PGA to be suspended. But for the war, Snead would have had four more opportunities to win an Open, which he never did, Hogan might have blossomed earlier, and Nelson could have won everything. Turned down for military service because of a blood condition, he played throughout the war, and in 1945 won nineteen tournaments, eleven of them in succession, but while he was at the peak of his career, there was no U.S. Open or Masters Tournament. What might he have accomplished?

He left one other enduring record. At a time when tournaments paid perhaps fifteen places (the 1942 Masters paid only twelve, and the 1941

Open, the last until 1946, paid just twenty-five), Nelson finished in the money in 113 in succession.

Byron Nelson

Walter Hagen won the last of his five PGA Championships in 1927, beating Joe Turnesa by one hole at the Cedar Crest Country Club, in Dallas. On his way to the final, Hagen met Al Espinosa in the semi-final. Late in the afternoon, as Hagen faced a shot into the setting sun, he squinted into the glare and held his hands across his brow to shield his eyes.

A young man standing next to him wearing his school cap asked, "Would you like to borrow my cap?" Hagen turned and said, "Yes." He set the cap on his head low enough to block the sun, then played a stunning shot within eight feet of the hole. Beaming, he handed the cap back to fifteen-year-old Byron Nelson.

Hagen holed the putt to pull even with Espinosa, then beat him on the first extra hole.

Shortly after Nelson began playing amateur golf, he was chosen to play in a caddie match against the Bob-O-Link Golf Club, in Dallas. The caddies would play thirty-six holes at Bob-O-Link and thirty-six at Katy Lake, in Fort Worth. As his opponent Nelson drew Ralph Guldahl, who had grown up in Dallas. Both boys were sixteen at the time.

After two rounds at Bob-O-Link, Guldahl stood twelve holes ahead of Nelson. The match shifted to Katy Lake the next day. After the morning round Guldahl had added six more holes to his lead and stood eighteen up with eighteen to play. They didn't bother to play the final round. It was the worst beating Nelson ever took. Years later Guldahl won both the 1937 and 1938 U.S. Opens, and Nelson won it in 1939.

Nelson turned pro on a bus. On his way from Fort Worth to play in an open tournament in Texarkana, Texas, in November of 1932, shortly after losing his job with a magazine that collapsed during the early years of the Great Depression, Nelson began thinking about the $500 prize money. He believed he was good enough to win money playing golf.

When he arrived at the club he asked tournament officials what he must do to become a professional. They told him, "Pay five dollars and say you're playing for the money." Nelson did, placed third, and won $75. Ted Longworth, the Texarkana Country Club professional, won, and Ky Laffoon finished second. Hogan, Jimmy Demaret, and Dick Metz finished out of the money.

No one played more deadly irons than Nelson. On his way to winning the 1939 U.S. Open, he hit the pin six times during the regulation seventy-two holes at the Spring Mill Course of the Philadelphia Country Club, each time with a different club—a wedge, a 9-iron, 6-iron, 4-iron, 1-iron, and a driver.

With 284, he tied Denny Shute and Craig Wood for first place, setting up a playoff. Both Wood and Nelson shot 68 in the first playoff round, but Shute slipped to 76 and dropped out. Nelson and Wood would have to play again the following day.

Nelson effectively ended the playoff in the early holes. From a downhill lie on the third, he pitched next to the hole for a birdie, and on the fourth—a long, sweeping dogleg to the right—he drove past a nest of bunkers where the fairway bends to the right and drilled a low 1-iron directly at the flagstick. The ball carried to the green, rolled past the hole, curled back, drifted downhill, and wedged itself between the pin and the lip of the cup. An eagle 2. Nelson shot 70 and Wood 73.

Nelson's Year

No other golfer has ever had a year equal to Nelson's year of 1945. Most record books give him credit for winning eighteen tournaments, but he actually won nineteen. In addition to those normally listed, he won a tournament at Spring Lake, New Jersey, that wasn't part of the tour. Most sensational of all, though, Nelson won eleven tournaments in succession and set a scoring average of 68.33 that has never been matched.

Nelson's eleven-tournament streak began in March when he and Jug McSpaden won the Miami Four-Ball. Next he won the Charlotte (North Carolina) Open, where he beat Sam Snead in an eighteen-hole playoff. The following week he won by eight strokes at Greensboro, North Carolina, and a week later at Durham he shot 65 in the last round and beat Toney Penna by five strokes.

By winning four consecutive tournaments, Nelson had tied the record Johnny Farrell had set in 1927. A week later, in the Iron Lung Open, in Atlanta, Nelson bogeyed the last hole but still set a tour scoring record of 263, and of course won the tournament—his fifth in succession.

Now the tour took a two-month break—there simply weren't any tournaments scheduled until early June.

Nelson returned sharper than ever, breaking par at the Islesmere Country Club, in Montreal, by twenty strokes and winning the Canadian PGA by ten strokes over McSpaden. A week later McSpaden had gone into the last round of the Philadelphia *Inquirer* Invitational leading Nelson by two strokes and closed with 66. Byron learned of McSpaden's score as he was about to drive on the thirteenth and figured he'd have to birdie five of the remaining six holes to win. He did; he shot 63 and beat McSpaden by a stroke. Now he had won seven in succession. Nelson took a week off, then won the Chicago Victory Open by seven strokes over McSpaden once again, shooting 275 and beating par by thirteen strokes. Next Nelson drove to Dayton, Ohio, for the PGA Championship at the Moraine Country Club. After winning the qualifying medal, which returned a prize of $125, Byron beat, in order, Gene Sarazen; Mike Turnesa in a very close match; Denny Shute, whom he'd beaten in a playoff for the 1939 Open; Claude Harmon in a very good match with Nelson shooting 65 against Harmon's 68 in the semi-finals; and in another spurt of sensational golf, played the fifth through the fourteenth holes in six under par and came from three down to beat Sam Byrd, a former New York Yankees outfielder, 4 and 3 in the final.

After a layoff of two weeks, Nelson closed with 32 on the last nine holes, shot 269 at the Tam O'Shanter Country Club, in Chicago, won a $100 bet with George May, the club owner and tournament sponsor, and won the All-America Open by eleven strokes over Sarazen and Hogan. This was Nelson's tenth consecutive victory.

The next week, over a very difficult Thornhill Country Club course, in Toronto, Nelson shot 68–72–72–68—280 and won the Canadian Open by four strokes over Herman Barron. He had won eleven consecutive tournaments.

His streak ended with the Memphis Invitational, where he played some loose golf and placed fourth. The tournament was won by Fred Haas, an amateur who had won the Intercollegiate championship as a student at Louisiana State University.

Losing didn't last long. The following week he won at Spring Lake and again at Knoxville, and he played his best golf of the year at Seattle. Hogan had beaten him by fourteen strokes at the Portland Invitational the last week of September by playing the best golf of his career—Ben shot 65–69–63–64 for 261. Nelson placed second with 275.

Asked how long he thought Hogan's record might hold up, Nelson answered, "You can never tell in this game. It might last forever or it might be broken next week." Two weeks later Nelson shot 62–68–63–66—259 at the Broadmoor Golf Club and broke Hogan's record by two strokes. McSpaden and Harry Givan, a former Walker Cupper, shared second place at 272, a highly respectable score, but thirteen strokes behind Nelson's.

A Swap

Nelson became acquainted with the game by caddying at the Glen Garden Country Club, in Fort Worth. Each year at Christmastime the club staged a tournament for the caddies. The 1927 tournament was to be played over nine holes. Glen Garden's first nine carried a par of 37. Nelson shot 40 and tied fellow caddie Ben Hogan.

Since the weather was good and there was plenty of daylight left, the members decided to extend the playoff to nine holes. Nelson beat Hogan by one stroke. As prizes the club offered both the winner and the runner-up a golf club: Nelson was given a 5-iron and Hogan a 2-iron. Nelson already owned a 5-iron and Hogan a 2-iron, so they swapped.

Wifely Criticism

Disappointed in his first practice round leading up to the 1935 U.S. Open, at Oakmont, Nelson brooded over his driving that night while his wife,

Louise, a compulsive knitter, did her needlework. "Louise," he said, "I have to buy another driver; I'm driving terrible." Louise, a non-golfer, continued to knit and purl for a moment, then said, "Byron, we've been married over a year now. I haven't bought a new dress or a new pair of shoes or anything for myself in all that time, but you've bought four drivers and you're not happy with any of them. That means one of two things: either you don't know what kind of driver you want or you don't know how to drive."

When the message sank in, Nelson agreed Louise was right. Instead of buying still another driver, he took one he already had to the Oakmont shop and worked on reshaping the face to the contour he wanted. Not only did he never worry about his driving again, he developed into one of the two or three best drivers the game has known.

The Hagen Treatment

Leading the 1935 General Brock Hotel Open, in Font Hill, Canada, after fifty-four holes, twenty-three-year-old Byron Nelson was thrilled to be paired with Walter Hagen in the final round. Notorious for his gamesmanship, Hagen didn't show up for the appointed starting time, but in those days no one was about to disqualify Walter Hagen. So Nelson waited. Then he waited some more. Hagen finally turned about about an hour late, and they began their round. Upset and nervous because of the delay, Nelson shot 42 on the first nine and lost the tournament by one stroke to Tony Manero.

Anatomy Lesson

While he held the professional's job at the Reading Country Club, in Reading, Pennsylvania, Nelson was giving lessons to a rather buxom woman named Ann Metzger, the wife of a local dentist. Nelson kept telling her how he wanted her to swing the club, but for the life of her she couldn't do it as Nelson taught. Finally she approached Louise and said, "I wish you'd explain to Byron how we women are made, because I get in my own way, and that's why I can't swing the way he wants me to swing."

At Mrs. Metzger's next lesson, Nelson said he had changed his plans for her and told her, "Bend over more, stick your back end out more, and swing your arms out a little farther from your body." After listening to the new approach, Mrs. Metzger said, "Louise must have talked to you."

Nelson vs. Hogan

Nelson and Hogan tied for first place in the 1942 Masters, setting up an eighteen-hole playoff the next day.

Always nervous the morning of a difficult match, Nelson lost his breakfast, which was not uncommon for him. Hearing that Nelson felt sick,

Hogan went to Byron's room and offered to postpone the playoff. Realizing his sickness often signaled a good round, Nelson declined Hogan's offer and said, "No; we'll play."

In an unusual testament to the great playing ability of both Nelson and Hogan, about twenty-five players stayed over in Augusta to watch the playoff. Nelson started badly, losing two strokes to par on the first hole, and stood three strokes behind Hogan after five holes. Hogan played the next eleven holes, from the sixth through the sixteenth, in one under par and lost five strokes.

Nelson picked up two strokes with a birdie on the sixth, another par 3, where Hogan bogeyed, then eagled the eighth, an uphill par 5, where Hogan birdied.

At the dreaded Amen Corner, Nelson birdied the eleventh, twelfth, and thirteenth. He had played the eight holes from the sixth through the thirteenth in six under par. Leading Ben by two strokes coming to the eighteenth, he played it safe, leaving his second shot short of the green in order to avoid the bunker on the right side of the green, and bogeyed. He shot 69 against Hogan's 70.

Years later as he reminisced about Nelson's golf that day, Hogan said, "To see him play those eight holes in six under par on that golf course was something to behold."

Clumsy Caddie

Nelson stood two strokes behind Hogan and Vic Ghezzi going into the third round of the 1946 U.S. Open, at the Canterbury Golf Club, near Cleveland. About 12,000 fans swarmed unchecked over Canterbury's hills the last day, restrained only by ropes carried by marshals. Players and caddies had to push and shove through the crowds, and the players hit their shots through narrow funnels of spectators.

After Byron played his second shot on the thirteenth hole, a par 5, the gallery rushed ahead and crowded close behind his ball. The marshals strung their ropes, hoping to hold back the fans, but they had squeezed so close to his ball that when his caddie, Eddie Martin, so recently discharged from the military he still wore his Army uniform, ducked under the ropes, he lost his balance and stumbled into Byron's ball. The accident cost Nelson a penalty stroke. Eventually he and Ghezzi tied with Lloyd Mangrum, and Mangrum won the playoff.

Donald Ross

Donald Ross, the immigrant Scot who became the leading golf course architect of his time, rose to the top levels of administration of the Pinehurst complex, in Pinehurst, North Carolina. At the same time he managed the Pinehurst Country Club, he was the club professional, clubmaker, caddie master, and occasional labor negotiator as well.

Sometime during the 1940s, when the caddies declared a strike, Ross handled the negotiations. He had a novel approach. Picking up a 5-iron, he walked into the caddie yard and asked who was their leader. One of the bigger and older caddies stepped forward and said, "I am, Mr. Ross. We're striking."

Holding the grip of the club, Ross bounced the clubhead off the caddie's shoulder a few times and hissed, "If you don't get back to work, I'll bounce this 5-iron off your head."

The strike ended.

In designing his courses, Ross often worked off topographical maps and then visited the site to oversee the actual construction. The Canadian Pacific Railroad hired him to lay out a course in the Canadian Rockies in 1910, when his reputation as an outstanding designer had just begun to spread.

Ross worked out his design on the maps, and then asked for a $50-a-day fee for inspection visits. Cornelius Van Horne, Canadian Pacific's president, had pushed the line through the Rockies against overwhelming obstacles, building an impossible trestle across a deep chasm called Kicking Horse Pass, and drilling tunnels through the mountains. The result was a rail line that spiraled thousands of feet upwards and then eased back down to sea level for its final run to the Pacific. The railroad ranks among the world's astounding engineering achievements.

Told of Ross's request for paid inspection visits, Van Horne turned him down, claiming none of the engineers who created the railroad's wonders were paid such high fees. He should have agreed. Ross had calibrated the slopes and gullies he wanted to create in inches—a three-inch rise on the left of this green, and a slight 20-inch depression in front. They were all marked on his blueprint. Van Horne's engineers weren't accustomed to such subtleties; after all, they tamed mountains rising to the clouds. When they read the blueprints, they interpreted the measurements in feet. The slight three-inch rise became a three-foot-high hillock, and the twenty-inch depression plunged downwards twenty feet.

By the time someone suspected all was not as it should be, they had completed nine unplayable holes. Van Horne abandoned the project, leaving the ground to be reclaimed by nature.

Ben Hogan

"You'll never get anywhere fooling around at those golf courses." Clara Hogan to her sixteen-year-old son, Ben.

Listening to her husband complaining he couldn't seem to hole those twenty-foot putts, Valerie Hogan told Ben, "Hit the ball closer to the hole."

Hogan was never easy to please. He began his golf career as a caddie, principally to help support his family, but at the same time he became fascinated by the game. He seldom had enough money to buy clubs, but when he did he would spend an hour or so at a barrel of old, used clubs at the W. T. Grant store. Later, when he and Byron Nelson became affiliated with the MacGregor Sporting Goods Company, Hogan would spend two or three days assembling a set of clubs, watching as the technicians ground and milled his irons, creating just the right look to the top edge and shaping the sole just the way he liked it. Nelson was different; he could assemble a set of clubs in half an hour out of inventory.

Hogan had an older brother, Royal, who managed a business supply store in Fort Worth. Finally realizing what the game meant to Ben, Clara Hogan gave Royal $40 and told him to buy a set of clubs for Ben. They would be a Christmas present.

Hogan turned pro in 1931—the year after Bobby Jones retired—he was eighteen. He entered the U.S. Open for the first time that year, shot 165 over thirty-six holes in the qualifying rounds at Dallas, impressed no one, and was eliminated.

Hogan tried twice to make a living on the tour in 1932 and 1933, but he failed both times. Still he wouldn't give up his dream. He tried again.

He married Valerie Fox in 1935. Valerie traveled with him. It was a hard life. By January of 1938 they had reached the point of utter despair, living on little more than oranges for a month while the tour moved through California.

Hogan stayed in contention through the first three rounds of the Oakland Open, but the morning of the last round when Ben left his hotel to pick up his old, used car that he had parked on a gravel lot, he saw it jacked up. Someone had stolen the wheels. He hitched a ride to the Claremont Country Club, and that day, playing harder than he ever played, he shot 67, finished five strokes behind Harry Cooper, and won $285. He was never close to giving up again.

Hogan had just lost a playoff to Jimmy Demaret for the Phoenix Open and had decided to take a few weeks off. On the morning of February 2, 1949, he was driving his new black Cadillac through a heavy fog outside the small town of Van Horn, Texas, with Valerie alongside him when out of the gloom he saw four headlights dead ahead. A Greyhound bus was speeding directly at him.

Held up for seven miles and falling behind schedule because of a lumbering truck up ahead, the bus driver tried to pass. He picked the wrong time; both vehicles were crossing a bridge bordered by concrete guard rails. There was no escape; the collision couldn't be avoided.

When Alvin Logan, the driver, realized he was bearing down on a car, he tried to swerve right, Hogan swerved right as well, but he couldn't get off the highway. With a 19,000-pound Greyhound bus about to crash into his 3900-pound car, Ben flung himself across the seat to protect Valerie.

By trying to save his wife he saved his own life. The force of the impact drove the steering column through the front seat and into the back. Had Hogan not dived away, he would have been crushed. Even so, the wheel crashed into his left shoulder, and broke his collarbone. The engine was driven back into the car's cabin, mangling Hogan's left leg and smashing into his midsection. Luggage flew from the back seat into the front, and Hogan's clubs lay scattered on the road.

When help arrived, it took an hour to free the Hogans. When Valerie saw Ben's clubs strewn about, she asked a highway patrolman if he would pick them up. The patrolman looked at Hogan lying beside the road, covered by a thin blanket, then looked back at Valerie. His expression left no doubt about his feelings: this man would never use those clubs again.

Hogan was taken to a hospital in El Paso, where doctors found that in addition to the obvious injuries, he had sustained a broken rib, broken pelvis, broken ankle, bladder injuries, and deep and massive bruises on his left leg. He lay in bed with a cast covering him to his waist.

Slim and wiry at 138 pounds, Hogan progressed so rapidly over the first few days that doctors expected to release him by February 16, two weeks after the accident. Within the next few days, however, the prognosis changed. Hogan's life was threatened. A blood clot had formed in his bruised leg, passed through his pulmonary artery, and lodged in his lung. Doctors were afraid a larger clot might clog the artery completely and kill him. He lay seriously ill; he might die at any minute. Doctors decided that the vena cava, the large vein that leads from the leg to the heart, would have to be tied off in the abdomen.

On the night of March 2 doctors found a second clot. The operation would have to be done quickly. Doctors at the Hotel Dieu Hospital, in El Paso, contacted Dr. Alton Oscher, professor of surgery at Tulane University and chief of staff at the acclaimed Oscher Clinic in New Orleans. They asked Oscher if he would perform the operation. Oscher agreed to.

There were further complications. A storm had grounded all commercial flights from New Orleans. Now Valerie acted. She telephoned General David W. Hutchinson, the commanding officer of Forbes Air Force Base in Topeka, Kansas. Hutchinson had been commanding officer at Biggs Field, in El Paso, at the time of the accident. She reminded him that Hogan had served in the Army Air Forces during the Second World War. The general diverted a B-29 bomber flying a practice mission to New Orleans, where it picked up Dr. Oscher and flew him to El Paso. The doctor arrived at eight the next morning.

When Oscher explained the operation to Hogan, Ben raised his head and asked, "Will I be able to use my legs and play golf?" The doctor said he was certain he would. Hogan said, "All right." The operation was successful.

Hogan had won both the 1948 Open and the PGA Championship, the first man to win both in the same year since Sarazen in 1922. He was to defend his Open championship at the Medinah Country Club, near Chicago, in 1949, but with the accident it didn't seem likely. Still, he had hopes.

Recovering at home, he wrote a letter to the United States Golf Association and included his entry for the Open. Addressed to Joe Dey, the USGA's executive secretary, the note read:

> I am getting along great just now, up all day and walking as much as possible. The doctor tells me walking is the only cure for my legs, so that's my daily thought and effort.
>
> Enclosed is my entry for the Open, with the hope that I will be able to play. Up to now I haven't taken a swing, but miracles may happen. Would you please do me a favor and not release my entry? If I can play I should like it to be a surprise. I hope and pray that I may see you in June.

He didn't play; he withdrew before the qualifying rounds.

Hogan returned to competitive golf in the Los Angeles Open, in January of 1950. The LA Open was played at the Riviera Country Club, where Ben had won both the 1947 and 1948 LA Opens along with the 1948 U.S. Open. Riviera, in fact, had become known as Hogan's Alley.

Limping on his aching legs, Hogan opened with 73 and finished the first round five strokes behind Ed Furgol. Most fans felt he had performed a miracle by finishing. It wasn't quite good enough for Hogan; he played to win. The next day he shot 69, added two more 69s in the next two rounds, and not only did he finish seventy-two holes, he looked like the winner. Only Sam Snead could catch him, but Snead would have to make up four strokes over the last five holes.

Performing a miracle almost comparable to Hogan's, Snead birdied four of the last five holes and tied Ben. They took a week off to play the Crosby, then returned to Riviera for the playoff. Ben had nothing more to give. Tired beyond reason and his body aching, Hogan shot 76, Snead 72, and Sam won the playoff. Hogan had won the moral victory.

Hogan returned to the Open in June of 1950. This was the fiftieth Open conducted by the USGA, and the association took it to the Merion Golf Club, near Philadelphia, one of its favorite golfing grounds.

Becoming stronger every week, and leading by three strokes after eleven holes of the final round, Hogan very nearly didn't finish. Forced to soak his legs in a tub, then wrap them in Ace bandages, he struggled to play thirty-six holes in one day, one of the requirements of the Open before conditions were changed in 1965.

Ben had already played eighteen holes in the morning and seemed to be dragging himself around Merion in the afternoon. He hadn't played thirty-six holes in one day since he had won the 1948 Open at Riviera two years earlier, and the strain was telling. Still, he held a three-stroke lead as he stepped onto the twelfth tee. It nearly ended there. Moving into his drive, Ben's legs locked. He nearly fell. Harry Radix, a rather well-known golf official and a friend of Hogan's, stood by the gallery ropes. Hogan stumbled toward him, wrapped an arm around his shoulders, and said, "Let me hang on to you for a little bit. My, God, I don't think I can finish. My legs have turned to stone."

Standing on Merion's eighteenth tee, Hogan figured he had lost the championship by bogeying two of the previous three holes. A perfect drive over an abandoned stone quarry put him in the fairway, but now he faced a dilemma. The hole measured 458 yards, a very long par 4 for those times. If he could tie for first place with a par 4 he'd play a safe shot to the left front of the green. If he needed a birdie 3, he'd play a 4-wood to the right rear, where the hole had been cut. It would be a dangerous shot because the green fell away at the back and too strong a shot would zip off into the high rough beyond. Hogan wasn't playing for second place.

Spotting Jimmy Hines, who had missed the thirty-six-hole cut, Hogan asked, "What's low." Grossly misinformed, Hines said, "Two eighty-six. Bill Nary, I think." Nary had actually shot 290.

Fred Corcoran, who at one time had run the pro tour, knew better. Actually George Fazio and Lloyd Mangrum shared the low score at 287. Standing alongside Hines he spoke up. "No," he said, "287 is low."

Hogan froze Corcoran with his icy stare. In a slow, measured, emphatic tone, Corcoran repeated, "287 is low."

Hogan thought for a moment, then drew his 1-iron. He'd play the safe shot. He drilled a lovely shot that flew straight at the front left of the big green, hit, and braked itself perhaps forty feet from the flagstick. His first putt ran five or six feet past the hole. Dreadfully weary and thoroughly discouraged because he had thrown away so many strokes over the last six holes, Hogan took no time at all with the second putt. It fell nonetheless. He shot 287 as well. He would be in the playoff with Fazio and Mangrum.

Hogan won the next day, but he won the playoff without his 1-iron. Sometime after he played his shot into the eighteenth and before he stepped off the green, someone stole it from his bag. He never saw it again during his playing career, nor did he replace it; he did not use a 1-iron again.

Thirty-three years later the club was returned to him through a tortuous route that no one wanted to talk about. Evidently it was seen in a shop by someone who suspected its origin. The club was a MacGregor Ben Hogan Personal model with a circular worn spot near the heel where its owner had hit a countless number of balls, consistently, each on that one spot. It is also likely it wasn't hidden away all those years; the grip had been replaced.

When it was returned, Hogan stood it in a corner of his office in Fort Worth and took the occasional tentative swing trying to decide if this was or was not his long-lost club. Finally making up his mind that this was indeed his old club, Hogan restored a grip of the type he had used in 1950 and sent it to the museum of the United States Golf Association, where it was put on display near his portrait

Defending his Open championship in 1951, Hogan seemed despondent after the first thirty-six holes. Robert Trent Jones had toughened the Oakland Hills course to such a degree that only two men had shot a round in par 70 over the first thirty-six holes. With two rounds to go, Hogan felt he couldn't win, saying, "I'd have to be Houdini to win now. I'd need 140, and how can anybody shoot 140 on this course?"

The next morning Hogan had a great round going, shot 32 on the first nine, but then let it slip away and came back in 39. He shot a disappointing 71, but even so he passed ten others and had climbed into a tie for fifth place, two strokes behind Jimmy Demaret and Clayton Heafner.

Hogan had been angry when he finished the third round. After a quick lunch he stepped back onto the first tee to begin the fourth round and said to Ike Grainger, the referee, "I'm going to burn it up." He did. He went out in 35, even par, then came back in 32. His 67 may have been the best round of competitive golf he ever shot, considering the difficulty of the golf course. It was the lowest score of the week and the second lowest closing round ever shot by the champion. Gene Sarazen had finished with 66 at Fresh Meadow in 1932. With 287 for the seventy-two holes, Hogan won by two strokes over Heafner, whose 69 was the only other sub-par round shot in the tournament. Accepting the trophy a few minutes after the final putt had fallen, Hogan told the gallery, "I'm glad I brought this course, this monster, to its knees."

Deception

Before he began making clubs under his own name, Hogan was a member of the MacGregor Company's playing staff. Toney Penna, who headed the staff that designed the company's clubs, remembers a special design problem he had to solve for Hogan.

Ben had a very fast swing; he'd whip the club back, then whip it through almost faster than the eye could follow. To slow it down Ben wanted a driver with a swing weight of D-4 or D-5, rather heavy of itself, and a dead weight of 13¾ ounces. That is a very heavy club; only a very strong and very good golfer could control it. A normal club with that heavy dead weight would swing into the E scale—which is heavier than the D scale. Penna solved the problem by counter-balancing. He added weight under Hogan's grip, which increased the dead weight and lightened the swing weight. Chuckling, Penna said, "He didn't even know it."

Hogan at Seminole

A member of the Seminole Golf Club, a wonderful Donald Ross-designed course in North Palm Beach, Florida, Hogan often played casual rounds with Chris Dunphy, for years the irascible president of the club. During one round Hogan complained the greens were too slow. Never one to take criticism lightly, even from Hogan, Dunphy snapped, "If you didn't take so much time to putt, the grass wouldn't grow that long."

Hogan usually spent a month at Seminole getting ready for the Masters. Most often he stayed as Dunphy's house guest. Dunphy had invited the Duke of Windsor to a dinner party at his house one year, and at the same time asked a number of high powered executives, men like Robert R. Young, who, among his other achievements, had brought about the merger of the B&O and the C&O railroads.

With all the places at the dinner table taken, Dunphy told Hogan he would have to dine out. When the Duke showed up the business executives were in place, but neither they nor Dunphy were prepared for the duke's reaction.

The duke, a *bona fide* golf nut who had spent at least part of his youth following golf pros around British courses and inviting Walter Hagen into clubhouses where he ordinarily wouldn't have been allowed, shook hands with all those introduced to him, then looked around for the one person he wanted to talk to. With a frown on his face he turned to Dunphy and asked, "Where's Ben?"

No Help Needed

Ken Venturi claimed Hogan could say more in fewer words than any man who ever lived. Max Faulkner, who had won the 1951 British Open, watched Hogan playing his precise, controlled fade on Maniac Hill, Pinehurst's practice ground, a few days before the 1951 Ryder Cup Match. As each ball hit the ground, the caddie would take a step to the right, or a step to the left, or else simply reach out and catch the ball on the first or second bounce.

Not realizing Hogan deliberately shaped his shots to drift right, Faulkner finally spoke up. "Ben," he said, "I think I can help you get rid of that fade." Startled, Hogan snapped, "You don't see the caddie moving, do you?"

Growing weary of questions from reporters about how he played a round at the Masters, Hogan snapped, "One day a deaf mute will win this thing and you guys won't be able to write a word."

A Club—and a Neck—Saved

Midway through the final round of the 1953 British Open, Hogan studied his approach to the seventh hole at Carnoustie, a fairly straightaway par 4 of 397 yards, considering which club to play. Hogan stood with his hand on his 3-iron, but, with the wind coming directly at them, Cecil Timms, his veteran caddie, felt he needed a 2-iron. After a brief discussion, Hogan fixed Timms with his steely glare and warned, "If this shot goes over the green, I'll wrap this 2-iron around your neck." The 2-iron soared onto the green, Hogan made his 4, and the 2-iron went back into his bag unbent.

Happenstance

Who knows what might have happened to the dog that almost stepped into the path of a Hogan drive at Carnoustie during the 1953 British Open. As Hogan took his stance on the tee, he saw a big black dog walk across about ten yards in front of him. He thought he saw the dog walk into the crowd, but as he swung into the shot, the dog walked across again. The ball didn't miss him by more than two inches. Had it hit him, it might have killed him and probably changed the outcome of the tournament.

A Trophy for Golf House

When Hogan shot 287 at the Olympic Club, in San Francisco, almost everyone believed he had won the 1955 Open, his fifth. No one had ever won five; until then only Hogan, Bobby Jones, and Willie Anderson had won four. Looking over who hadn't finished and figuring what they would have to shoot to beat him, Gene Sarazen concluded Hogan had won and said as much to a nation-wide radio audience. When Hogan reminded him that a number of players still on the course had a chance, Gene brushed it aside, saying no one could possibly catch him.

Even though he had refused to acknowledge publicly he believed he had won, Hogan was fairly certain he had. Knowing that many items of historic significance to golf were on display at Golf House, the headquarters of the United States Golf Association, in New York City, Hogan approached Joe Dey, the USGA's executive director, handed him the ball he had used to hole out on the final green, and said, "This is for Golf House."

Later in the day Hogan was tied by Jack Fleck, an unknown golfer who in forty-one tournaments over three years had won only $7400. The next day Fleck beat him in a playoff.

Hogan finished the playoff against Fleck by pulling his drive into knee-high rough lining the left side of the eighteenth fairway. He moved his ball only a foot or so with his second shot, tried again and moved it perhaps three feet, then chopped it out with his fourth. He pitched on with his fifth, then holed a slippery downhill putt for a 6.

Since the championship was played on the West Coast, reporters for Eastern newspapers found themselves close to deadline and didn't have time to question the players. Since he hadn't seen a ball move, Bob Drum, writing for the *Pittsburgh Press*, believed Hogan's first two swings had been practice swings and reported that Hogan had finished with a par 4.

Later he and Hogan met in a corridor of Olympic's clubhouse. After offering his condolences, Drum added it was good to finish with a 4. Hogan explained that he had scored 6, not 4. Drum disagreed, saying he had watched every shot from the path leading up the steep incline to the clubhouse, and began ticking them off on his fingers:

Drive into rough, second shot on the fairway . . . Hogan interrupted, explaining that he had swung three times to get back to the fairway. Surprised, Drum said, "Weren't those practice swings?" Furious, Hogan barked, "I was trying to hit the ball, you damn fool."

Drum claimed later that Hogan didn't speak to him for five years. "Ben doesn't talk to idiots," he explained.

Course Analysis

Partnered by Sam Snead in the 1955 Canada Cup, Hogan had played one practice round at the Wentworth Golf Club, in England, when Snead joined him for their second practice round. Standing on the tee of a par-4 hole, Snead pulled out his driver. Seeing his choice of club, Hogan said, "Use your 3-wood, Sam." Puzzled, Snead said, "Why? It's a par 4 isn't it?" "Yes," Hogan replied, "but use your 3-wood." Snead did and hit a good shot down the middle, but as he studied his approach he complained to Hogan, "I need a 5-iron from here, but I hit a 7-iron this morning." Hogan agreed that was probably true, "But," he went on, "you played it from a downhill lie." "I never thought of that," Snead agreed.

Hogan and Snead had arrived at London's Heathrow Airport on the same flight, but Hogan had thought ahead and ordered a limousine to take him and his wife, Valerie, to their hotel. Traveling alone, Snead hadn't. As the Hogan car arrived, a man arranging transportation for arriving players asked if Hogan would take Snead with him to their hotel.

Hogan said, "No," climbed into his car, and was driven away.

Your Choice

Sitting in front of his locker after the second round of the 1956 U.S. Open, Hogan fielded questions from the press. He had shot 140 for thirty-six holes and stood a stroke behind Peter Thomson, the young Australian who had won the previous two British Opens. As the interview progressed, a reporter asked Hogan, "Ben would you rather be one stroke behind right now or one stroke ahead?" With an icy stare, meant to freeze his questioner, Hogan snarled, "Would you rather be rich or would you rather be poor."

No Help Needed

On the practice tee a few days before the 1957 U.S. Open began, another professional watched Cary Middlecoff struggling with his irons and said he thought he would wander over and offer help. Hogan said, "Leave Middlecoff alone."

Ask Elsewhere

Hogan had given Gary Player a Hogan driver with a sole plate bearing the Hogan Company's distinctive seal, but Player was under contract to another manufacturer, so he replaced the plate. Some time later, Player developed trouble with his swing. While he was playing in Brazil, he telephoned Hogan at his home, told him of his problem, and asked if he could fly to Fort Worth and have Ben take a look as his swing. Hogan said, "I'm going to be real curt with you. What clubs are you playing?" Player said he was using Dunlop clubs. "Well, Gary," Hogan said, "ask Mr. Dunlop." Then he hung up.

Cherry Hills

Hogan had gone into the last thirty-six holes of the 1960 Open in 11th place with a score of 142, even par. Beginning the last day of thirty-six holes, he hit every green in the morning round, holed nothing to speak of, shot 69, two over par, and climbed into a tie for fifth place, at 211, three strokes behind Mike Souchak, the leader.

Continuing to play the most consistently precise tee-to-green golf of his career, Hogan hit the first nine greens of the afternoon round without making one birdie. His luck changed slightly on the second nine. With birdies on the twelfth and fifteenth (his approach to the thirteenth hit the flagstick but bounded nine feet away and Ben missed the putt), Hogan stood four under par, tied with Arnold Palmer and Jack Fleck. Three more pars would give him 280.

He hit the sixteenth green but missed from twelve feet, then moved on to the seventeenth, a par 5 of 548 yards with an island green separated from the fairway by a twelve-foot wide band of water. The water forced the hole to be played with a drive, a layup, and a short pitch. Hogan drove nicely and followed up with a 3-iron just short of the pond.

Figuring both Palmer and Fleck would birdie there, Hogan felt he had to go for a birdie of his own. With his putting so bad, he also felt his best chance lay in dropping his approach close to the hole, which sat on the front of the green, barely clear of the water.

With his ball sitting nicely, about an eighth of an inch off the ground, Hogan played his wedge. As soon as the ball took flight, Phil Strubing of the USGA, the referee, cried, "Oh, no."

The ball didn't carry quite far enough. It hit near the top of the bank and spun back into the pond. Hogan took off a shoe and sock and played from the water, but be bogeyed there; then drove into the water on the eighteenth. He finished with 73, and shot 284 for the championship. He tied for ninth place, four strokes behind Palmer. Looking back,

Hogan said, "When will I learn not to go for birdies in situations like that?"

A little more than ten years later, Hogan's portrait was unveiled at Golf House, the USGA's headquarters, located at the time in New York City, Phil Strubing mentioned his crying out, "Oh, no," when Hogan made contact. Then he confessed that he wouldn't have known what to do if Ben's club had touched the water on his backswing. Strubing had often asked himself if he would have had the nerve to call the penalty. Hogan answered, "You wouldn't have had to. I'd have called it myself."

Strubing then told of a conversation he had had with Clarence W. (Gus) Benedict, a former USGA president. "You know what to do in situations like that?" Benedict counseled. "Stand where you can't see."

A middle-aged engineer, Karsten Solheim stood in the gallery watching Hogan and Nicklaus play the last round. As Hogan's ball splashed into the water, he thought to himself, "That ball would have cleared the water if Hogan had been using my clubs." Solheim had just begun work on his Ping irons—the most successful clubs in terms of sales in the history of the game.

Hogan had been given a special exemption into the 1966 Open, played once again at Olympic, in recognition of his nearly winning a fifth Open there in 1955.

Bruce Devlin arranged a practice round with him a day or so before the first round. When they stepped onto the sixth tee, Hogan pointed at a bunker on the left of the drive zone and told Devlin he was looking at Olympic's only fairway bunker.

"Is it in play, Ben?" Devlin asked.

"No," Hogan answered. "You hit to the right of it."

Hogan's eyesight began to fail in the 1960s, after he had retired from competitive golf. He played only socially, with friends, but he was never able take the game casually, and never once did he relax his standards. He played often at Brook Hollow, a fashionable club in Dallas, mostly with Shelley Mayfield, a former tour player who had become the club's professional.

Hogan continually amazed Mayfield with his precise ball striking and his assessment of distance. Time after time Hogan fired his shots directly at the flagstick, and Mayfield stood by shaking his head as Ben's ball braked itself ten or fifteen feet to the right or to the left, seldom short, and never long.

One of Brook Hollow's par-4 holes called for a blind approach to a green set well below the fairway. To indicate the line of play, the club had raised a tall stanchion, about eighteen feet high, behind the green. Hogan played his iron in the perfect direction. It took off low, climbed to its apogee, then floated downward toward the green, never varying its line. As they walked onto the green together, Mayfield was shocked. Hogan's ball had carried to the back of the green, bounced over and down a hillside, and stopped yards beyond. Seeing his ball deep in the woods, Hogan turned to Mayfield, and with a bashful smile said, "I thought that marker was the flagstick."

The Voice of Doom

Dave Marr had never played with Ben Hogan until the first round of the 1964 Masters. Arriving at the eighth hole, Dave drove into the right fairway bunker, set at the base of a gradual rise and out of sight of the green, well beyond the crest of the hill. Since he was away, Marr would play before Hogan, who was in the fairway. While Hogan stood with hands on hips glaring toward him, Marr waited for the forecaddie at the top of the rise to signal that the group ahead had left the green, which indeed was well out of his range. Finally reaching the end of his patience, Hogan growled, "You think you can get there from here?" Startled and, he admitted later, scared to death, Marr leaped into the bunker and slashed his ball out.

His Portrait

Hogan's portrait was done by Anthony Wills, who painted the portrait of Lyndon Johnson that hangs in the White House. Wills visited Hogan in his home in Fort Worth, posed him standing alongside his golf bag as if he were choosing a club as he gazed toward a green. Satisfied with the arrangement, Wills shot a Polaroid picture and began painting. After painting Hogan's figure, Wills filled in a background. Wills was a member of Champions Golf Club, in Houston, a course covered with pine trees. Hogan was a member of Shady Oaks, in Fort Worth, given its name for obvious reasons.

The painting done, Wills showed it to Hogan. Ben looked at it and gave Wills his critique. "There are no pine trees at Shady Oaks." Wills went back to his studio and painted oak trees over the pines.

A Rare Compliment

A young mother from Fort Worth took her six-year-old son to her golf club to give him a taste of the game. As they enjoyed themselves knocking the ball around, they noticed a man on a nearby practice tee. When they walked past, the man approached and said, "Young lady, I want to compliment you for taking the time to teach your son to play golf. You'll never regret your effort. Your son will learn a game he can enjoy the rest of his life." Hogan tipped his white cap and walked back to his practice.

Iron Master

Hogan never clubbed himself by yardage; he depended on how he felt, how the ball was carrying, and the kind of shot he wanted to play. He might hit anything. Like all the good players, he could finesse a shot as well. Nor did he like anyone checking his club. In a round at Shady Oaks he noticed Gene Smyers, a good player, sneaking peeks in his bag. When they came to the thirteenth, a par three of modest length, Hogan played a 5-iron about ten or fifteen feet from the cup. A 5-iron seemed quite strong for this hole, but Smyers, figuring Hogan knew something he didn't, played a 5-iron as well. When he last saw his ball it was passing over the green, still climbing.

Squelched

Hogan was playing an informal round with R. H. Sikes, a journeyman professional during the 1960s and 1970s. Sikes played rather loose tee-to-green golf, but hole after hole he saved himself with remarkable recoveries from around the green. Throughout the round Sikes asked Hogan for help with his game, saying he'd appreciate any little bit of help Hogan might be able to give him. Hogan remained quiet, but Sikes continued to ask.

The round over, Sikes added up his score and found that despite a shabby performance from tee to green, where he constantly drove off the fairways and missed more than half the greens, he had turned in a score of 67, five strokes under par.

Bracing himself, once again he asked Hogan for advice, saying, "As you could see, Mr. Hogan, my short game is pretty good, but my long game isn't. If you say anything—anything at all—that might help I'd really appreciate it, because my long game is pretty bad." Hogan turned to him and said, "Your short game isn't so hot, either."

Try Practice

A young pro eager to improve his game asked Hogan if he could give him any advice.

"Do you have any practice balls?" Hogan asked. Expecting to be led to the practice tee for a lesson, the young man assured him he did indeed have practice balls. Hogan said, "Then use them."

Accuracy Pays

Another young pro came to Hogan for help, saying he was having trouble with thirty- and forty-foot putts. Could Hogan offer him advice?

Hogan offered the young man the same advice Valerie had given him years earlier: "Hit the ball closer to the hole."

Last Effort

Hogan played his last tournament in 1971. He hadn't played in a tournament since the 1967 Open, but he entered the 1970 Houston Open at the Champions Golf Club, in Houston, owned by Jimmy Demaret and Jackie Burke, two of his closest friends. While a number of other players walked in his gallery, he shot 287 and tied for ninth place. A year later, approaching his fifty-ninth birthday, he entered again. He treated the tournament as he had every important event he ever played. He arrived a week early, practiced every day, and shot superb golf in practice. In one round he had seventeen pars and a birdie, and in two others he shot 65 and 67.

Once the tournament began, though, he was in trouble from the start. He stepped onto the fourth tee two over par and faced a big par 3 of 228 yards with a deep ravine running along the left. He tried to fade a 3-iron close to the hole, but the shot didn't work. The ball flew straight and disappeared into the ravine. Climbing down the steep hill to see if he could play the ball and save a stroke, he strained his left knee, which had been hurt so badly in the car accident so long ago. When he found the ball it wasn't playable. Since the ball had crossed the margin of the hazard so close to the tee, he had to return to the tee and play another shot. With the gallery hushed and willing him to play well, he hit two more into the ravine, finished the hole with a 9, then murmured to Dick Lotz and Charley Coody, his partners, "I'm sorry, fellows."

Ben lasted through the eleventh hole. Embarrassed by then, and in pain, he couldn't go on. He called for a cart to take him back to the clubhouse. As he drove away, he said to a friend, "Don't ever get old."

Thanks But No Thanks

Hogan and Lanny Wadkins were to play at Shady Oaks with two men who called to say they couldn't keep the date. Wadkins and Hogan went off alone. When they reached the twelfth hole, a third man rode up in a cart and announced, "I'm going to join you guys." Wadkins looked at Hogan, Hogan looked at Wadkins and asked, "You ready to go in?" They left the stranger alone on the tee and drove back to the clubhouse.

Souvenir

Wadkins had been ahead in their match. After a few days the mail brought a check from Hogan for $15. Rather than cash it, Wadkins put it away as a souvenir. After bank statements arrived, Hogan's secretary called Lanny and asked that he cash the check. He didn't. After her final plea, Wadkins told her, "Clarabelle, you can forget that check. It's never gonna be cashed. It's in my file and I'm never gonna send that check back."

The Big Sting

Hogan had been taking a beating in money matches with his regular group of eight Shady Oaks members on Wednesday afternoons during 1964. He was fifty-two then and still playing wonderful golf; four years earlier he had nearly won the 1960 U.S. Open. No one at Shady Oaks could touch him, of course, but in club matches he played off a handicap of plus five (which meant he added five strokes to his score). Against his regular four-ball group, whose handicaps ranged from one stroke to ten, he had been giving up from six to fifteen strokes in every round. It's tough to beat anybody giving up that many strokes. They weren't playing one dollar nassaus, either; real money was involved.

After losing all kinds of ways for three or four weeks, he was disgusted. Then, in a match that became known as The Big Sting, he got his revenge.

Hogan arrived at Shady Oaks about 6:45 one cold and miserable April Wednesday, rapped on the golf shop door, and told Art Hall, the club's professional, to find Henry Martin, the greenkeeper, and have him come to the shop. Hogan, meanwhile, marched into the locker room and changed clothes.

When Martin showed up, he and Hogan headed out to the course, Hogan in a standard cart, Martin in a Cushman scooter with a bunch of flagsticks hanging out the back; he'd been interrupted setting up the course for the day's play.

About three hours later Hogan came back to the shop, winked at Hall as he walked by and said, "I wasn't here," changed back into his business suit, then drove off to his office at the Hogan factory.

At about 11:30 the whole group, including Hogan, began strolling into the Shady Oaks clubhouse for lunch and to arrange the day's teams and terms. Lunch over, Hogan announced he would take Earl Baldridge, the weakest player in the group at a ten handicap, as his partner, play everybody every way, and double the bet.

After a lot of needling about Hogan's annoyance at losing the last few times, they all agreed, changed clothes, and went out to the first tee.

When they arrived they found all three sets of tee markers, including those for the ladies, stacked at the far back right corner of the farthest back tee. Not suspecting trouble just yet, Dr. Harvey Small, one of the better players, asked, "What are they doing, Ben, mowing the tees?"

"That's right," Hogan smirked. "Hit it."

Reaching their drives, the group looked at the green and saw the flagstick tucked into the front right corner dangerously close to a bunker.

By then a number of members had already played the course, knew what was afoot, and spread the word. A gallery of about twenty or thirty came out to watch. They saw that, sure enough, all the tee markers on every hole had been stacked in the farthest back right corner of the farthest back tee, and every flagstick set in the front right, close to a bunker if there was one.

Since most of the others played a draw-hook, in the standard Texas tradition, Hogan cleaned up. On a grim and overcast day, with the temperature hovering in the 50s and the chilly wind blowing at about twenty-five miles an hour, he shot 64, which at the time might have been the lowest score ever shot at Shady Oaks. With Hogan telling him what to do, Baldridge, his partner, shot 81, the second lowest score of the remaining seven. Hogan and Baldridge were unbeatable.

Later, when everyone returned to the grill and Hogan sat at his table collecting the money, the walls of Shady Oaks trembled from the roar of very unhappy losers. Words that hadn't been heard in years ricocheted down the hallways, and it is said that one enraged loser slammed his check in front of Hogan and drove a penknife through it, pinning it to the table.

No one will admit how much money Hogan and Baldridge won, but it certainly covered the lunch bill. For the next two months Hogan couldn't get a game on Wednesday afternoons.

Try Work

Gary McCord, best known for doing commentary for televised golf tournaments, played the pro tour before turning to television. Sitting around a table with a number of others, McCord was introduced to Hogan. After the usual introductory remarks, Hogan asked McCord what business he was in.

McCord answered, "I'm a professional golfer."

"How many tournaments have you won?" Hogan asked.

"None," McCord answered.

"How long have you been on tour?"

"Ten years," McCord admitted.

Stunned, Hogan said, "What the hell are you doing out there? You ought to find yourself a job."

It's No Secret

After playing in the 1992 Grand Slam of Golf, a competition limited to the winners of the U.S. Open, the British Open, the PGA championship, and the Masters Tournament, Nick Faldo flew to Fort Worth to meet with Hogan.

Like everyone else, Faldo held him in awe. They spent an hour talking. After lunch at Shady Oaks, Faldo asked if he would watch him play some practice shots. Hogan declined. Then Nick asked if Hogan would share a secret with him. Hogan asked, "What secret?"

"I really want to win the U.S. Open," Faldo said, "and I'd like you to tell me the secret to it."

Frowning, Hogan said, "Shoot a lower score than anybody else."

Thinking Hogan was joking, Faldo laughed a little, then persisted, "No, really; what is the secret?" Hogan said he wasn't joking. "Just score lower than anybody else." Then he excused himself and walked away.

Sam Snead

As a raw rookie, Sam Snead played his first tournament in Hershey, Pennsylvania, in September of 1936. Arriving in time for a practice round, he spotted George Fazio, a total stranger, and asked if he might join him and two other pros waiting to tee off. A kind-hearted man, Fazio agreed.

Snead dropped his worn golf bag on the tee, scrounged in one of the pockets for a ball fit to play, and while the others in the group scowled, drove his first ball almost into the lobby of the Hershey Chocolate Company.

Again he rooted about in his bag for a playable ball and drove it in the other direction, out of bounds. While the others growled at Fazio for inviting this incompetent rookie to join them, Snead teed up another ball and drove it straight down the middle about 300 yards. No one said another word.

Snead shot 291 and tied Ralph Guldahl for sixth place, four strokes behind Henry Picard, the winner. No one ever again laughed at his golf game.

Snead's Driver

Snead was known primarily as the longest straight driver of his time. Actually, he was much better with his pitching clubs. Put him within 100 yards of the green and he'd consistently get down in two strokes. He was a superb putter as well, a terrific long iron player, and he could recover from the sand with anyone.

His driving actually caused him trouble. He had an unpredictable and uncontrollable hook at first; when it was bad, the ball never rose more than a few feet off the ground and swerved sharply left. It had grown so bad he was ready to go home early in January of 1937, but just before the Los Angeles Open, Henry Picard handed him a George Izett driver that weighed 14.5 ounces, heavier than the normal driver, with a swing weight of E-5, again quite heavy. It had a loft of eight degrees, three degrees less than normal, which made it almost straight-faced. The shaft was stiffer as well.

A week later Picard was leading the Oakland (California) Open after three rounds, but using his new driver, Sam shot 67 and won by two strokes. First prize was $1200. Sam paid Picard $5.50 for the club.

Curl Your Toes

Snead often told Fred Corcoran, his manager, of his early days in western Virginia, where the kids took off their shoes the first of May and didn't put them on again until the first frost. A great promoter, Corcoran embellished the story and claimed Sam could play golf barefoot. His audience scoffed at Fred, saying anyone who played barefoot would break his toes.

His reputation at stake Fred pleaded with Sam, beseeching him to play at least a couple of holes barefoot. Sam agreed, took off his shoes and socks, and while a sizable gallery tagged along, he played the first and ninth holes. He birdied the first and barely missed another on the ninth. Explaining the technique to his gallery, Snead said, "You just curl up your toes and pivot off the inside of your foot.

Buried Treasure

Jimmy Demaret helped feed Snead's reputation as an uneducated hillbilly, claiming that Snead stuffed his money in tomato cans buried in his yard. According to the legends somebody evidently believed it. Snead is said to have come home unexpectedly one night and found a man with a pick and shovel digging up his lawn.

These Modern Inventions

After Snead won the 1937 Oakland Open, his first victory on the tour, Corcoran showed Sam his photograph in the *New York Times*. Jokingly, Sam asked Fred, "How'd they get my picture? I've never been in New York." For years Corcoran passed it off as the remark of a backward countryman.

Snead and the Open

Although Snead never won it, no one meant more to the Open. No other golfer won—and broke—so many hearts trying. He had so many chances, and the stories of his losses are enough to wrench anyone's emotions.

Snead played his first Open at Oakland Hills, in 1937, and came into it as the favorite. He shot 283, one stroke over the Open record, and stood around accepting congratulations while Ralph Guldahl raced around the first nine in 33 and came back in 36. With 281, Guldahl broke the year-old Open record and beat Snead by two strokes. This was the first in a series of disappointments.

Two years later Snead was set up for the greatest failure of his career. Coming to the seventy-first hole at the Spring Mill course of the Philadelphia Country Club, Sam apparently had won yet another Open. He needed two pars for 282, and since 284 was the best score in, it seemed obvious the Open was his.

Then Snead three-putted the seventy-first. One stroke gone. Now he needed a par 5 on the seventy-second to win, a bogey 6 to tie. Drawing back his driver in that languid, fluid motion, Sam pulled his drive into the left rough. Now he had a problem.

The primitive communications of those days left Snead wondering what score he needed to win. Byron Nelson had already finished with 284; but Craig Wood and Denny Shute had started behind Snead, and Sam didn't know how they were playing. He felt he probably needed a birdie. His ball sat in a tight lie 275 yards from the green, but Sam believed he might get there with a good brassie. Even if he didn't catch the ball flush, he'd still have an easy pitch left.

His decision was a disaster. Sam topped the shot. The ball rolled into a steep-faced bunker about 110 yards short of the green. The bunker face rose about five feet, low enough to clear with a wedge, but Sam tried to reach the green with an 8-iron. Off on too low a trajectory, the ball slammed into the bunker face and wedged between cracks of freshly laid sod.

From an awkward stance, Snead chopped the ball free, but it rolled into another bunker alongside the green. With his feet on the grass outside the bunker, Sam played his fifth shot forty feet past the hole. He three-putted and scored 8. With 286 he finished two strokes behind Nelson, Wood, and Shute.

Snead had holed an eighteen-foot putt on the seventy-second hole at the St. Louis Country Club to tie Lew Worsham and force another playoff for the 1947 Open.

Three times he led by two strokes, and he still led by two with three holes to play. He looked like a certain winner. Then Worsham birdied the sixteenth, and Snead dropped a stroke at the seventeenth. Now they were even with only the eighteenth to play.

Both men played their approach shots past the hole. Snead's ball lay in almost the same position as it had the previous day, and Worsham's ball had run over the green. Worsham played a bold chip that ran true to the hole, nipped a corner of the cup and spun away. Playing a more timid shot, Snead left his putt short. Both balls lay about two and a half feet from the cup, Worsham's below and to the right, Snead's above and to the left.

Snead stepped up briskly to continue putting. Suddenly Worsham called, "Whoa, Sam. Are you sure you're away?" With the easier putt, Worsham wanted his ball in the hole first to put more pressure on Snead.

Snead stopped. Not sure of the proper procedure, they called to Ike Grainger, the chairman of the USGA's rules committee, the final arbiter on the rules of play. When Grainger reached them, Snead insisted he had the right to continue putting. Grainger said he hadn't, reminding him, "Not unless you're away, Sam."

Using a steel tape, Grainger measured each ball's distance from the cup and saw that Snead's ball lay 30½ inches from the cup, Worsham's 29½ inches away. Snead could continue.

Sam's putt would break left to right. He tapped his ball a touch too softly; the crowd groaned as the ball turned away from the cup and stopped two inches outside the hole.

Worsham gave his putt a firm rap. It dived into the cup for the par 4. Lew shot 69, Sam shot 70. Snead was frustrated once again.

Snead had a coarse sense of humor that often led others to turn away in embarrassment, even though they snickered at some of Sam's expressions. The afternoon before the first round of the 1956 Open, Merrell Whittlesey, a reporter for the Washington *Evening Star*, who knew Snead well, asked what score Sam felt he'd need to win. Without hesitating, Snead said, "Oh, I'll take 280, sit in the clubhouse, eat hot dogs, drink Cokes, and fart."

The conversation turned later to the 1953 Open, when Snead went into the last round just one stroke behind Hogan, then shot 76 and lost by six. Whittlesey asked, "Were you tight, Sam?" With a look of disgust, Snead grumbled, "Tight? I was so tight you couldn't a drove flax seed up my ass with a knot maul."

Tricks of the Trade

Snead and Dave Marr were paired together at Medinah Country Club, near Chicago, during a Western Open. They reached the tee of the fourteenth hole, a par 3 across a lake. Up first, Marr figured the shot called for a 6-iron by him and a 7-iron by Snead, but he noticed Snead had positioned himself alongside Dave's bag so he could see which club Marr chose.

Instead of the 6-iron, Marr drew his 5-iron, hit the ball a little softly, and watched it settle nicely on the green.

Knowing he hit the ball a club longer than Marr, Sam played a 6-iron. His ball flew so far the crowd behind the green didn't have to move as the ball soared over their heads.

Keep Your Perspective

Snead was in golf for more than the joy of playing, although hardly anyone ever enjoyed it more. He was in it for the money.

Sometime during the 1960s, the artist Leroy Nieman did an abstract painting that featured a number of great players spanning several eras—Ben Hogan, Jack Nicklaus, Arnold Palmer, Lee Trevino, Gary Player, and Sam Snead. Years later a fan named Jim Trimpe, who lived in Atlanta, took a print of that painting to the 1981 PGA Championship, at the Atlanta Athletic Club, and asked each of the players there to sign under his likeness.

He found Nicklaus and Palmer, and then, strolling through the clubhouse he saw Snead sitting at a table eating an ice cream cone. Approaching Snead, Trimpe said, "Hi, Sam," as if he might have been an old friend, "how about signing this for me—right there under your picture."

Frowning, Snead looked at the picture and then at Trimpe. Ever one to keep matters in perspective, Snead signed his name in the appropriate spot and then snarled, "You know, none of us got a dime out of this."

Ice Plant

Paired with Ed Tutwiler, an outstanding amateur from West Virginia, Snead overshot the fifth hole at Spyglass Hill during a Crosby tournament. His ball landed in a cluster of ice plant, a devilish growth with thick, squishy, nearly cylindrical stalks rising four or five inches above ground. It looks like a field of shiny green beans, but it's nearly impossible to play from; a golf ball tends to sink among the stalks. Sam's ball sat on top of the stalks, almost as if it were on a tee. Seeing a heaven-sent lie, Sam drew his 8-iron, ready to pop the ball onto the green and maybe save his par. As he moved into position, Sam's nephew, J. C. Snead, stopped him.

"Unkie," he said, "have you ever played out of this stuff?"

Sam said he hadn't, and so J. C. advised him, "You better just take a drop out of there, or just take your sand wedge and blast it."

Sam stopped in his tracks and glared at J. C., as if saying, "Boy, do you know who you're talking to?," then flicked at the ball. It just sat there. He tried again, and again didn't move the ball. Then he switched to his wedge, chopped the ball onto the green, and made 6, a triple bogey.

The round over with Sam having birdied three of the last four holes, J. C. said, "Unkie, a bogey on that par 3 would look pretty good right now, wouldn't it?"

Perils of Flight

Flying out of Greenfield, Iowa, in 1959, Snead and his pilot did everything but take off. Racing down the runway at full throttle, their plane scraped a parked aircraft, spun around and wiped out two billboards, ran out of runway and crashed through a fence of barbed wire, then plunged into a ditch. Pieces of propeller, meantime, whizzed past Sam and beheaded cornstalks in the surrounding fields.

Frantic to escape, Snead cut several fingers of his right hand at the base and injured his right thumb. A rescue party claimed that if the plane had flipped once more, Sam could have lost a few fingers. For the next few weeks, while his injuries healed, Sam couldn't hit a shot.

South American Fling

The State Department organized a tour to sixteen South and Central American countries by leading American golfers, both amateur and professional. Preparing to play from a bunker during an exhibition in Brazil, Snead was attacked by a rhea, a big, flightless bird resembling an ostrich. Noticing Sam's trademark Panama hat, the bird thought he had spotted lunch and went for it. Trying to protect himself, Sam threw up his right hand. The rhea wasn't picky; a little red meat would go nicely with the salad, so he stabbed with his beak. Sam became a one-handed golfer for a time.

On to Argentina and the San Andreas Country Club. On a clear day, Sam faced a fifteen-foot putt for a birdie he just knew he could make. He tapped his ball and watched it run true to the cup. Just as it seemed it would fall, the ball suddenly bounded backwards.

Stunned, Sam knelt at the cup for a closer look. There he found something he hadn't noticed before. Someone had planted green-colored toothpicks all around the hole.

Gamblers evidently had set up the barrier to un-nerve Snead, the betting favorite, and prevent him from beating the local players.

Late in his career, during the middle 1950s, Sam was offered a $5000 appearance-fee to play in the Brazilian national championship, in São Paulo. Never reluctant to accept money, Snead agreed, but when he arrived he was told he would not be eligible for the prize money. The tournament promoters wanted the money to go to South Americans, such as Roberto De Vicenzo and Antonio Cerda.

Sam agreed and played anyway. Opening with 71, he found himself well behind the home-growns, who were scoring in the middle 60s. Seeing how far behind Snead had fallen, the promoters changed heart and ruled him eligible. When he heard money was at stake, Snead's eyes lit and his clubs became hot. He shot 64 in the second round, setting the Brazilian record. In one stretch he holed chip shots from 40, 25, and an estimated 100 feet. The record lasted one day. Sam shot 63 in the third round, then followed with 69. He not only won, he finished eleven strokes ahead of the second-place finisher. The money was all his. "Best jinx I ever beat," he crowed.

Royal Singapore

As a prisoner of war during the Second World War, William Herbert Day, once president of the Malayan Golf Association, lined up for roll call one morning and began thinking his life might suddenly become more pleasant. The prisoner of war camp was just across a road from the Royal Singapore Golf Club. Shortly after Day was captured, the officer in charge marched all the prisoners to the club, lined them up, and asked everyone who had been a club member to step forward. Day thought, "Well, those little fellows want to play some golf. I'll be happy to oblige," and stepped forward with about five others. The commandant obviously felt six men weren't enough, so he asked for everyone who played golf to fall out, and a number of others stepped forward. "That's better," the commandant said.

Day was all set for some golf, but the Japanese weren't. They were evidently fond of sweet potatoes, so the prisoners were marched onto the golf course and told to dig up the greens and plant potatoes. Day tried to outwit the Japanese by simply scratching a shallow trench and sticking in the plants, but an officer inspected the work and said, "No good. Dig deep."

The sweet potatoes thrived, but all that tenderly nurtured prize sod was tossed into the river.

Disqualified

According to weather reports, a storm was moving in from Lake Erie toward the Canterbury Golf Club, near Cleveland, on the last day of the 1940 U.S. Open. Six players who had finished their third rounds decided to start their fourth earlier than their scheduled times, hoping to avoid the threatening weather. Leland Gibson, Ky Laffoon, and Claude Harmon started first; Dutch Harrison, Johnny Bulla, and Porky Oliver followed.

By the time the USGA's executive committee heard of it, all six men had played at least one stroke. The USGA ruled they had violated the rules, and all six were disqualified. They heard of it when they finished.

Their disqualification didn't matter to five of them because they were out of the hunt, but Oliver had 287, matching Lawson Little and Gene Sarazen. They would play off the next day. Because Oliver had been disqualified, he was not eligible for the playoff. Little beat Sarazen.

Oliver was heartbroken, of course. The United States had not fully recovered from the Great Depression, and money was scarce. Porky sat at his locker for a long time with his head bowed and an occasional tear running down his cheek. "It's not just the honor of having a chance to win the Open," he said, "I need the money, and I need it badly."

Letters and telegrams flooded to the USGA, nearly every one protesting Oliver's disqualification, some of them in forceful language. One correspondent wrote, "I feel certain that your committee stands indicted on a charge of complete lack of sportsmanship. . . . To me it is a great shame that it is not possible to disqualify those who have had anything to do with this decision from all further participation in USGA activities."

Another wrote, "If the association continues to conduct its affairs in such an autocratic, dictatorial, and unsportsmanlike manner, thousands of us average golfers feel the sooner the governing body is replaced, the better it will be for golf." Still another claimed, "There is a limit to poor sportsmanship and you have reached it. You are a moldy lot, and replacements are urgent."

The next day Oliver rushed to Detroit as a replacement in a charity exhibition match for the Red Cross. Gene Sarazen had been scheduled to play, but he couldn't make it; he was held over in Cleveland for the playoff.

Aerial Architecture

Fleeing from an angry Spitfire during the Battle of Britain, in 1940, a German bomber jettisoned its cargo over the Sunningdale Golf Club, on the outskirts of London. The bomb fell alongside the eighteenth green, leaving a deep furrow. Using remarkable foresight, club officials ignored the furrow, and in later years it grew into an interesting grass bunker.

Across the Schuylkill

During the 1940s, golfers in Philadelphia had a favorite challenge when they weren't actually on the golf course. They tried to drive golf balls across the Schuylkill River, which flows through the city. As a sporting ritual it held overtones of George Washington's throwing a silver dollar across the Rappahannock. The money involved in the Philadelphia challenge most often exceeded Washington's silver dollar, though. Since everyone knew a solidly hit 3-wood would clear the far bank of the Schuylkill, the betting centered on pulling off the trick with another club.

Woody Platt had grown up in the Roxborough neighborhood, which overlooked East Falls, where the challenge was set up. A man who could regularly clear the river, Platt somehow was backed into claiming he could drive a ball across the river with a putter. All his bets were covered.

A wily man, Platt had something in mind. Taking a mallet-headed Braid-Mills putter, made with an aluminum head, he ground the face to give it more loft and wound the hickory shaft with the strong twine used to bind the heads of wooden clubs to the shaft. Wrapped in twine, the shaft became firm as an iron pipe. A smooth swinger, Woody ripped the ball across the Schuylkill and picked up all the bets.

Wartime Rules

During the Battle of Britain early in the Second World War, the St. Mellons Golf and Country Club, located in Monmouthshire, adopted a set of unusual rules for unusual circumstances. Written by G. L. Edsell, the club secretary, they read:

1–Players are asked to collect the bomb and shrapnel splinters to prevent their causing damage to the mowing machines.

2–In competition, during gunfire or while bombs are falling, players may take shelter without penalty for ceasing play.

3–The positions of known delayed-action bombs are marked by red flags at a reasonable but not guaranteed safe distance therefrom.

4–Shrapnel and/or bomb splinters on the fairways or in bunkers within a club's length of a ball may be moved without penalty, and no penalty shall be incurred if a ball is thereby caused to move accidentally.

5–A ball moved by enemy action may be replaced, or if lost or destroyed, a ball may be dropped without penalty, not nearer the hole.

6–A ball lying in a crater may be lifted and dropped not nearer the hole, preserving the line to the hole, without penalty.

7–A player whose stroke is affected by the simultaneous explosion of a bomb may play another ball under penalty of one stroke.

Open Trophy Lost

Lloyd Mangrum won the 1946 U.S. Open in a playoff with Vic Ghezzi and Byron Nelson. At the time he held the job of playing pro at the Tam

O'Shanter Country Club, in Chicago. Lloyd gave the trophy to George S. May, the owner of the club, to put on display in the clubhouse. No one ever showed it again. That winter the clubhouse burned to the ground and the trophy melted.

Snead at St. Andrews

Peering through the window as his train approached St. Andrews for the 1946 British Open, Snead asked, "What's that over there? It looks like an old abandoned golf course." It was, of course, the Old Course.

Snead hadn't wanted to play in the British Open, but he hadn't much choice since he was under contract to Wilson Sporting Goods, and I. C. Icely, the company president, insisted he enter. Except for the outcome, the trip was a disaster from the start.

First, seated in his aircraft in New York while the pilot warmed up engines of the propeller-driven Constellation, his plane caught fire. Smoke poured into the cabin, the passengers panicked, and, Snead remembered, "Us passengers came popping out of there like ants, afraid of an explosion."

Reaching London, Snead couldn't find a place to stay; still ravaged by wartime bombing, the city had lost so many hotels, rooms were scarce. Snead slept a few hours on a bench, then boarded the train for St. Andrews. When he arrived, he learned he couldn't use his favorite center-shafted putter. The style had been banned since Walter Travis had won the 1904 British Amateur using his center-shafted Schenectady putter.

The problems continued to mount. His caddie, said to be among the canniest at the Old Course, showed up for work drunk. Snead fired him. His replacement was no help; he whistled through his teeth while Sam played his shots.

Finally, as Sam walked onto the final tee, a man stepped from the crowd and handed him a new ball. Under contract to play the Wilson ball, Sam had played another brand through seventy-one holes, but he used the Wilson on the seventy-second.

Along with Icely, Johnny Bulla, an old friend who palled around with Sam when they were both young professionals, had prodded him into going over for the tournament. So what happened? Snead, Bulla, and Welshman Dai Rees tied at 215 after 54 holes, but Sam finished with 75, Bulla shot 79, and Snead won by four strokes. Sam shot 290 for the seventy-two holes.

Rees was doing all right until a friend rushed out to him on the fifth hole and told him, "For God's sake keep your head. Everybody else is playing badly, and you only need to play your game to win." Rees began playing so badly he shot 80 and tied for fourth place, five strokes behind Snead.

Cary Middlecoff

After he won the 1945 North and South Open as an amateur, Cary Middlecoff, a graduate of dental college, was selected for the 1947 United

States Walker Cup team. This was to be a historic match. Because of the Second World War, the Walker Cup hadn't been played since 1938, when the British and Irish had won for the first time.

Middlecoff declined the invitation, though, because he had a personal agenda. In a letter to the USGA he explained:

> Needless to say I am gratified beyond words about being selected for the Walker Cup team. I know it is the highest honor than can be bestowed upon an amateur golfer in this country, but a few things have come up that make my participation impossible.
>
> First, I am going to be married March 4.
>
> Next, I have planned to turn pro at that time and play tournaments until I have proved to myself one way or the other if I am good enough to make golf playing a life work.
>
> Ever since I can remember I have wanted to play golf without being worried about one thing or another but have never had the chance. I know I would never be happy practicing dentistry without knowing for sure if I were a good player or a great one, and dentistry is too confining ever to offer me that opportunity.
>
> My decision was reached because I love the game of golf and I can see no other suitable out."

Middlecoff gave himself two years to prove himself. In 1949 he won the United States Open Championship.

Short Fuse

One of the great players of his time, Middlecoff occasionally showed flashes of temper. He was especially sensitive to late starting times, which forced him to play after greens had been spiked up by the early starters.

In 1953, when those who organize the U.S. Open were still grouping players by whim rather than by their scores, Middlecoff found himself with a late starting time for the fourth consecutive year. This annoyed him.

He had begun by shooting 76–73—149, which left him ten strokes behind Ben Hogan, the leader, then added 38 on the first nine of the third round, one over par. Clearly headed nowhere without an abrupt turnaround, Cary drove into a fairway bunker from the tenth tee. After playing a safe shot back to the fairway he hit his third shot into a greenside bunker.

In a vile mood staring at a bogey at best, Middlecoff glared at his ball, and in a moment of clarity, decided he didn't have to put up with this. Oakmont's tenth green stands close to the Pennsylvania Turnpike, which slices through the course and separates the second through the eighth holes from the rest. Suddenly wheeling away from the green, Middlecoff drew back his wedge, slammed into his ball, and flew it into the racing traffic on the turnpike.

Amazed, Walter Burkemo, his playing partner, said, "It went into the turnpike like a dart." Without a word, Middlecoff climbed from the sand and walked in. Reminded of it later, Middlecoff said, "I must not have been in a very good mood."

They Took It with Them

With Britain still rationing food several years after the Second World War ended, the American Walker Cup team shipped their rations to St. Andrews for their 1947 match. Aside from his baggage and clubs, each team member carried along 25 pounds of food. In addition, the USGA shipped 300 pounds of fresh and smoked meats—steaks, hams, and beef roasts.

Among other things, the 25-pound packages contained canned lamb stew, boned chicken, codfish cakes, ham spread, sausage and tongue, powdered milk and eggs, cereal, soup mix, dried fruits, orange juice, tomato juice, jam, jelly, butter, and chocolate.

Don't Be a Stranger

Colonel Brian Evans-Lombe, an imperious retired cavalry officer, was appointed secretary of the Honourable Company of Edinburgh Golfers (Muirfield) in 1947, one of a series of overbearing autocrats to hold this position. He reigned for seventeen years.

Early one Sunday morning he noticed an unfamiliar figure sitting alone in Muirfield's smoking room. Asked whose guest the chap might be, Evans-Lombe was told rather sharply that he was talking to a member of some twenty years' standing. Not one to give away an advantage, Evans-Lombe snorted and parried, "In that event you should come more often; then I might recognize you."

Lloyd Mangrum

Opening the 1948 Masters with 69, good enough to lead the field, Lloyd Mangrum continued to play well and hang onto the lead through the first seven holes of the second round. A little too quick with his hands playing his drive on the uphill par-5 eighth hole, Lloyd hooked into the woods. As he took his stance to pitch back to the fairway, Mangrum stepped on a twig. The twig nudged against the ball, and the ball turned over. Stepping out of the woods after playing his shot, Mangrum told Byron Nelson, who was paired with him, that he had taken a penalty stroke. Mangrum double-bogeyed the hole, shot 73 for the round, and eventually tied for fourth place behind Claude Harmon.

Tall and slender, with dark, slicked-back hair, a thin mustache, and a lighted cigarette usually dangling from his lips, Mangrum fulfilled everyone's notion of the Mississippi riverboat gambler. That he did a little betting on the golf course only heightened the image. He was tough. A sergeant with a reconnaissance unit in Europe during the Second World War, he was awarded two Purple Hearts for wounds sustained in combat.

Born in Texas in 1914, Mangrum played the professional tour in the late 1930s into the 1950s, leaving school when he was only fifteen so he could play for money. Since his older brother Ray Mangrum had been mildly successful, Lloyd asked Ray to back him in tournament golf. Ray pulled a thick bankroll from his pocket, peeled off two one-dollar bills and handed them to Lloyd. "You ain't good enough," Ray is supposed to have said. "Go find a job."

Lloyd went on to win the United States Open without help from Ray.

Trailing Hogan by one stroke after fifteen holes of their 1950 U.S. Open playoff. Mangrum sprayed his drive into heavy grass bordering the right side of the sixteenth, leaving his approach blocked by trees. After a safe pitch back to the fairway, he played a terrific shot no more than eight feet from the cup. Hogan's ball, meanwhile, lay six feet from the hole in two.

George Fazio, the third man in the playoff, asked Mangrum to mark his ball, since it lay on his line. The rules in those days forbade a player from either lifting his ball unless he had been asked by another player, or from cleaning it even if he had been asked to pick it up. He held the ball gingerly between his thumb and middle finger.

After Fazio putted, Mangrum replaced his ball and lined up his putt, but as he was about stroke the putt he saw a bug crawling around on top. Without thinking, he picked up his ball and blew the bug away. The bug gone, he replaced his ball and rolled it into the hole for his 4. Hogan missed his birdie, and they walked toward the seventeenth believing Hogan still led by one stroke.

Since he had birdied the fifteenth, Mangrum held the honor, but as he stepped onto the tee to play his shot, Ike Grainger, the referee, slipped under the gallery ropes and stopped him. Grainger explained that by blowing away the bug he had incurred a two-stroke penalty.

Mangrum held an iron in his hand, ready to play to the seventeenth green, a par-3 hole. He slammed the club back into his bag, fixed Grainger with a cold stare, and snarled, "You mean I had a 6 instead of a 4?" Grainger agreed that was indeed the case. It was a tense moment. Finally Mangrum's expression eased, and he said, "Well, I guess we'll all eat tomorrow."

At the presentation ceremony later in the afternoon, after Hogan had won the playoff by shooting 69 against Mangrum's 73 and Fazio's 75, Jim Standish, the president of the USGA, mistakenly referred to the Merion Golf Club by its former name of Merion Cricket Club. As he accepted his second place check, Mangrum shot back, "The brass may not know where they are, but they sure do know the rules."

A photographer had taken a shot of Mangrum walking with Grainger. Some years later Mangrum sent it to Ike with the inscription, "Love and kisses, Lloyd."

Mangrum opened the 1951 St. Paul Open with 67, 67, and 62. Leading by a wide margin, he took a telephone call the night before the final round. The caller threatened to kill him if he won. Not certain it was a hoax, Mangrum called the police. The authorities wanted him to withdraw, but Mangrum refused. "Withdraw!" he cried. "What do you think I'm playing for, tin cups?"

Asked what he felt the police could do, Mangrum told them, "Give me some bodyguards who'll keep their heads up. In my racket you have to keep your head down." They did. With uniformed policemen standing beside him all the way around, he shot 70 and won the tournament.

Mangrum survived eleven heart attacks; the twelfth killed him, in 1973. While he was recovering from his seventh, President Eisenhower had his second. Thinking he could cheer up the President, Mangrum sent him a telegram. "Dear Ike," it read, "I'm five up on you."

Harmon at Muirfield

Just as he was about to tee off in the first round of the British Open, at Muirfield, in 1948, Claude Harmon learned that the man he had been paired with had withdrawn. A hasty search for a marker turned up only one Major W. H. Callender, retired from the Royal Scots Greys. Callender promptly said he would be "delighted to give the fellow a game."

He did indeed give Harmon rather more than he expected, although not in the sense of competition. Harmon had won the Masters Tournament only a few months earlier, but that didn't impress the good major. Reporters following in Harmon's wake noticed a heated discussion, perhaps an argument raging between the two. Hurrying to them, they heard Callender giving Harmon a lesson, along with a demonstration, on the proper way to swing a club. "Gripped farther down the club," the major said, " a short swing controls the ball more easily in the east wind."

When he was driven from Holyrood Castle, near Edinburgh, to Muirfield for the 1948 championship, King George VI, a former captain of the R and A, became the first reigning British sovereign to watch a British Open. Henry Cotton put on an exhibition worthy of a king. Cotton shot 66 in the third round, on his way to an aggregate of 284 in winning his third and final Open. He won by five strokes, playing the ultimate in precision driving. Along the way he hit fifty-three of the fifty-six fairways on driving holes.

Joan Hammond

In addition to developing into a first-class golfer, Joan Hammond, an operatic and concert soprano, was an outstanding athlete in her native Sydney. She reached the final of the Australian women's squash championship, indulged in yachting and riding, won a gold medal for life-saving, and at one time held the lowest handicap in Australian women's golf.

A sportswriter with a Sydney newspaper as well, Miss Hammond studied the violin, but she had to give it up when she injured her left arm in an automobile accident. With the violin gone, she concentrated on her singing. She became so exceptional that before the outbreak of the Second World War, women golfers of New South Wales raised funds to send her to Europe to study voice.

During the London blitz, Miss Hammond alternated between driving an ambulance and singing in air-raid shelters, subways, and service camps.

The war over, Miss Hammond remained in Britain and began playing golf once again, while at the same time advancing with her operatic career. She entered the 1948 British Ladies championship, played that year at Royal Lytham and St. Annes, on the west coast of England, and drew as her first round opponent Louise Suggs, one of the finest women golfers ever developed in the United States. Playing with borrowed clubs, Miss Hammond putted like a demon and took Miss Suggs to the sixteenth green before, predictably, losing.

Miss Suggs went on to win the championship, but as soon as the match ended Miss Hammond leaped into a car and sped forty miles to Manchester, where that night she sang the title role in *Tosca*, and of course lost again, taking that final plunge over the castle wall.

Harry Bradshaw

Harry Bradshaw, a florid-faced Irishman, opened the 1949 British Open by shooting 69 at Sandwich, on England's southeastern coast, then began the second round with four steady 4s. On the fifth, a strong par 4 of 451 yards over a hill, Bradshaw cut his drive into the right rough. The ball hit into the heavy grass, bounced, and settled in the bottom half of a broken beer bottle.

Finding his ball in the bottle, Bradshaw didn't know his options and he couldn't find anyone to advise him. In those days players didn't send word to the clubhouse for help, and the R and A hadn't yet established the practice of stationing rules officials throughout the course.

Since the broken bottle qualified as a movable obstruction, Bradshaw could have taken the ball from the bottle and dropped it without penalty. At the same time, the rules also stated, "The ball is unplayable if the player considers he cannot make a stroke at it and dislodge it into a playable position." Not sure of the rule, Bradshaw closed his eyes and chopped at the bottle. The glass shattered, and the ball carried about twenty-five yards ahead. Shaken, Bradshaw missed the green and made 6. He was never the same again. He shot 77 that day, then added 68 and 70, tied Bobby Locke, of South Africa, at 283, then lost to Locke in the playoff.

Larry Smith, Bradshaw's caddie, had reached the ball first, and like Bradshaw he had no idea what to do. A native of Belfast, Smith eventually emigrated to Canada. On a visit to his son, who still lived in Belfast in 1986, Smith couldn't miss the opportunity to see Bradshaw again. He drove to Portmarnock Golf Club, near Dublin, where Bradshaw had been the golf pro for twenty-seven years. During their conversation, Smith said, "If I had known what was going to happen, the ball wouldn't have been in that bottle when you got there."

Locke shot an incandescent 67–68—135 against Bradshaw's 74–73—147, but not until he threatened to refuse to play unless the cups were moved. They had been left in the same positions over the four rounds of the championship proper.

Good Luck Ball

Lee Mackey, a twenty-one-year-old golfer from Birmingham, Alabama, started the 1950 U.S. Open with a round of 64, at that time the lowest single round ever shot in an Open. He created a sensation. Here was a practically unknown player shooting the lowest single round ever in the national championship and leading by four strokes over the greatest players in the game—Ben Hogan, Sam Snead, Cary Middlecoff, Lloyd Mangrum *et al*. Newspaper reporters surrounded him asking for details of his round and his life history. It was heady stuff.

Mackey, though, had a firm grip on reality. Standing on the first tee the next day, he sidled over to Joe Dey, the USGA's executive secretary, and quietly asked, "What do you think will make the cut?"

After shooting the 64, Mackey had given his golf ball to his caddie, whom he described as "an elderly fellow, probably about 45."

Mackey had lost some of his edge overnight. He was on his way to shooting 81 when his caddie tapped him on the shoulder, handed him the ball he had used to shoot the 64, and said, "Better try this one again."

His 145 did indeed survive the thirty-six-hole cut. When he followed with 75 and 77 on Saturday, the last day, he put up a score of 297 and tied for twenty-sixth place. He won $100.

Mackey eventually drifted out of golf and became sales manager of a concrete firm in Birmingham. Many years after his historic round, he became reinstated as an amateur.

It's My Honor

A few years after he had won the 1948 Masters, Claude Harmon was paired with Ben Hogan in the first round at Augusta. Holding the honor on the twelfth tee, Hogan lofted a nice pitch within ten or twelve feet of the hole. Up next, Harmon hit what he described as the best 6-iron of his life. It climbed up the flagstick, down the flagstick, and darted into the hole. A hole-in-one.

The crowd went wild as the two men left the tee and crossed the little bridge leading across the creek flowing in front, and stepped onto the green. Hogan, however, said nothing, not even when Harmon strode toward the hole, lifted his ball from the cup, and waved to the crowd. It was then that Hogan said his first word. "Caddie," he said, "leave the flag out." With the crowd hushed, Hogan holed his putt for the birdie 2.

Once again the crowd cheered, but Hogan still said nothing about Harmon's ace. With the crowd still cheering, Harmon teed up his ball on the thirteenth, then looked up as Hogan began walking toward him. Harmon thought to himself, "Now the Master is finally going to congratulate me." Hogan looked Harmon straight in the eye and said, "You know, Claude, I've been waiting three years to make a 2 on that hole."

Patty Berg

Patty Berg had grown up as a tomboy in Minneapolis playing football with the Fiftieth Street Tigers, a neighborhood team organized by Bud Wilkinson, who grew up to become the football coach at the University of Oklahoma. Miss Berg said the Tigers had one signal—22, and everybody ran every which way—but never lost a game, "Only teeth."

When she began hurting her legs and tearing her clothes before they were paid for, she became a speed skater for the Minneapolis Powderhorn Club, then turned to golf, taking lessons from Les Bolstad, who had played in the 1930 United States Open, the third leg of Bobby Jones's Grand Slam.

In 1951 she led a team of American women professionals against a team of first-class British male Amateurs, most of them veterans of international competitions. They played at the Wentworth Golf Club, near London, playing on level terms from the men's tees. Against the British, the Americans sent Babe Zaharias, Betsy Rawls, Peggy Kirk, Betty Jameson, Betty Bush, and Patty. The British team was led by Leonard Crawley, the man who dented the Walker Cup with an over-ambitious shot during the 1932 Walker Cup Match. Crawley was a commanding figure, partly because of his glorious mustache. He vowed to shave it off if his team should lose.

The match began badly for the American women. They lost two of the three foursomes and halved only one, trailing by 2½ to ½ at the luncheon break. They were to play six singles matches in the afternoon. Sitting around the table bedecked with American flags, the women lunched in grim silence. The meal over, Patty shot to her feet and announced, "All those who expect to win their singles, follow me." The Babe smiled, then said, "C'mon. Follow Napoleon." The women swept the singles and beat the British, 6½ to 2½.

Without a word, Crawley climbed into his car and drove away. He was never seen sans mustache.

Patty and Babe played at least one other match against British men, meeting Edward Davenport and John Beck in foursomes. With the women a hole down on the seventeenth tee, Babe flew their approach into the heather. Unconcerned, she said to Patty, "We'll halve this hole and eagle the next. All you have to do is put this one on the green." "Fine," Patty answered, "but I've got to *find* the ball first."

She did indeed find the ball and somehow coaxed it onto the green. Determined to putt first, Babe said, "When I sink this they'll tighten up and blow it." Davenport, on the other hand, was convinced he was away, and he crouched over his ball, ready to putt. "Time out," Mrs. Zaharias shrieked, a tactic hitherto unknown to the game. "Whaddaya mean, time out?" Patty said in a fierce whisper. "I want a measurement," she told Davenport.

A tape measure showed the men were indeed away; Davenport holed his putt, and the Babe made hers. On to the final hole, still a hole down. Babe hit the longest drive Patty had ever seen a woman hit. They eagled the hole, halved the match, and the Babe lifted Patty off the ground and carried her off the green.

Babe Zaharias

Mildred (Babe) Didrikson Zaharias may have been the greatest woman athlete who ever lived. She won gold medals in three track and field events in the 1932 Olympic Games, played baseball against the likes of Babe Ruth, and dominated women's golf in the late 1940s and 1950s, first as an amateur, then as a professional. Asked if there was nothing she didn't play, she snarled, "Yes. Dolls."

Playing the tour, Babe occasionally found the rules of golf interfering with her game, and she could be rather casual about them. Two large stone lions stood on the site of one Florida tournament and, predictably, the Babe's ball rolled behind one of them. Looking over her position, she saw the lion blocked a clear shot to the green, but because the lion didn't interfere with her stance or swing, she couldn't move her ball. She knew enough about the rules, though, to know she could move the lion if it could be moved easily.

The Babe was married to George Zaharias, a very big and strong man who performed as a professional wrestler. Seeing him in the gallery, she called to him, "George, that's a movable obstruction. Move it." Zaharias strained and huffed, but he couldn't move it. Babe had to play around it.

At another tournament, she was approached by her playing partner, who had driven into a boggy area in the rough. Pointing to her ball, she asked Babe, "This is casual water, isn't it? I get a free drop, don't I?" "Honey," the Babe purred, "I don't care if you send it out and get it dry cleaned."

Babe entered the hospital with cancer in 1953 and had a tumor removed. It looked for a time as if her golf career had ended, but as she lay in bed she prayed, Please, God, let me play golf again.

Her prayers evidently answered, Babe was back on the golf course the following year just as good as ever. She entered the 1954 Women's Open, at the Salem Country Club, in Salem, Massachusetts, played absolutely superb golf, and completely outclassed the greatest women golfers of the time.

Over a difficult and trying course, she shot 72 and 71 in the first two rounds. As an indication of just how good that was, at the end of thirty-six holes she led the field by seven strokes.

The women, like the men, played thirty-six holes on the last day. Babe shot 73 in the morning, increasing her lead to ten strokes, and followed with a closing 75. She shot 291 for the seventy-two holes. Betty Hicks placed second. She shot 303 and finshed twelve strokes back. This was one of the finest exhibitions of overpowering golf a woman had ever shot. It was also the Babe's last moment of glory.

In July of 1955, doctors at the John Sealy Hospital, in Galveston, Texas, examined the Babe and diagnosed inoperable and terminal cancer. In December, Babe and George visited Bertha Bowen, an old friend who lived in Fort Worth, to spend Christmas together. Growing progressively weaker and realizing her time was running out, Babe asked to see a golf course one more time.

Mrs. Bowen and George helped her into the car and drove to Colonial Country Club, where the second green lies close to the road. Bertha parked the car and helped Babe out. She could barely walk, but dressed in pajamas and a robe, she struggled to the green, knelt down, and laid the palm of her hand flat on the ground. Then she climbed back in the car, and they drove away.

Six months after doctors severed her spinal cord to relieve unbearable pain, Babe lay on her hospital bed, looked up at George and said, "Honey, I ain't gonna die." Those were her last words. On September 26, 1956, Babe died.

Wet Socks

As the 1932 British Amateur champion, Johnny DeForest, a titled nobleman—Count de Bendern—had an annual invitation to the Masters before the rules of eligibility changed. Playing the home nine during the third round of the 1953 tournament, he misplayed his approach to the thirteenth into the creek in front of the green. Walking up to the creek, he saw his ball lying in shallow water just a foot or so short of the far bank. He had three options: He could take a penalty stroke and drop the ball behind the creek, he could play the ball as it lay, or he could go back and play the original shot again. After carefully studying his lie, DeForest chose to play the shot.

He carefully slipped off his left shoe and sock, rolled up his left trouser leg above the knee, then stepped into the creek with his still shod *right* foot. In a classic case of *savoir faire*, he held his ground and played a nice shot onto the green.

Sic Semper Tyrannis

Those close to British golf often wondered why Leslie McInally had never been chosen for the Walker Cup team. The answer may have lain in a match he played with Gerald Micklem, a pillar of British golf and a man who ruled the Sunningdale Golf Club, near London, with an iron fist inside a mailed glove.

Micklem and McInally met in the 1953 British Amateur, played that year at Hoylake, in western England. Micklem could be rather abrupt in dealing with others, and McInally was a rough sort, not at the forefront with social graces. McInally knocked a drive into the rough on the fourth hole and searched frantically for his ball. Time flew by. Eventually, Micklem called to him in an imperious tone, "McInally, you've had five minutes. My hole"; then stalked off to the fifth tee.

Later in the round, as they reached the thirteenth hole, Micklem let a shot get away from him and began beating through the undergrowth with the clock again ticking away the minutes. The search became futile, Micklem and his caddie thrashing about and McInally standing in one spot watching them. As the time limit passed, McInally said nothing and still Micklem and his party searched. Finally McInally decided he'd had enough. Calling loudly to Gerald he said, "Micklem, you've had seven minutes," and then pointing downward where he stood, he added, "and here's your ball."

Micklem won the match, and McInally went through life without earning a place on a Walker Cup team.

One Pro to Another

Seminole Golf Club held a two-day pro-amateur tournament during the 1950s that attracted most of the big-name golfers. Claude Harmon was the club professional. One day while the tournament was going on he had a call from the gate. Jimmy Demaret and Vic Ghezzi had arrived with two female acquaintances, but, they complained to Harmon, the guards wouldn't pass their friends through the gate. Not knowing the guests were women, Harmon asked, "Are they pros?" After a full two-beat stop, Demaret said, "Yes. They're pros."

Worsham's Eagle

The World Championship of Golf was the creation of George S. May, a Barnum-like former Bible salesman who grew into one of the game's flamboyant promoters and bought the Tam O'Shanter Country Club, in suburban Chicago.

At a time when most professional tournaments offered prize funds between $15,000 and $20,000, May's World Championship offered $75,000; the winner took home $25,000. With that much money at stake, the tournament drew the best players in the game.

Chandler Harper, a former PGA champion, had finished the 1953 tournament with a 72-hole score of 279 and stood behind the eighteenth green accepting congratulations as Lew Worsham stepped onto the final tee. Lew needed a birdie to catch him.

Worsham's drive left him 100 yards from the green, just a wedge, the same club he had played in his seven previous rounds over Tam. Lew drew back his club, but as he moved into the ball he took a touch too much turf and hit the shot a little fat. The ball had no backspin to speak of, which turned out to be the luckiest break he ever had. The ball hit short of the green, bounced on, rolled directly at the hole, and tumbled into the cup—an eagle 2. He had won the tournament.

More people saw Worsham hole that wedge shot than had ever before seen a golf shot played. May had arranged for a nation-wide telecast, the first coast-to-coast hookup of a golf tournament. Jimmy Demaret was doing the television commentary. As the ball dived into the hole, Demaret blurted over the air, "The son-of-a-bitch holed it."

Rank and Privilege

President Eisenhower took occasional license with the rules of golf. During his presidency he played Burning Tree Club every Wednesday he could, quickly becoming known for his novel method of identifying his ball. Rather than stoop over for a closer look, he would use a club and roll his ball over until the trademark became visible. If his ball moved into a better lie, well, Ike *was* the President.

One afternoon Eisenhower's ball settled in the rough. When he tried to roll it over, it lodged against a rock. Startled, the President glared at his caddie and snapped, "What happened?" The caddie answered, "Mr. President, I'm afraid you have overidentified your ball."

Invited to play a round with the President at Burning Tree, Sam Snead kept his thoughts to himself through seventeen holes, but as they walked onto the eighteenth tee he said to the President, "Mind if I tell you one thing?"

Pleased to take advice from one the greatest golfers the game has known, Eisenhower said, "No. Not at all." Snead looked him in the eye and said, "Stick out your fanny, Mr. President." His fanny in the proper position, Ike then whacked a 220-yard drive that split the fairway.

Throughout his years as a member, of the Augusta National Golf Club, Eisenhower was bedeviled by the broad oak on the left of the seventeenth fairway; his drive seemed always to land behind it. He often asked the club to cut it down.

Recovering from a heart attack, Mr. Eisenhower attended a membership meeting at which Clifford Roberts, the club chairman, rose and said, "Mr. President, we're so pleased to have you back, we'll make any change to the golf course you'd like."

Face alight with the prospect he'd finally have his way, Eisenhower cried, "Cut down that tree on seventeen." Nonplused, Roberts calmly replied, "I'm sorry, Mr. President, but we can't do that." And it kept on growing.

A man who had risen to command the invasion of Europe during the Second World War and to serve eight years as President of the United States, Eisenhower claimed his happiest moment occurred on February 6, 1968, when he had reached the age of seventy-seven. On that day, playing the Seven Lakes Country Club, in Palm Springs, Ike reached the thirteenth hole, a 104-yard par 3. There he hit a 9-iron into the cup for the only hole-in-one of his career. That, he said, was the thrill of a lifetime.

No Help

Before he turned pro, Julius Boros played lots of golf with Bob Toski, Sam Snead, and George Fazio, and he once spent most of a summer with Tommy Armour, the professional at Boros's club in Connecticut. Julius took a few lessons from Armour as well.

Armour always taught "Hit the hell out of the ball with your right hand," which became the subtitle of his book *How To Play Your Best Golf All the Time*. It didn't work with Boros. "That was the only time I ever shanked the ball," he claimed. Not able to hit the ball as Armour wanted, Boros finally told him, "Tommy, don't give me any more tips."

When Boros told Armour he was turning pro, Tommy wasn't impressed, and predicted Boros would never succeed. Admitting that Boros was exceptional from sand bunkers, Armour gave him more advice. "Aim for the bunkers and you might make it."

Boros developed his touch from sand early by playing bunker shots with an R. T. Jones 9-iron with a fairly wide flange. He used it when he finished second in the North and South Open. Seeing Boros using a 9-iron when every self-respecting golfer owned a sand wedge, Vic Ghezzi, the 1941 PGA champion, told Julius, "I'll send you a sand iron."

Clayton Heafner

Blunt and grouchy, Clayton Heafner, a big man who stood six feet-two and weighed 225 pounds, may be remembered best as the only man other than Ben Hogan to break 70 at Oakland Hills in the 1951 U.S. Open. Hogan shot 67 and Heafner shot 69. Hogan won by two strokes.

Heafner, with a crown of reddish-blond hair, is remembered almost as well for his flaming temper. There was a story that he became so angry he tore off the steering wheel of his car and had to drive with his big hands gripped around the column. Jimmy Demaret called him the most even-tempered man he'd ever seen—"He was mad all the time."

Heafner played the tour during the 1940s and 1950s. He often shared a room with Skip Alexander, like Heafner, a North Carolinian. Clayton didn't care much for people. Alexander awoke one morning and saw Heafner leaning out the window glaring at a noisy crowd on the street below. Sensing Skip had awakened, Heafner growled, "Do you see all those people down there? They're the same people who'll come out to the golf tournament tomorrow and bother me all day long."

Heafner was also known to walk off a course before he had even struck a blow. He represented a club in Linville, North Carolina, but he liked to be introduced as playing from Charlotte, where he lived. When the starter at the Oakland Open introduced him as Clayton Heefner, from Linville, North Carolina, and added he hoped Clayton wouldn't have the same trouble with the trees he'd had a year earlier—Heafner had hauled his bulk up one to play a shot from the fork of a branch. Annoyed by the introduction, Heafner turned purple. In measured tones he told the official, "My name is Heafner, not Heefner. I come from Charlotte, not Linville. And as for the trees, I'm not allowing myself the chance to get into your goddam trees." He spun around, motioned for his golf bag and told his caddie, "Put that stuff in the car," and drove away.

Another time he hooked his opening drive badly, sent his caddie after the ball, and again walked away without playing. He was also known to leave in the middle of a round. Playing indifferent golf one day, Clayton hit a particularly bad shot, erupted, and told his caddie to pick up his ball, he was through. Hearing him, a woman in the gallery raced up and confronted him face-to-face. She told him he couldn't pick up his ball—she had him in a Calcutta pool. Heafner said, "Okay," turned to his caddie and told him, "Leave the ball." *Then* he walked in.

Ed Furgol

As a boy of twelve romping around in a playground, Ed Furgol tried to leap from one parallel bar to another. He missed and fell onto the cinder-block base and landed on his left elbow. The bone broke and popped through the skin—a compound fracture.

The pain must have been beyond bearing, but gripping his broken arm with his good right hand, Furgol walked a mile before someone picked him up and drove him to a hospital. Doctors performed three operations but couldn't restore Furgol's arm to normal condition. After the last operation, he wore a cast that held his arm over his head for six weeks. When doctors removed the cast, he couldn't lower his arm.

Because of the accident, his left arm bent permanently at a 45-degree angle, and eventually some muscles withered. Still, he was determined to become a professional golfer. He fulfilled his dream and, in fact, became a very good golfer. From 1945 through 1952, he averaged 71.4 strokes in more than 1000 rounds.

He also played one of the more intelligent and unorthodox shots the Open had ever seen. Standing on the eighteenth tee of the Lower Course at Baltusrol, Furgol was in position to win the 1954 U.S. Open. The closing holes of both the Lower Course and the Upper Course run parallel, both of them par 5s separated by a stream and a line of tall trees. Furgol tried to drive close along the left side, hoping to shorten the hole, but he turned his shot a little and yanked it into the woods.

Looking over his position, Furgol realized that if he played a safe shot back to the Lower Course, he wouldn't have a shot at the green. As he pondered his decision, he noticed an opening in the trees that led to the eighteenth fairway of the Upper Course.

Checking first to be sure the Upper Course was indeed in play and not considered out of bounds, Furgol played an 8-iron from a scruffy lie and threaded his ball through the trees into open ground on the adjoining course. Left with a 7-iron shot, he misjudged the distance and left his approach four or five feet short, chipped to five feet, and although he had averaged 35 putts in each of the earlier rounds, he holed it, beating Gene Littler by one stroke.

Ingenuity Will Find a Way

In the years immediately following the Second World War, Ted Rigney ranked among the better amateur golfers in Australia. He might have been better still had he taken the game a bit more seriously.

Rigney opened the 1954 Ampol Open, at the Lakes Golf Club, in Sydney, by shooting 69. He led quite a strong field, which included Dutch Harrison, Porky Oliver, and Tommy Bolt, three of the best American pros. So signal an achievement should not go unrecognized, Rigney proclaimed, and so he embarked on a one-man celebration that lasted well beyond darkness.

Even after he had reached his considerable limit, Rigney was still in close enough touch with reason to remember he was leading the tournament and certainly didn't want to sleep through his starting time for the second round, which by then was approaching with speed. Rigney was equal to the problem. Rather than head for his bed, he returned to the Lakes, found himself an ample lantana bush alongside the first tee, curled up under it, and went to sleep. He emerged in the morning and shot 76. While he didn't win the tournament, he finished as the low amateur.

Night Fighter

Rufus Stewart, the golf professional at the Kooyonga Golf Club, in Adelaide, Australia, once shot 77 at his home course, which is nothing to write home about except:

He played in the middle of the night under only the light of the moon and a flashlight held by a friend. The round took him only an hour and a quarter to play; over a very fine course with its share of hazards and matty rough, Stewart played the entire round with the same ball. Undamaged and unscarred, the ball was mounted on the wall of the Kooyonga clubhouse.

Jack Fleck

Jack Fleck, who ran a municipal course in Davenport, Iowa, had played wonderful golf through the early holes of the fourth round of the 1955 U.S. Open. Out in 33, two under par, he stood on the tenth tee just as Hogan finished and heard the roar of the great crowd. Hogan had shot 287 and apparently won his fifth Open. Only one stroke behind, Fleck was the only man still on the course who had a remote chance of catching Ben.

As he stood waiting to play, a friend trying to urge him on reminded Jack he needed only one more birdie to tie.

Fleck was paired with Gene Littler, a droll man with an impish sense of humor. Hearing the comment, he reminded Fleck's friend, "He'll need a few pars, too."

Using one club too much, Fleck overshot the fourteenth green and bogeyed, dropping two strokes behind Hogan. Now he needed two birdies to catch up. He made one with a marvelous 6-iron into the fifteenth, a little par 3, barely missed two others at the sixteenth and seventeenth, and came to the eighteenth hole just one stroke behind.

While Fleck struggled home, Hogan sat in front of his locker accepting congratulations from the press and from the other players. He had packed his gear and he was ready to leave for home as soon as he could. Meantime reports on Fleck's progress filtered in. Fleck had driven into the rough on the eighteenth. He'd played a 7-iron onto the green about seven feet from the cup. A roar and then, "The kid's holed it," someone said.

Hogan froze. Barely above a whisper he said, "I was hoping he'd either make a two or a four. I was wishing it was all over." Then he called the locker-room attendant and said, "Put those sticks back in the locker; it looks like I'll be playing tomorrow."

The following day as Hogan prepared to play his tee shot to the third, a 220-yard par 3 from an elevated tee, a frightened rabbit darted through the crowd and raced across the tee. At first it looked like a good omen for Hogan when he played a laser-guided 2-iron within four feet of the hole and Fleck got away with a sloppy shot that hit the top of a bunker and jumped onto the green.

Fleck saved his par, and with what looked like a certain birdie, Hogan missed. The good omen worked for Fleck, not Hogan. With six holes left in the playoff, Fleck had opened a three-stroke lead, but, fighting back, Hogan had pulled within one stroke after the seventeenth. Ben's fourteen-foot putt grazed the lip of the cup; had it fallen they would have been even. Now Hogan needed a birdie on the home hole.

Olympic's eighteenth tee had been freshly top-dressed and the soil lay loose. Ben squirmed his feet into as firm a stance as he could, but as he moved into his drive, his right foot slipped. He yanked his ball so badly into knee-high rough he lost any hope of a birdie.

Finding the ball he slashed at it once and moved it a foot or so. He swung again and moved it perhaps three feet. He chopped out to the fairway with his fourth, and he pitched to the back of the green with his fifth. Game to the end he holed a slippery downhill thirty-foot putt for a double-bogey 6. He had shot 72 and had fallen to Fleck's 69.

Ironic Delivery

Shortly before the 1955 Open, Hogan had opened his own manufacturing company and played his clubs in the championship. So did Fleck. After hearing of Hogan's new company, Fleck sent him his specifications. He had received all but two wedges by the time he headed for the Open. Hogan brought them to San Francisco and handed them to Fleck himself.

At the birth of his son, Fleck proposed to his wife that they name him Snead Hogan Fleck. Lynn Fleck absolutely refused that combination of names, so they compromised. Jack gave her the names of every Open champion. Mrs. Fleck chose Craig Wood Fleck.

Lost Ball

An amateur hadn't won the U.S. Open in twenty-two years when Harvie Ward, the 1952 British Amateur champion, shot 144 in the first two rounds in 1955. Halfway through, Ward had tied Tommy Bolt for the thirty-six-hole lead, and he still ranked among the leaders after seven holes of the third round. His challenge ended on the eighth, a dinky little par 3 of only 139 yards.

While the eighth might have been short, the tee shot had to thread a narrow opening between trees. Ward pushed his shot toward a palm tree growing to the right of the green. Everyone there saw the ball come down among the palm fronds, but no one saw it fall to the ground. It was a lost ball.

Teeing another ball, Ward scored a double-bogey 5, shot 76 in the third round, added another 76 in the fourth, and tied for seventh place at 296, nine strokes behind Hogan and Fleck.

During the first round, Lawson Little drove into a copse of cypress trees bordering the fifth fairway. After Little, his caddie, and the spectators searched for a time, one of the fans said to Little, "Wait a minute, I'll help you."

He turned away and shinnied partway up one of the trees, gave it a healthy shake, and saw three balls tumble out. Little looked them over, saw that his ball wasn't among them, and encouraged the fan to shake again. He did, but nothing happened. Little then walked back to the tee to play a second drive. Fans picked up the balls and strolled off.

College Pranksters

College students behave pretty much the same the world over. On the evening before the first day of the 1955 Walker Cup, the students at St. Andrews University:

1–Removed all the flagsticks from the holes and laid them out on the first fairway spelling BRITAIN;

2–Carried a scoreboard designed for television 300 yards and replaced it on the edge of the Swilcan Burn, in front of the first green;

3–Tore down a "No Cycling" notice and tossed it into the burn;

4–Took down a "No Passing" traffic sign and stuck it into the eighteenth cup;

5–Carried away an officials' tent and set it up 400 yards from the course;

6–Took down a British Railways information tent and erected its nine-foot sign over the entrance to MacIntosh Hall, the university's women's residence.

The grounds crew took three hours to put everything back together again.

Nicknames

Nicknames are part of Walker Cup ritual—everyone gets one. As a member of the 1955 American team, Joe Conrad, a slightly built Texan, was assigned two. He was known first as Smokey Joe, then Gaylord. Why Gaylord? Because, as one teammate explained, he was so unlike a Mississippi riverboat gambler.

Rules Bonanza

Those responsible for such things at a U.S. Open wore out their rules book during the first day of the 1956 championship. At least four situations strained their powers of reasoning and had effects that caused problems the following season.

First—Walker Inman, a young professional, arrived at Oak Hill early enough to be sure to make his 9:40 starting time. Waiting for the appointed hour he hit practice shots for a time, then wandered onto the practice putting green and dallied until his time drew close. When he arrived at the first tee he found he had a problem.

In a dyslexic reading of the pairings, Inman had mistaken 9:04 for 9:40. Not only had Julius Boros and the amateur Bill Hyndman, the other two members of his group, left the first green by then, but the following group was about to putt as well.

By missing his starting time, Inman was ripe for disqualification, but he had caught the rules officials on one of their lenient days. Instead of disqualification, he was penalized two strokes and allowed to play late in the day, alone, after the final pairing had teed off.

With his two-stroke penalty attached to the par 4 he scored on the first hole, Inman shot 77. He finished the Open with 304, twenty-three strokes behind Middlecoff, the winner.

Second—Playing his second shot to the tenth, a straightaway par 4 of 420 yards, Doug Ford hit his ball so badly it splashed into a creek knifing across the fairway yards short of the green. Ford didn't know it but it also splashed right out again.

Believing his ball had sunk into the water, Ford dropped another ball. Before he could play a shot with the second ball, though, someone shouted that his first ball had indeed jumped up the stream bank onto the other side. Ford picked up the second ball and holed out with the first.

This brought up some more interesting questions in the occasionally mind-boggling rules of golf. Some strict constructionists argued that the second ball became the ball in play as soon as Doug dropped it, which meant that when he picked it up and played the original ball he had played a wrong ball, and that after he holed out with his first ball he had only one avenue to salvation: he should have counted all the strokes he played with that ball, gone back and played the second ball and added still two more strokes to his score.

Others felt differently. They argued that since Ford hadn't played a shot with the second ball he was within his rights to pick it up, add a two-stroke penalty for dropping it in the first place, and get on with it. The liberals carried the day.

Third—Now the rules connoisseurs found some real meat to chew on. Did Henry Cotton cheat or didn't he? Although he had won his third British Open in 1948, by 1956 he had reached the age of 49 and was past his prime. Nevertheless, he had played sixteen respectable holes and stood just three over par coming to the seventeenth, a murderous par 4 that doglegs right and whose green is almost sealed off by bunkers that pinch the narrow entrance. It measured 463 yards, stretching the definition of a par 4 hole.

Like a great many players that week, Cotton missed the green. He chipped on and stroked a nice putt that skimmed the edge of the cup and ran a few inches past. He had a literal tap-in left. Reaching across the hole to rake the ball toward him, he either did or did not stub his putter on the ground and miss contact entirely.

Henry had been grouped with Middlecoff and Demaret. Both of them swore that is exactly what happened. Cotton said they were wrong. He claimed that as he reached across the cup he lost his balance and jabbed the putter into the ground deliberately. No one could be sure.

Middlecoff and Demaret believed he had stabbed at the ball, but the rules officials could do nothing other than accept the man's word that he hadn't. Even though neither Demaret nor Middlecoff would attest his score by signing his card, his 5 on the hole stood. He shot 74. The following morning one Rochester newspaper carried the headline "Britannia Waives the Rules."

In the final accounting, Cotton tied for seventeenth place, Demaret missed the cut, and Middlecoff won the championship, his second.

Fourth—In the incident that had everlasting effects, Jackie Burke made a bogey 5 on the last hole, but his marker, the person who kept his score, posted a 4 on his scorecard. Burke didn't check closely enough, signed it, and turned it in.

He had gone off at about 1:30. Checking over the scoreboard later in the day he noticed the 4 in the block for the eighteenth hole and realized he had committed suicide—in this case disqualification. He of course reported his crime.

Alas the liberals still controlled the debate. Instead of being sent packing, Burke was given a reprieve in the form of a two-stroke penalty, which that day seemed to cover all sins.

The next morning Gil Cavanaugh, a club professional from Long Island, reported that he, too, had made a mistake on his scorecard. The rules zealots ground their molars and slapped another two-stroke penalty on the unwitting and careless; by then they could do no more.

Eventually, however, the strict constructionists overcame the wafflers. During winter meetings the USGA said, in effect, never again. Hereafter the rules book will mean what is says.

The following summer Jackie Pung apparently won the Women's Open, but she and Ruth Jessen, who played the last round together, made the same kind of mistake Burke and Cavanaugh had. The USGA disqualified them both.

A Grand Match

Played for absolutely nothing except bragging rights, a four-ball match set up at a cocktail party sent two of the greatest professionals the game has ever known against two of the best amateurs the United States ever bred. The idea was born a few days before the opening round of the 1956 Bing Crosby Pro-Amateur. It was played at the Cypress Point Golf Club, one of the game's great courses, befitting the contestants.

It came about when Eddie Lowery, Francis Ouimet's caddie in the 1913 Open, held a discussion with George Coleman, a wealthy oil baron from the Southwest and later president of Seminole.

Both Harvie Ward, the 1952 British and 1955 U.S. Amateur champion, and Ken Venturi, who would come so close to winning the Masters a few months later, were on Lowery's payroll as salesmen at his automobile agency. Lowery believed that as a team they could beat anyone, and he said as much to Coleman. Lowery offered to back his boys against anyone Coleman chose. Coleman chose Ben Hogan and Byron Nelson.

Nelson and Hogan agreed with the provision that no one was to know of the match except those immediately involved. To throw everyone off the scent, they arranged a starting time for Hogan and Nelson the next morning at Pebble Beach.

The match began with only a handful of spectators. They were awed by what they saw. Ward and Venturi birdied nine of the first ten holes. They stood one hole down. Hogan and Nelson had matched them stroke for stroke through the first nine, and then Hogan had holed an eighty-five-yard pitch on the tenth, an uphill par 5 of just under 500 yards.

Still the battle went on. Perhaps the greatest player of long irons who ever lived, Nelson rifled a 2-iron within twenty feet of the flagstick on the eleventh, a strong 450-yard downhill par 4, and holed the putt. Venturi matched him.

On and on they struggled, the amateurs never able to make up that one hole. At the fifteenth, one of two consecutive par 3s, both Hogan and Venturi birdied. On the picturesque sixteenth, a long par 3 of 235 yards across a rocky chasm where the waves of the Pacific Ocean churn below, both Nelson and Ward hit drivers. They halved the hole in birdies. On the seventeenth, 376 yards along the ocean, Ward and Nelson halved again in birdies.

The match then hinged on the eighteenth, an uphill par 4 of 342 yards that is the weakest hole at Cypress. Venturi pitched to twelve feet and Hogan placed his shot inside him by two feet. As cool as ever he was, Venturi rolled his putt into the bottom of the hole. Now the result was up to Hogan, as tough a competitor as the game has known. Hogan crouched over his putt, and through gritted teeth he was heard murmuring, "I'm not about to be tied by two goddamned amateurs." He ran the putt home; he and Nelson had won by one hole.

Their medal scores were sensational. Both Ward and Nelson had shot 67s, Venturi had 65, and Hogan had played Cypress Point in 63. Ward and Venturi had posted a better-ball score of 59, thirteen strokes under par, and yet had lost to a 58. Talking about the match many years later, Venturi said, "I hate to lose but I just had to love it when Ben drained it at eighteen. We played our best, but at least on that day we got beat by the best."

An Unbeatable Record

Before a battery of cameras, and with former British Walker Cuppers Leonard Crawley and John Beck standing by, Ralph A. Kennedy, a New

Yorker, teed off at the Old Course in St. Andrews, Scotland, one sunny day in 1957. He was beginning a round over the 3000th golf course he'd played in a forty-three-year quest to set a record. It wasn't his last.

Before he gave it up at the age of seventy-two, Kennedy had played 3,165 courses all over the world. He played throughout the United States, in Europe, in South America, wherever golf courses could be found. His trail began at the Van Cortlandt Park golf course in Manhattan, believed to be the first public golf course in the United States, and ended at the Hamilton Inn Golf Course, in Lake Pleasant, New York.

Kennedy kept a dated, numbered, and signed scorecard from every one of those courses locked in four safety-deposit boxes in a bank vault. They recorded his every round in every state and in thirteen other countries.

He was indefatigable. Five cards grouped together represented one day's golf in Peoria, Illinois. Four from different counties in Arkansas reminded him of an all-day dash that began at 5:00 a.m. and ended in pitch darkness with six forecaddies straining to locate flying golf balls. He ended the day by holing a chip shot in total darkness from the edge of the eighteenth green for a par 5. He returned the next morning to play the hole in daylight and scored 8.

In 1933 he played eight courses on Bermuda in two days. On a seven-day visit to Chicago the following year he played twenty-one more. Visiting Maine in 1935, Kennedy played thirty-one courses in nine days. He also persevered. Visiting two courses an oil company had commissioned that were laid out over sandy wastes in Peru, Kennedy sailed into a bay and had to be ferried ashore by lighter. From there he traveled by truck through a tunnel gouged into a mountain and on to neighboring Nefretis.

He rode a donkey to the first tee at the Uniontown Country Club, in Uniontown, Pennsylvania. Playing in Canada, he found a bear in Jasper Park sniffing his ball, and he stood silently at Cypress Park while stags reared their antlers until their does and fawns had slipped into the brush.

As he stepped onto the tee at the Old Course and noticed the army of photographers and prominent people gathering behind him, Kennedy assumed they were waiting for a celebrity and attempted to step aside and wait. Amused, Crawley assured him the crowd had gathered solely to witness the start of his round, much as they assemble to see the new R and A Captain drive in. With shaking knees and trembling hands he slammed a solid drive down the middle.

By Any Other Name

Randolph Scott, a courtly Virginian who starred in many a forgettable movie, played perhaps the best golf of any of the better-known Hollywood stars. He also lusted to become a member of the Los Angeles Country Club, whose North Course is regarded the best in the city, better even than Riviera, which had been the site of the 1948 U.S. Open championship.

LACC, however, didn't want Randolph Scott; as a matter of fact it didn't want anyone associated with the movie business. One actor wanted membership so badly he sent critics' reviews of his work to prove he was no actor. Scott eventually left the movies and went into the oil business. He also bought a house alongside LACC's fifth fairway. At about that time the club accepted a new member named George Randolph Crane. The members learned later that this was Scott's real name. Too late, though; Scott was now a member. He was asked, however, one favor. As he and his wife were leaving the club after dinner one evening shortly after he had been accepted, a member of the club's board of governors approached and told him, "Mr. Scott, the board would greatly appreciate your doing your best to keep your old movies off television."

Stuffiness Lives

Stuffiness seems to abound everywhere. The Rumson Country Club, close to the New Jersey shore, dates back to 1908, making it one of the old-line clubs. It is also stocked with scions of old money. In 1955 the club hired Dave Marr, then a young club professional from Texas.

One Saturday morning when Marr had been at the club barely a month, three members invited him to join them for a round of golf. Marr complied and everyone had an enjoyable round. The game over, they retired to the grill for a club sandwich, a Rumson special, and a few glasses of beer. All seemed well with the world.

Three days later Marr received a letter from the club's president telling him no longer would he fraternize socially with the members. Furthermore, since he did not hold a membership and was merely one of the help, he would hereafter take his meals in the kitchen. A thoroughly likable man, Marr remained through the season, then resigned. Later he joined Claude Harmon's staff at the Winged Foot Golf Club, in Mamaroneck, New York, eventually joined the tour, won the 1965 PGA Championship, and developed into one of the more entertaining television commentators. He's welcome everywhere.

Ward and Kocsis

During the middle 1950s, while Ben Hogan was still a power and Sam Snead could still play shots others only dreamed of, there were many who believed the amateur Harvie Ward was the best player in the game. A North Carolinian, Ward had won the 1952 British Amateur and the 1955 U.S. Amateur. No one had won consecutive Amateurs since Lawson Little, in 1934 and 1935, but if anyone had stood a chance to do it it was Ward. He was a dream golfer, solid in every phase of the game, but he excelled with his irons. Consequently, he was favored to win again in 1956.

After fighting his way through the early rounds, he met Chuck Kocsis in the final. A member of a prominent family of golfers from Detroit,

Kocsis had played in the 1930 Amateur as a teen-aged prodigy. He was forty-three in 1956, Ward was thirty. Ward shot 72 in the morning of the final, but Kocsis threw 71 at him and led by one hole at lunch. He would go no further. Ward played untouchable golf, rifling his drives down the middle and firing irons that covered the flagsticks, and moved relentlessly ahead. As they stood on the thirty-first tee with Ward well ahead, Kocsis sidled up to him and said, "Take it easy on an old man, Harvie." Smiling, Ward said, "I see your lips moving, but I've turned off my hearing aid." He polished off Kocsis by 5 and 4.

Accuracy Be Damned

Always among the most accurate of iron players (he had hit the flagsticks on six holes during the 1939 U.S. Open), Byron Nelson lived to regret it during the 1957 Masters. Byron mis-played his tee shot to the sixteenth, a par 3 across water, and hooked it into the pond to the left of the green. Teeing another ball, he drilled it directly at the flagstick. It hit the stick and ricocheted into the pond once again.

Press Gang

Finishing off two consecutive rounds of 68 on the last day of the 1957 U.S. Open, Cary Middlecoff holed a curling nine-foot putt on the seventy-second hole at Inverness and tied Dick Mayer, forcing a playoff the next afternoon. Those last two rounds must have cost Middlecoff more than he had to give. Far off the standard he had shown the previous day, he fell three strokes behind Mayer after nine holes.

Since the walk from the ninth green to the tenth tee passes the clubhouse, Middlecoff ducked inside to answer nature's call, with Des Sullivan, who was covering the Open for the Newark (New Jersey) *Evening News*. As they stood side-by-side, Sullivan said to Middlecoff, "Doc, are you going to press on the back nine?"

Costly Error

Jacqueline Pung, a slightly overweight, good-natured Hawaiian, shot 72 over the East Course of the Winged Foot Golf Club, in Mamaroneck, New York, in the last round of the 1957 Women's Open. She had played the last thirty-six holes in 145 strokes, one under a tough par, superb golf in a stiff crosswind sweeping across most of the holes.

Her 298 had nipped Betsy Rawls by one stroke, and she had evidently won. As she stepped off the green, the gallery cheered, and crying with joy, her fifteen-year-old daughter, Barnette, rushed to embrace her. But she hadn't won. A scoring error on the fourth hole not only cost her first place, it disqualified her as well. In an odd coincidence, both Mrs. Pung and Betty Jameson, who played together, wrote down 5s for their fourth hole, even though both had actually scored 6s. The round over, they had both signed their scorecards, again oddly, both with the correct eighteen-hole scores, and turned them in. That doomed them. It doesn't matter that they turned in the correct total score, only the hole-by-hole scores matter. Under the rules of golf, a player is disqualified for submitting a score for any hole lower than the actual score. With Mrs. Pung out of the way, Betsy Rawls was declared the champion.

Hearing of the disqualification, Winged Foot members collected a prize fund for Mrs. Pung. She left Winged Foot with $3000. Miss Rawls won $1800.

Bobby Locke

Although he finished well in the 1936 *Golf Illustrated* Gold Vase, his first, Bobby Locke left behind a bad impression. The correspondent for *Golf Illustrated* wrote, "We were disappointed in Mr. Locke. There is a certain crudeness in his game that must prevent him from becoming a consistent winner of major championships." The magazine's comments no doubt ringing in his ears, Locke won four British Opens.

Born in South Africa, Locke was christened Arthur D'Arcy but his father called him Bobby because he idolized Bobby Jones. So did Locke. Even though he took the club back in a languid tempo, looped the clubhead at the top of his backswing, and hooked every shot he played, he insisted he modeled his swing after Jones's.

Unlike Jones, though, Locke avoided practice. As his starting time for the 1957 British Open approached, he strolled to the practice ground where he ran into Peter Alliss, with whom he was to play an exhibition match two days later. After discussing arrangements for a few minutes, Locke glanced at his watch and said, "Must be going." He hadn't hit a single shot. He strolled to the first tee, took two practice swings, and drove off. Locke shot 70 and won by three strokes over Peter Thomson.

Locke came to the United States in 1947, won eleven tournaments and very few friends. Dressed in baggy plus-fours and a white shirt and tie, Locke looked like a ten-handicapper. With his wristy swing, he hooked every shot; some say he hooked his putts as well. Having seen him play, Clayton Heafner told Jimmy Demaret he'd bet Locke against Demaret and Hogan the rest of the summer. Demaret had been the leading money-winner at the time. He accepted the bet, and lost a lot of money.

The sponsor of one tournament didn't like Locke, and confided to a couple of players that he wouldn't want him to win his tournament. One of the foreign professionals told the sponsor he knew how to stop him. "Put all the pins on the right side of the green," because Locke was a hooker and couldn't play his approaches anywhere near the holes. Locke finished twelfth.

In the first British Open shown on live television, Locke won the 1957 championship at St. Andrews under what later became controversial circumstances and might have been the first rules infraction called by a television viewer.

Leading by many strokes, Locke pitched no more than four feet from the cup on the eighteenth hole and now found himself in the exquisite position of being able to four-putt and still win. As Locke marked his ball with a coin, Bruce Crampton, asked him to move his coin a clubhead's length away. After Crampton putted out, Locke replaced his ball at the coin and holed out. He had forgotten he had moved his marker off Crampton's line.

No one running the championship noticed Locke's mistake, which would have called for a two-stroke penalty, but Gerald Micklem, an influential R and A member, had seen it on television and telephoned St. Andrews. Word reached Peter Thomson, who had placed second, three strokes behind Locke. He checked films of the incident with the BBC and found Micklem had been right. The Open, however, was over by then, and no penalty could be called. Locke remained the champion.

Locke became a father in February of 1960, a few months after his forty-second birthday. On his way to visit his wife in the hospital, his car was hit by a train at a crossing. The accident left him unconscious for several days and cost him most of the sight of his left eye. He was such a complete golfer, though, that four months later he shot 289 in the Canada Cup and, with Gary Player as his teammate, helped South Africa to fifth place.

Staggering Finish

Leslie McClue, a five-foot-six Scot from a small town near Glasgow, came up against the American Walker Cup player Dale Morey in the fourth round of the 1957 British Amateur. With McClue up by two holes, the two men were walking across the eleventh tee when Morey suddenly felt the urge to take a practice swing with his driver. McClue paused, then thinking Morey had finished, he began walking ahead once more. Just then Morey once more drew his club back and struck McClue a frightening blow to the right side of his head. McClue dropped to the ground and lay stunned. Morey rushed to him and helped revive him. Blood streamed down McClue's face and onto his clothes, but he insisted on finishing the match. Morey, of course, was perhaps more upset than McClue, who went on to win the match.

Morey's blow had torn McClue's ear so badly it needed nine stitches. McClue pleaded with the doctor, a friend of his, not to do anything that would interfere with his playing the next day. The stitching was done therefore without anesthesia. Still suffering from a mild concussion, McClue collapsed later that evening. Worried about McClue's condition, the doctor sat up with him all night. After a restless night, McClue awoke about 9.30, asked for breakfast, and went back to the golf course for his meeting with Harold Ridgley, another American.

Pale and shaken, his right ear encased in its dressing, McClue wasn't the player he had been the previous day. He fell to Ridgley by 6 and 5. McClue left England the next day and drove to Glasgow, where he shot 78 and 74 in a thirty-six-hole tournament.

Snake in the Cup

Frank Phillips, an Australian, spotted a blacksnake in the woods alongside Augusta National's sixth green during a practice round before the 1958 Masters. He drew out his sand wedge and bludgeoned the poor thing to death. Then he stuffed the dead snake into the cup.

Playing with the mischievous Phillips, Mike Souchak holed his putt, but when he walked up to the hole and saw a snake curled around his ball, and not caring if it were alive or dead, he refused to take his ball from the cup. A spectator slipped under the ropes and picked it up for him.

Tommy Bolt

Always quick with a quip, Tommy Bolt occasionally wore black and gold saddle shoes on the golf course. They of course attracted comments from the gallery, who in those days walked along the fairways with the players during the course of tournaments. They often asked where he had those special shoes cleaned. "At the jeweler's," Bolt riposted.

Two Clubs Broken . . .

Bolt was known not only to throw a club or two, he also broke the occasional offending stick. He had hired a caddie named Hagan, who knew the Riviera Country Club, in Los Angeles, better than anyone. Studying an approach shot, Bolt asked Hagan, "What do you think?"

"A 6-iron," Hagan answered. Bolt snapped his head toward Hagan and said, "No; it's a 5-iron." Hagan said, "No, it's a 6, and when you hit it, just hit it firm and don't press."

"You're crazy," Bolt insisted.

Tommy hit the 5-iron twenty yards over the green, turned red, and broke the club over his knee. Hagan, meanwhile, had been holding the 6-iron. When he saw Bolt snap the shaft of the 5-iron, Hagan broke the 6-iron over *his* knee, dropped Bolt's bag, and walked away.

One Drowned . . .

Bolt was paired with Porky Oliver during the Colonial National Invitational Tournament, a big event during the 1950s. Tommy was having a terrible day, missing putts that should have fallen and threatening to snap his putter over his knee. Oliver spent a considerable part of the afternoon trying to calm him down and save the putter.

Sensing that as soon as the round ended Bolt would try for a new distance record with a thrown putter, Porky said to Bolt, "Tommy, let me see that putter." Bolt handed it over, and Porky tossed it into the lake.

. . . and One That Got Away

Bolt had a decent round going in the first round of the 1960 Open, at Cherry Hills, but his game began unraveling at the eleventh, where he drove out of bounds, and continued onto the twelfth, where he dumped his approach into a pond, then fell into an impassioned argument with a USGA official over where he could drop his ball. Bolt lost the argument. Upset, he three-putted the thirteenth, bogeyed the fourteenth, and then stepped onto the eighteenth tee in a black mood.

The drive on the home hole calls for a carry across a broad pond. Steaming, Bolt hooked his first shot into the pond, teed up another ball and hooked it into the pond as well. Now he had reached the point where he could no longer control himself. Teeth bared in crazed rage, he drew back his driver with a perfect pivot and exceptionally fine hand position at the top of the backswing, took one step forward, and while Claude Harmon, his playing partner, ducked out of the way, he flung his driver into the pond.

As Bolt stared at the widening ripples while Harmon cowered because he wasn't certain what Bolt might do next, a small boy slipped under the gallery ropes, waded into the pond, held his breath as he ducked under the waves, and climbed out holding Tommy's driver overhead.

The gallery cheered, and even Bolt began to smile a little as he walked toward the boy with his hand held out for the driver, a favorite club. The boy saw Bolt headed his way, then began a broken-field run. He dodged Bolt, dashed across the fairway toward a chain link fence separating the club property from a line of houses beyond, and when a spectator gave him a leg up, he leaped over the fence and disappeared, Bolt's driver still waving aloft.

Bolt vs. Brown

When Bolt met the Englishman Eric Brown in the 1957 Ryder Cup match, at Lindrick, England, both men made it clear they'd much rather have been somewhere else. They played in an atmosphere of pure dislike.

Bolt and Dick Mayer had beaten Brown and Christy O'Connor by 7 and 5 in foursomes, but Brown turned around and beat Bolt, 4 and 3, in the first of the singles matches, leading Great Britain and Ireland to a 7½–4½ victory, their first since 1933.

Both men practiced gamesmanship. Knowing Brown to be a fast player,

Bolt slowed to an agonizing pace. His strategy backfired, though. Brown whispered something to his caddie, who sped off toward the clubhouse. Brown carried his own bag until his caddie reappeared, bringing with him a lounge chair. Brown rested while Bolt fiddled.

At the end of their match, Bolt snarled, "I can't say I enjoyed the game." Brown snorted, "That's because you don't like getting stuffed."

It's Only a Typo

Leading the 1958 U.S Open, Tommy Bolt at first refused to come to the press tent for the customary post-round interview. Asked why, Bolt said, "Did you see that item in the paper this morning?" he sizzled while he painstakingly combed his hair in the locker room. His questioner thought for a moment, then realized Bolt was referring to a sentence that described him as "forty-year-old Tommy Bolt," whereas Bolt claimed he was just thirty-nine (he was actually forty-one).

"But, Tommy," the questioner said, "that was only a typographical error."

"Typographical error, my ass," Bolt bellowed, "it was a perfect four and a perfect zero."

Talking with Bolt twenty years later, the writer Charles Price asked, "How old are you, Tommy?" Bolt muttered, "I was sixty last January." Remembering the Southern Hills incident and capable of elementary mathematics, Price shot back, "Sixty! If you were sixty in January, then in June of 1958 you were forty, not thirty-nine." Bolt stared at the ceiling deep in thought and admitted, "Yeah, I guess I was," obviously pleased he still held a year in reserve.

A Patriot

Philomena Garvey won the 1957 British Women's championship and of course was selected for the Curtis Cup side that would meet the United States the following year at Brae Burn Country Club, near Boston. Nothing unusual in that; Miss Garvey had represented Britain and Ireland in the previous five Matches, every Curtis Cup since the end of the Second World War.

Now, though, the British added something different. Before the team left for the United States, every member was told she must wear a badge depicting the British Union Jack. Miss Garvey, though, was Irish. She refused. The Ladies Golf Union dropped her from the team.

Stumbling Finish

Although they became standard later in the century, such amenities as scoreboards hadn't been scattered throughout British courses by the time of the 1959 British Open.

Eight strokes behind Frank Bullock beginning the last day of thirty-six holes, Gary Player approached the seventy-second hole at Muirfield convinced that to win he needed a par 4 on this demanding 432-yard hole. He had shot 70 in the morning round, and the 4 would have given him 66 in the afternoon. He wasn't aware that Bullock had shot 74 in the morning and was on his way to another 74 in the afternoon. Instead of the 4, though, Player made 6 and, dejected, slunk off the green in a morose mood, believing he had thrown away his opportunity. Then he learned he could have played the hole in 7 and still won. He beat both Bullock and Flory Van Donck of Belgium.

Things To Come

In the days leading up to the 1959 Walker Cup match, at Muirfield, the hopes of Great Britain and Ireland ran high. On their side they had such players of proved quality as Michael Bonallack, who had won five British Amateurs, Joe Carr, with two, Michael Lunt, who in 1963 had won the British Amateur against the mass invasion of the American Walker Cup team—the first Briton to beat back such a threat in forty years—and a number of others of high achievement.

The United States, on the other hand, leaned heavily on a group of inexperienced college boys. No one in Britain had heard much of Jack Nicklaus, Tommy Aaron, or Deane Beman. A day or so after the Americans arrived, a platoon of British supporters strolled out to see for themselves how well these youngsters played. They all headed for the second hole, a 349-yard par 4.

Scotland's weather had been quite dry for some time; the unaccustomed warm sunshine had baked the fairways hard, and the Americans were testing the 1.62-inch diameter ball, which flies farther than the 1.68-inch American ball.

The Brits arrived at the second green just in time to see Nicklaus launch one of his missiles. They watched in wonder as the ball arched against leaden sky, cracked against the brick-hard fairway and bounded onto the green. Like a flock of birds reacting to a mysterious signal, they turned as one and marched smartly and silently back to the clubhouse and straight into the bar. At the end of the week the Americans had won once again, 9–3.

Explosion Shot

High winds swept across Royal Birkdale throughout the 1961 British Open, flattening the tented village, confounding club selection and flinging balls off course, and assuring that sub-par rounds would be rare. Battling for the lead, Arnold Palmer lost his drive to the right of the fifteenth fairway. His ball swung viciously off line and dived into deep rough at the base of a vertical grassy bank about 150 yards from the green.

Surveying the shot, Palmer saw his ball almost obscured by grass tendrils hanging down from above. First he asked his caddie for a 7-iron, obviously to play a safe shot. Suddenly he changed his mind, switched to a 6-iron, and slashed into rough the gallery figured he'd need a wedge to escape. As he scythed through the grass and finished his swing, so much grass clung to the club shaft it looked as if he had reaped a bale of hay. Meantime, the ball flew onto the green and rolled within easy two-putt distance of the cup. Arnold had won the championship.

Immensely impressed by the shot, club officials embedded a small plaque at the site. At Royal Troon the following year, Palmer won his second consecutive British Open.

Years later a newspaper photographer approached Palmer while he was playing the eleventh hole at Troon during a practice round for the 1989 Open and asked him to pose alongside the plaque.

Palmer thrashed around in the gorse without finding it, then called to Alfie Fyles, his caddie, waiting some distance away, "Hey, Alfie, where's that plaque?" Fyles's jaw dropped. After a momentary silence he called back, "About 600 miles away. You're on the wrong course."

Following Palmer through the early holes in 1962, a crowd stood in awe watching Arnold draw a 1-iron from his bag on the sixth hole at Royal Troon, a long par 5 of 577 yards. As Arnold tore into the shot and the ball cut through the air with a loud buzzing sound, a man from Glasgow turned to his wife and said, "You see, he hits it farther than we go for our holidays."

Sam Snead's Streak Snapped

In seven Ryder Cups before 1961, Sam Snead had won ten of thirteen matches, and he was in the running for a place on the next team, even though he was forty-nine at the time. He didn't make it. A month before the scheduled start of the match, Sam played in what he believed was an exhibition match in Cincinnati. Late on the first day he learned he was playing in a tournament in conflict with the Portland Open, an official PGA tournament. As soon as he found this out he called the sponsors of the Portland event and asked for permission to skip their tournament. They refused. Snead was handed a six months' suspension and fined $500; he forfeited his place on the Ryder Cup team.

Eventually the PGA reduced his suspension to forty-five days, the United States, with Billy Casper, Arnold Palmer, and Gene Littler leading the way, won the Ryder Cup, 14½ to 9½, and Snead never again played on the team. He did, however, serve as captain in 1969.

Speed Demons

Although they were well along in age and no longer able to compete in the Masters, Fred McLeod, the 1908 U.S. Open champion, and Jock

Hutchison, winner of the 1921 British Open, teed off first on opening day for many years and played a ceremonial round. Until they had advanced into their eighties, they played a full eighteen holes. In spite of their age, they didn't dawdle.

One year during the early 1960s they played eighteen holes in two hours and fifteen minutes. Speaking of it later, McLeod complained, "We had lunch and a couple of drinks before the next group finished. I don't know what they were doing."

Open Bunny Hutch

Leading up to the 1962 U.S. Open, at Oakmont, Fred Seitz, a robust, husky man with a man-sized appetite for the things men like, held the position of club manager. Blessed with an active sense of humor, Seitz delighted in telling the USGA man on the site that he planned to turn a section of the clubhouse basement into a nightclub during the championship and staff it with barely clothed girls recruited for the week from a downtown establishment resembling a Playboy Club.

Since he represented a rather staid and proper organization, the USGA type laughed along with the joke. It was a joke, wasn't it? Horrors, no, it was no joke. Seitz meant what he said. When the gates opened early in Open week, the nightclub was in full swing. Oscar Fraley, an irrepressible correspondent for United Press International, announced he would ask the president of the USGA to pose for a photograph with the bunnies the next day.

Alas, the next day was too late. By then the USGA had swooped down on Seitz, and the atmosphere had changed. For the remainder of the week, the waitresses were dressed in outfits someone likened to habits borrowed from the nearest nunnery.

Shell's Wonderful World of Golf

Shell's Wonderful World of Golf appeared on television for the first time on January 7, 1962, with a match between Jerry Baber, a small, gruff Californian who the summer before had holed every putt he looked at and won the PGA Championship, and Dai Rees, an equally small and gruff Welshman who, at forty-eight, had been runner-up to Arnold Palmer in the 1961 British Open. They played at the Wentworth Golf Club, near London. The last show appeared on February 28, 1970, with Dan Sikes playing Roberto De Vicenzo at the Olympic Club, in San Francisco.

Each week two golfers, either men or women, played a match at one of the world's great courses. It was recorded on film and shown on television, usually on wintry Sunday afternoons. In nine years the show took viewers to the six inhabited continents of the globe, to most of the countries where golf was played then, to all but a few of the world's great courses, and showed every one of the great players of the time, and some from earlier times.

Shell's Wonderful World of Golf was the best filmed golf show ever produced. It was a miracle it ever appeared at all.

Although the Barber-Rees match appeared in television first, the first match that was filmed was done at Pine Valley, although not before John Arthur Brown, the club's president, could be persuaded to agree. The assignment of winning Brown's heart was turned over to Fred Raphael, who was to produce the show. Raphael was a strange choice. He produced shows for Filmways, an organization best known for the television shows "Petticoat Junction" and "Green Acres." He'd never been on a golf course in his life. When he was handed the assignment he spent three days in the New York Public Library taking a crash course on the game by reading the books of Herbert Warren Wind, who wrote about golf for *The New Yorker*.

Raphael made an appointment to meet Brown at Pine Valley. When he arrived he was shocked. "I looked around and said to myself, 'This has got to be some kind of disaster.' All I could see was sand and water. How far can these guys hit a ball, I wondered. I didn't know a thing about the game."

Raphael had a director with him named Lee Goodman who was also a good salesman. Brown went around the course with Goodman and Raphael, and Goodman kept telling Brown how great the show was going to be. Finally Brown told Raphael, "Okay. You can come down here and film a show provided you can do it in two days, as you say you can, if it is in color, and if this gentleman (Goodman) will be the director."

Raphael accepted the conditions, then raced back to New York and into the Filmways office and told Marty Ransahoff, who had the over-all responsibility for the show, the result of his visit with Brown, and said, "We've got a little problem here, Marty, because Lee doesn't know anything at all about directing a golf show. He doesn't know as much about it as I do."

Ransohoff glared at Raphael and snapped, "Teach him."

Raphael and Goodman went to Wind for ideas, since he had been working on the project for Shell. Not understanding how little he knew, Goodman began telling Wind how he wanted to place the cameras, because he had done this whole thing before with Sam Snead. Feeling a little better because he believed Goodman knew something about filming golf, Wind said, "Tell me, where was this match you filmed with Snead?" Astonished, Goodman said, "What do you mean, match? It was an Alka Seltzer commercial."

Raphael hired a backup director.

The first match filmed, between Byron Nelson and Gene Littler, began with an uneasy feeling. It turned out that those involved had every reason to feel worried.

Nelson had the honor and played his drive precisely as it should be played on the first hole at Pine Valley—long and down the left side, opening the green to the approach. As Nelson picked up his tee peg, Raphael looked down the fairway and saw the cameraman climb off his truck and walk to the ball. Then, as Raphael and everyone else cringed in horror, the cameraman picked up the ball, threw it back toward the tee, and called over his radio, "Ask him to hit it again. We missed it."

His playing career over, Jimmy Demaret joined the show as a commentator. During its last years, a match was being done in Faro, Portugal, between Doug Sanders, an American, and Peter Alliss, an Englishman.

The film crew arrived in Portugal just after a big bank robbery. As a matter of routine, an accountant with each crew deposited $25,000 in a bank as soon as he arrived in any foreign country and converted the money into local currency, because the Shell Oil Company didn't want its crew leaving unpaid bills behind. On arriving in Lisbon, the accountant deposited the money and put the Portuguese currency in an attaché case, then climbed into a car with Raphael and Demaret for the drive to Faro.

Along the way they came to a roadblock where every car was being stopped and searched by a horde of policemen looking for the stolen money. Here they were, strangers in the country, carrying a briefcase stuffed with $25,000 worth of Portuguese escudos driving away from a city where a bank had just been robbed. They were worried.

Then it was their turn to be searched. As they pulled up to the roadblock they were ordered out of their car and told to open all their baggage. Demaret picked up the briefcase and stood around holding it while the policemen rooted through their baggage, tossing clothes onto the road, probed for secret compartments, and searched throughout the car.

Finally satisfied, the policemen told them to pick up their debris and go on their way. No one had bothered to search the briefcase.

Although the matches were played months before they were shown on television, the results somehow were kept secret. Only those involved knew who had won, and they generated a moderate amount of friendly betting.

Since Raphael didn't own a color television set, he'd go to the Tudor Hotel, on Forty-second Street, in New York, and watch the show in the hotel's bar. One Sunday afternoon, two men and a woman were watching the show and betting against the bartender. They won every hole.

When it ended, the couple left, and the bartender struck up a conversation with Raphael, not aware he produced the show.

"I don't believe what happened," he said. "They wiped me out."

"Yeah," Raphael said. "I saw it."

Then he made a suggestion. "Next week," he said, "Byron Nelson is playing Gerry DeWit at The Hague. Why don't you bet DeWit beats Nelson." The bartender stared at Raphael and said, "You're crazy."

"Why don't you really bet a lot of money that Nelson doesn't break 80."

"Mister," the bartender said, "I'm cutting you off. No more beer."

The following week Raphael was back in the Tudor Hotel bar. Nelson shot 80, DeWit won the match, and the bartender lost money again. When the bar emptied, the bartender came back to Raphael.

"Have a beer," he said, drawing a draft. "Tell me," he went on, "what do you know about this golf show?"

"I'm the producer," Raphael answered.

The bartender smiled as the possibilities dawned, and asked, "What about next week?"

"Forget it." Raphael said.

Tony Lema

Throughout most his career, Tony Lema, a tall, slender, handsome man with an elegant golf swing, was known as Champagne Tony, a designation that grew out of an incident in California at the Orange County Open of 1962. Played in late October, as the season ran toward its close, the Orange County was a small tournament with total prize money of $20,000. Most of the bigger events paid $50,000 or more.

Because of its limited importance, only a half-dozen or so reporters showed up, so few that the Mesa Verde Country Club needed only turn over the men's card room as a press room. For refreshment, the club kept a picnic cooler supplied with a few soft drinks and some beer.

On the evening of the third round, Lema answered reporters' questions while he sipped a beer. The interview over, he said, "Boys, if I win, we'll be drinking champagne tomorrow."

Donald (Doc) Giffin, the PGA's press secretary at the time, told the club manager to be sure he had a supply of champagne on hand. When Lema beat Bob Rosburg in a playoff, the champagne flowed.

Giffin wrote about it in the PGA's magazine. Those who distribute Moet champagne heard of it, and soon champagne appeared whenever Lema won a tournament.

Who's He?

While he worked as an assistant professional at the San Francisco Golf Club, principally cleaning and storing members' clubs, Lema was encouraged to enter the 1956 U.S. Open by Eddie Lowery, who had caddied for Francis Ouimet when he won the 1913 Open. Later, when he had become a Lincoln-Mercury dealer in San Francisco, Lowery went around town encouraging all the young golfers, both amateur and professional, to enter, hoping to swell the field and win more places for the city's golfers.

Hardly anyone had heard of Lema, but he won one of the seven available places reserved for San Francisco. When Lema's name appeared in the local papers the next day, identifying him as working at the San Francisco Golf Club, a dozen members called the sports editors and told them they had made a mistake. No one named Tony Lema worked at their club.

The 1956 Open was played at the Oak Hill Country Club, in Rochester, New York. Lema shot 77 in the first round, and on the last hole of the second round played a stunning 5-iron three feet from the hole. If he was looking for fan appreciation he was disappointed; he had gone off so late in the day, no spectators were left to applaud. He birdied, nevertheless, shot 71 and 148 for the first thirty-six holes.

Not aware of how things work in an Open, where only the fifty lowest scorers continued playing after the first two rounds, Lema returned to his hotel rather pleased with himself. Sitting in the lobby reading the local papers after dinner, he overheard a couple of spectators discussing the day's events and speculating over what score would qualify for the final thirty-six holes the next day. Stunned, Lema blurted, "What's this? Doesn't everybody play tomorrow?" Looking with pity at this naive young man, one of them answered, "No. There's a cut to the low fifty—149 will probably make it."

He was right; Lema did indeed qualify for the last two rounds, shot 79 and 81 on Saturday, and finished fiftieth, one stroke ahead of an amateur from Houston.

Lema also had a quick and biting sense of humor. During the 1964 PGA Championship, at the Columbus Country Club, in Columbus, Ohio, a reporter asked him to tell truthfully how much he drank.

Lema answered, "Not as much as you."

"How do you know that?" the reporter asked.

Lema riposted, "Because only a drunk would ask that question."

Lema and his wife were killed in the crash of a small plane in 1966.

Another Eruption

Steve Reid, who had a brief career on the professional tour in the 1960s, was another of those golfers blessed with an explosive temper. During a practice round leading up to the Philadelphia Open, he stood in the fairway preparing to play an approach shot when a ball driven from the tee behind him bounded into his group.

As it is with an overheated boiler, steam spewed from Reid's ears. He picked up the offending ball, stuck a peg in the ground, and drove it back toward the tee, sending the astonished group behind him diving for cover. Thinking back on that day, Reid said, "Best drive I ever hit."

Reid denies he once threw a network television executive into a swimming pool during a philosophical disagreement at a party. Offended when the subject was brought up, Reid raised his chin, looked down his nose, and said, "It was a lagoon."

Pardon the Interruption

Henry Cotton, the finest English golfer of his time, lined up his putt on the last hole of the Dunlop Masters tournament. As he stepped up to his ball and crouched over, about ready to tap his putt toward the hole a man stepped from the crowd and walked onto the green.

Startled by this breach of spectator etiquette, Cotton straightened up and watched in wonder as the man confronted Cotton's caddie, a man known simply as Barnes, and handed him a summons. It seems Barnes had fallen behind on alimony payments.

Bad Guess

The 1963 U.S. Open was played at The Country Club, in Brookline, Massachusetts, where Francis Ouimet had defeated Harry Vardon and Ted Ray fifty years earlier. A few days before the first round, Jack Fleck, the 1955 champion, approached Joe Dey, the USGA's executive director, and scolded him for picking what he believed would be an easy course. "There'll be a new Open record of 274 or 275," Fleck claimed. Seldom has anyone made such a bad prediction; Fleck missed the winning score by nearly twenty strokes. Julius Boros, Arnold Palmer, and Jacky Cupit tied at 293, the highest winning score since 1935, when Sam Parks won with 299.

Boros won the playoff the next day. After it ended, Mike Souchak said, "Someone could have won my house and my car if he had bet me I wouldn't beat Francis Ouimet's score from 1913. Mr. Ouimet played with hickory-shafted clubs and shot 304. I had fourteen clubs with steel shafts and a high-compression ball. I shot 307."

On his way to shooting 91 in the windswept third round of the 1963 Open, Tommy Aaron stood on the ninth tee, pulled out his white handkerchief, waved it overhead and called, "Where do I go to surrender?"

Locked in the playoff with Boros and Cupit, Arnold Palmer pulled his drive into the woods on the left of the eleventh hole, a terrifying 445-yard par 4. His ball caromed off a couple of trees, bounced along the ground, and somehow leaped into the rotted stump of a felled tree. Not alongside but actually inside the stump.

Already four strokes behind Boros and tied with Cupit, Palmer felt this was no time to give away another stroke by declaring his ball unplayable. Drawing back his wedge, Arnold chopped at the ball. It didn't move. He chopped again, and still the ball cowered within the stump. A third attempt dug it out, but instead of saving one stroke, Palmer had wasted two more. He scored 7 and fell seven strokes behind Boros. Boros shot 70, Cupit 73, and Palmer 76.

To judge from his practice rounds, Jack Nicklaus came into the Open at the top of his game. He had won in 1962, and the quality of his warm-up rounds indicated he might win a second. The Country Club has twenty-seven holes. On the eve of the first round, Nicklaus was given permission to practice on one of the holes that wouldn't be in play. Walking back to the clubhouse, he slipped onto the thirteenth tee of the Open course, which was closed, hit a massive drive and an 8-iron close to the hole, then went over to the fourteenth, a long par 5 with its green atop a rise. There he pounded a high drive that flew on and on, then drilled a 1-iron into the middle of the green. He had his Open game.

The next day he shot 77 and followed up with 76. With 153, he missed the cut.

Where There's a Will . . .

The championship course used at The Country Club for the 1963 and 1988 U.S. Opens combined holes from the club's three nines—Clyde, Squirrel, and Primrose. In setting up The Country Club for the Open, Ted Emerson, a club member, noticed the tee for the third hole of the Primrose nine, which played in the opposite direction and wouldn't be used for the Open, would make a sensational tee for the tenth hole of the championship course. There was one huge problem: A towering old oak just behind the tee would have blocked a drive down the tenth hole of the championship course.

A resourceful man not willing to give up a great idea easily, Emerson looked over the tree closely and spotted what he thought might be a diseased area. Furthermore, not willing to take unilateral action, Emerson called on the course superintendent, explained how he thought he could improve the tenth hole, and suggested the offending tree might have to be removed because he suspected it might be diseased. The superintendent looked and prodded, said "Hrrumph," a few times, and in the end agreed that, by golly, this indeed might be the case.

Down the tree came, and, evidently catching those concerned off guard, the trunk sure enough was diseased, and the tenth had a new tee for the Open.

Unrequited Love

In 1963 Michael Lunt became the first Briton to win a British Amateur in a year of a visiting American Walker Cup team since Roger Wethered in 1923. As the British Amateur champion and member of four British Walker Cup teams, Lunt received invitations to six Masters Tournaments. Strangely, he turned them down. His excuse? Evidently since the BBC hadn't felt it worth televising the Masters, few in Britain realized what a big event it had become. Speaking for Lunt, the secretary of Royal Mid-Surrey, his club, said, "We just didn't know what it was then. It wasn't until Joe Carr (winner of three British Amateurs) went and said, 'You have never seen anything like this place. You must go,' that we took any interest." For himself, Lunt said, "I wasn't invited again, and I've never been able to go subsequently."

Stay in the Rain

Sewsunker Sewgolum, a cross-handed golfer from India, stunned tournament sponsors by winning the 1963 Natal Open, in segregated South Africa. The tournament was not televised because of the South African Broadcasting Corporation's policy of never carrying multi-racial sporting events. Furthermore, Sewgolum had to accept his prize in the rain. Club officials were told that under South Africa's laws it was illegal for Sewgolum to enter the clubhouse.

Jackie on the Links

Since Democrats had mocked President Eisenhower's fascination with golf, those who knew President Kennedy played a fair game were cautioned to keep it mum. Then it turned out that Mrs. Kennedy was a golfer as well.

J. Walter Green, an Associated Press photographer based in Boston, fitted a high-powered lens onto his camera and shot a picture of Mrs. Kennedy, tastefully dressed in a golf outfit, hitting practice shots in Newport, Rhode Island. Spotting Green, Mrs. Kennedy called, "You're not going to use that picture, are you?" Like all good politicians, Green avoided answering by saying, "Mrs. Kennedy, you look lovely today."

Secret Servicemen closed in next, demanding Green turn over his film to them. Green stood fast, pointing out, "This picture was taken on public property." The next day signs littered the area where Green had stood proclaiming it off limits to photographers.

Snakes and Rakes

In a women's team match in Arizona, a player hit a misdirected shot into a sand bunker beside the green. Stepping into the bunker carrying her wedge, the woman spotted a rattlesnake sunning itself. Bent on self-preservation, she quickly scrambled out and refused to go back in, claiming the rules must provide relief from dangerous conditions like this.

Her opponent refused her relief, claiming she must play the ball because the snake qualified as a loose impediment and that a player may not touch or move a loose impediment in a hazard. The opponent made one concession to common sense, though. She offered to stand by, ready with a rake in case the snake made a move to attack.

Moment of Silence

Walking from the twelfth green to the thirteenth tee at Hazeltine, an older member was stricken with a heart attack and died instantly. As he lay on the path, another group arrived and saw the victim. Uncertain of what to do, the four players held a lengthy discussion and decided to remove their hats for a moment of silence, and then played through.

Others have died in similar accidents. After playing a poor drive on the thirteenth hole of the Pnoka Country Club, in Alberta, Canada, in 1993, Richard McCullough slammed his driver against his pullcart. The shaft snapped, whirled like a runaway helicopter and spun toward him. The jagged shaft drove itself into McCullough's neck, pierced his carotid artery, and he bled to death. He was just thirty-one.

Long Course

Floyd Rood played a version of cross-country golf during 1963 that ranks among the most ambitious ever taken on, a course that stretched 3,397 miles—not yards. He began his round, so to speak, on September 14 at the edge of the Pacific Ocean, continued across the United States, and ended October 3 at the Atlantic Ocean. He played the country in 114,737 strokes, and along the way lost 3,511 golf balls.

Violent Dispute

As Ken Venturi, bone weary and badly dehydrated, dragged himself toward the green of the seventy-second hole in the 1964 U.S. Open, at Congressional Country Club, near Washington, a cloud of dust rose from the parched ground a few yards short of the putting surface. Involved in a territorial dispute, two marshals assigned to control the galleries were rolling along the ground fighting one another.

Earlier in the day, a marshal assigned to walk with Venturi and Raymond Floyd had been ordered outside the gallery ropes by another marshal controlling the gallery at the eighteenth green. The walking marshal objected, but he complied. Then he want into the clubhouse for lunch, which he spiced with a few martinis.

Later in the hot and humid afternoon, when the temperature soared toward 100 degrees, nerves had rubbed raw, and the lunchtime alcohol had worked its magic, the walking marshal marched down the eighteenth fairway once again. This time the martinis erased his inhibitions.

Spotting the marshal who had ordered him outside the ropes, a man who stood at least six-foot-four and weighed well over 200 pounds, the walking marshal, three or four inches shorter, but broad as a World War II Sherman tank, approached him and said, "You're the son-of-a-bitch who threw me out this morning."

Without saying another word, the walking marshal delivered a sharp blow to the stationary marshal's chin that sent his hard-hat flying. The stationary marshal fought back, and while uniformed policemen and USGA officials struggled to pry them apart, they rolled around, raising dust, trying to batter one another to pulp. Make no mistake; this was a heavyweight brawl. The walking marshal had been a Congressional member for years. Naturally word of the incident, along with the identities of both men, spread through the club. When word reached the club's highest levels, the board of directors expelled the walking marshal from the club. The stationary marshal was a member of the Washington Golf and Country Club, across the Potomac River, in Virginia.

Ken Venturi

Venturi's 1964 victory ranks among the most unexpected in Open chronicles. A sensational amateur, he had nearly won the 1956 Masters Tournament, but he had shot 80 in the last round and lost to Jack Burke by one stroke. He joined the tour in 1957, won money in all but one of his starts that year, and had an incredible streak of twelve consecutive rounds in the 60s.

In subsequent years he had a series of disappointments, including two more close finishes in the Masters, in 1958 and 1960, losing both to Arnold Palmer. Then, in bending over to pick his ball from the cup during a pro-amateur tournament in Palm Springs, he felt a sharp pain in his back. When the pain wouldn't go away, his swing, once an elegant, compact, rhythmic motion, changed to a flat, crude, jerky movement you could see at the local club any Saturday morning. His game sank lower and lower. At thirty-two, he was washed up. No longer invited to the Masters, and no longer exempt for tour events, he had to plead for sponsors' exemptions.

He had to qualify for the 1964 Open, and nearly walked off the course when Larry Mowry played the first nine in three under par while Ken was shooting two over. Mowry, Billy Collins, and Mike Souchak talked him out of it. He played better the rest of the day and qualified.

After qualifying, Venturi received a letter from the Reverend Francis Murray, a parish priest in San Francisco. In hindsight the letter seemed prescient. It read:

> Dear Ken:
>
> For you to become the 1964 U.S. Open champion would be one of the greatest things that can possibly happen in our country this year.
>
> Should you win, the effect would be both a blessing and a tonic to so many people who desperately need encouragement and a reason for hope.
>
> Most people are in the midst of unremitting struggle, including their jobs, their family problems, their health, frustrations of various sorts, even the insecurity of life itself. For many there is a pressing temptation to give up, to quit trying. Life at times simply seems to be too much, its demands overpowering.

> If you should win, Ken, you would prove, I believe, to millions everywhere that they, too, can be victorious over doubt, misfortune, and despair. I'll be there with your mother and father and the children watching you on TV . . . your friend, Father Murray.

Venturi opened with 72 and 70 and followed with 66, a round that began with 30 on the first nine. He had tied Tommy Jacobs for first place. Wondering a few holes later if Venturi knew how he stood, Joe Dey, the referee, suggested he might want to look at the scoreboard as they walked past.

"Not interested," Venturi said. "I can't change what's up there, and I can't change what the other guys are doing. One shot at a time, that's all that interests me."

Since this was the last year of the double round on the last day of the Open, Venturi still faced another eighteen holes that afternoon. The weather had been oppressive, hot with high humidity. Exhausted by the weather and the tension, Venturi reached the clubhouse, where he was to rest and have lunch between rounds. He slumped on a locker-room bench, too drained to stand, and unable to speak. His face had no color, and his eyes were vacant and staring. He couldn't eat. He ordered tea with lemon, the first liquid he had taken since early morning, and swallowed salt tablets. Raymond Floyd, who had played with him, found his wife, Conni, and told her, "He's sick."

Dr. John Everett, a Congressional Country Club member, took him to one of the club's bedrooms and told him to lie down and rest. He also warned him he could endanger his health if he played. Venturi refused to quit.

After a fifty-minute break he was back on the first tee, apparently recovered. Dr. Everett was with him, carrying a plastic bag of ice cubes wrapped in a white cotton towel. He fed him salt tablets and periodically laid the ice against the back of his neck to cool him throughout the long and difficult afternoon.

Venturi shot 70. His total of 278 fell two strokes short of Ben Hogan's record 276, which he shot in 1948. He had won the Open by four strokes. As the last putt fell, Venturi dropped his putter, raised his hands to the sky, and groaned, "My God, I've won the Open."

A year later it was all over. A problem developed with blood circulation through Venturi's wrists. Some of the feeling in his hands was lost, and once again he sank to the bottom. At the Bellerive Country Club, near St. Louis, in 1965, he missed the cut. As he sat alone in the locker room straddling a bench with his head bowed, a reporter approached and told him how very sorry he felt that Venturi had played so badly in the first two rounds, "especially after last year." After a momentary pause, Venturi said quietly, "At least there was a last year."

Few people grasped how much winning the Open had meant to him, but he expressed his feelings five years later at a dinner the evening before the 1969 Open began. He was eloquent. His voice wavering with emotion, Venturi said to an audience, which included a number of players:

"This is my last year of exemption for the Open. How I won it I still don't know. At the end of 1964 I suffered a disease of the hands. I don't know why, but it happened, and I can't question it." His voice faltered; he paused, then, in control once again, he continued, "Unless a miracle happens, my golf days are limited. I may be forced to leave the game I love.

"There's Gary," he went on, nodding toward Gary Player, "a man with fortitude and determination. There's Jack," he said, nodding toward Jack Nicklaus. "He had a lot of early success, but it didn't spoil him. Gene Littler, one of the greatest talents I've ever seen. He won it with a great swing. And Arnold. Articles have been written that may have drawn us apart, but I'll always feel privileged to call Arnold a friend, a champion among champions."

Venturi turned to his left and looked down at Lee Trevino, the 1968 champion, and continued, "Lee has capitalized financially on his championship, but, Lee, the U.S. Open championship means more than money. It is something you will cherish forever."

Now he looked over the whole gathering, hushed by the strength of Venturi's emotion. "There's not much I fear, but I do fear leaving the game I love. I just thank God for the moment of glory He gave me. Whoever wins the Open next Sunday, whether it is an unknown or an established player, whoever wins it treat it well, because the U.S. Open is the greatest championship in golf."

Venturi never played in another.

Lighting the Way

I. S. Malik, who had risen high in the British civil service, which ran India, had been the first Indian to be accepted as a member of Royal Calcutta, until then an entirely British club. His son Ashok was an accomplished golfer who had played for India in a number of World Amateur Team championships. He also had a sucker bet. Led on by members in on the scam, someone offered to bet Ashok he couldn't par Royal Calcutta's first hole in the dark. Ashok accepted with one stipulation: somebody had to hold a flashlight over the hole for both his approach and his putts.

The night was pitch black, but he drilled his drive straight down the middle, lofted a 9-iron onto the green, holed his second putt, and made his par.

Player and Nagle

Player and Kel Nagle each shot 282 at the Bellerive Country Club, in St. Louis, and tied for first place in the 1965 Open, forcing a playoff the next day. On the morning of the playoff, after eating the same breakfast he had eaten every day of the previous week, Player tried to key himself up through psychological methods. Seated with Mark McCormack, his business agent, and George Blumberg, a friend from South Africa, Player said, "Make me mad."

McCormack tried first, asking, "Did you see that newspaper article that said you'd never win another major title?" (Player had already won the 1959 British Open and the 1962 PGA championship.)

"Yeah," Gary said."That's a good one. Make me madder."

"How about the guy who wrote you'd never win anything with those glass-shafted clubs?"

"That's a good one, too."

Sufficiently angered, Player won the playoff.

Just before Player and Nagle reached the thirteenth hole of the playoff, tournament officials found that vandals had damaged the green badly, gouging the turf with the base of the flagstick, creating a scar about eighteen inches long about a foot from the cup. Since there was no time to repair the ground properly, the two men were told that if the scar intervened between the hole and a ball lying on the green, the damaged ground could be treated as casual water, which would allow the player to move his ball enough to give him a clear line to the hole.

Bellerive's thirteenth is a 198-yard par 3. After playing his approach putt, Player found his ball in direct line with the scar. Even though he was entitled to move his ball under the local rule, he wouldn't do it.

"I hate to do that," he said. "It doesn't seem fair."

Instead, officials allowed him to repair a portion of the scar perhaps the size of a ball mark. He rolled his ball across it and holed the putt for his par. Player won the playoff.

Player shot 71 against Nagle's 74. As he reached for his ball after holing his last putt, he paused momentarily as if frozen in place, his hand still inside the cup. Asked later about the pause, Player said, "I was thanking God for letting me win."

Struggling to a final round 74 and a tie for sixth place in the 1962 Open, Player had confided to Joe Dey, the USGA's executive director, "I'm so disappointed not to win the Open. If I had, I planned to give the prize money to charities, especially cancer relief." Then he said, "Let's keep that a secret. One day I will win, and I'll turn back the money for good causes."

At the prize ceremony following his victory over Nagle, Player indeed gave back the money. Of his $26,000 prize, Player directed $5000 to be used for cancer relief in memory of his mother, who had died of cancer, and directed the USGA to spend $20,000 for junior golf. He gave the remaining $1000 to his caddie. Player said he made the donations because, "I am grateful to the American people, who have been so wonderful to me, a foreigner."

Optimism

Somehow worming his way into the qualifying rounds leading up to the 1965 British Open, Walter Danecki, an American from Milwaukee, shot 108 and 116 at Hillside Golf Club, whose course abuts Royal Birkdale, the site of the championship proper. Not so strong as Birkdale, Hillside is, nevertheless, a first-class course. Still, it isn't severe enough to cause a player of Open caliber to finish 80 over par for two rounds.

Commenting on his golf later, Danecki explained he was happy he had decided to use the larger American ball of 1.68 inches diameter rather than the smaller British ball of 1.62 inches. Otherwise, he said, "I'd have been all over the place."

Dave Marr

Dave Marr played in his first British Open in 1966, a year after he won the PGA Championship. The Open was played at Muirfield, where every blade of grass that ever grew is still there in its foot-high, thick-bladed rough.

Never having seen rough like this, Marr asked Roberto De Vicenzo, the delightful Argentinian golfer, "What do you think will win here?" Pondering for a moment, Roberto said, "Three hundred. If the wind blows, maybe 400." Jack Nicklaus won with 282.

What Game Am I Playing?

Reflecting on his nine-hole scores of 40-30 after stepping off the eighteenth green in the third round of the 1966 British, Phil Rodgers asked, "Anyone for tennis?"

Just Forgetful

It wasn't a case of nerves at all, simply a memory lapse. Anyhow, that was Ty Caplin's position before teeing off in the first round of the 1966 U.S. Open, at the Olympic Club. An assistant professional at Olympic until a few months before the Open began, Caplin looked down at his feet and insisted, "No, I'm not nervous—not at all. I just forgot to put on my shoes."

Punishment Fits the Crime

Following his press conference shortly after losing seven strokes to Billy Casper over the last nine holes of the 1966 U.S. Open, Arnold Palmer faced a difficult time forcing his way through a horde of his fans crowded around the entrance to the press facility. A USGA staff member suggested he slip out the back way and be driven to the clubhouse in a golf cart.

Palmer refused, saying he'd take his chances with the fans. He added, "Besides, I deserve whatever they do to me."

Practice Day

A man drove into the area set aside for public parking at Baltusrol, in New Jersey, early in the week of the 1967 U.S. Open. Watching him open his car trunk, take out a bag of clubs, and change into his spiked golf shoes, Matthew Glennon, the chairman of the club's parking committee, told him, "This is the back end of the course. If you haven't registered yet, just drive around to the main entrance and you'll find the registration desk. They take care of all the contestants."

Puzzled, the man told Glennon, "I'm not a contestant. I just came out for the practice."

"Practice?" Glennon asked.

"Sure," the man answered. "Here's my ticket."

Sure enough, the ticket read, "First Practice Day."

The Wisdom of Television

In the early years of ABC golf telecasts, Chris Schenkel, then the network's leading sports announcer, teamed with Byron Nelson on golf commentary. Noting the players' long drives during one of the winter tour tournaments, Schenkel led Nelson into explaining the reason for this phenomenon. Obviously a little nervous, since this was one of his early exposures to announcing, Nelson responded by expounding, "Chris, the boys are hitting the ball longer now because they're getting more distance."

A few years later, doing commentary during the 1983 Colonial National Invitational, Steve Melnyk explained, "Peter Jacobsen is in a position where a birdie will help him more than a bogey."

It was Melnyk, too, who said of Phil Mickelson while he was still an amateur, "His future is ahead of him."

One television commentator to another as a player prepared to play a very long shot:

First announcer: "Do you think he can get there?"

Second announcer: "Yes, if he can hit it far enough."

A Man's World

Among Britain's most exclusive clubs, Royal St. George's, on England's southeastern coast, has a decidedly male flavor. At one time the club displayed a sign declaring, "No dogs; no women."

Times change; women won gradual recognition, but still St. George's would have its way. A later sign, while acknowledging women were welcome in the clubhouse, proclaimed, "Ladies wearing trousers are requested to remove them before entering the clubhouse."

Easy Hole

The golf course designer Robert Trent Jones had revised the Baltusrol Golf Club, in Springfield, New Jersey, for the 1954 United States Open championship, turning it into a severe test of the game. Some Baltusrol members claimed it was too hard, especially the fourth hole, a demanding par 3 of 194 yards, all carry, across a broad pond.

Since Jones's business depended on cordial relations with clubs, an unhappy membership was the last thing he wanted. As a young man, Jones had been a very good golfer, and so to ease the members' strain he arranged a match with Johnny Farrell, Baltusrol's golf pro, who had beaten Bobby Jones in a playoff for the 1928 U.S. Open, C. P. Burgess, the tournament chairman, and another member.

Arriving at the fourth, Farrell and the club members carried their tee shots onto the green, and then Trent stood up. Swinging smoothly with a long iron, he lofted a lovely shot that flew straight at the flag, hit a few yards short of the pin, then dived into the cup. A hole-in-one. Turning to the others, Jones said, "Gentlemen, I believe this hole is eminently fair."

The Rules Prevail

During the early 1950s, spectators crowding around Augusta's eighteenth hole during the Masters inched almost to the green's edge. As Gene Sarazen played his approach, the gallery included an overweight young woman sitting on the grass in the first row of the gallery. Sarazen played his iron with a touch too much force. His ball skipped across the back of the green toward the young women, who sat petrified as the ball ran straight under her skirt. Stationed nearby, Ike Grainger, an Augusta member on duty as a rules official, watched the ball's path and gulped. He knew he had to handle the ruling.

First, he had to find where the ball lay. In real life chairman of the board of the Chemical National Bank, in New York City, one of the nation's most formidable financial institutions, Grainger threw himself prone on the ground and, and while the crowd roared in laughter, stared up the girl's skirt. He saw that the ball had gone as far as it could go. Seeming to think she had done something wrong, she asked Grainger, "What shall I do?" Grainger said, "Wait for Mr. Sarazen."

Arriving at the green, Sarazen asked, "Where's my ball, Ike?" "Have a look," Grainger answered. Now Sarazen and Grainger both dropped to their knees and looked at the ball.

Under normal circumstances Sarazen would have marked the position of his ball so that when the girl moved he would know exactly where to place the ball, but this was obviously out of the question. With a hint of a smile, Sarazen asked, "What do I do now?" Grainger told him, "Ask the young lady to stand up, and when the ball falls, you drop it as near as possible to where it was."

She stood, but the ball had wedged between her upper thighs and it wouldn't drop. She turned red and the gallery was out of control. Now she had to shake herself. Finally the ball broke free and tumbled to the ground. By then both Grainger and Sarazen were fighting so hard to hold back their laughter that tears rolled down their cheeks and they could hardly see. Calling on all his will power, Sarazen controlled himself, dropped his ball, chipped close, and holed his putt for a par 4.

Just Too Sentimental

Australian Peter Thomson had the 1954 British Open within reach after the fifteenth hole at Royal Birkdale, but he felt pressure from Tony Cerda, from Argentina, and the Welshman Dai Rees, who were closing in. Reaching the sixteenth hole, a long par-5 (in a revision of the golf course, this hole subsequently became the seventeenth), Thomson hooked his second shot. As he walked up to his ball he saw it resting in sand high on a steep upslope of a bunker twenty yards or so short of the green. To play his next shot, Thomson assumed an awkward stance, his left foot high above his right. Compounding the problem, he faced a delicate shot. Dig too deeply into the sand and the ball would carry only halfway there; hit it thin and the ball would fly miles over the green.

Taking his sand wedge, Thomson squirmed his feet into the sand and took a brutal swing. In a shower of sand the ball rose fifty feet high, carried to the front of the green, and rolled four inches from the cup for an easy birdie 4. Thomson won; it was the first of his five British Opens.

What happened to that magnificent wedge? "I don't know," Thomson admitted some time later. "It wasn't a very good club; it was too damned heavy. They didn't design good clubs in those days."

Arnold Palmer

No one had a more positive influence on golf than Arnold Palmer. He was at least as popular as Bobby Jones, and he came along just when television became an important part of the game. The galleries reacted to him because he reacted to them. He had a warm, friendly, tight-lipped smile he aimed directly at *you*.

All his emotions showed. When he hit a bad shot he screwed up his face as if he were in pain. When he hit a good shot, he glowed. He played golf for the joy of it, and by winning tournaments he became a public figure. He enjoyed his role.

Palmer grew up as the oldest of five children of Doris and Milfred (Deacon) Palmer, at first the greenkeeper and later greenkeeper and professional at the Latrobe Country Club, about forty miles east of Pittsburgh.

Arnold began playing golf when he was three, using junior clubs. By five he was playing full eighteen-hole rounds accompanied by his father and using his mother's clubs. By seven he was breaking 100 on the rare occasions he was allowed to play Latrobe.

Palmer grew up under a list of restrictions. He could not use the club swimming pool; he could not play with the members' children; he could not play golf by himself except very early in the morning or late in the evening with no one around. He could not play in a club tournament, but he could play in the annual caddie tournament. He won it four times, but he didn't accept a prize. Deacon told him he wasn't eligible for that, either.

What Deacon Palmer told Arnold stuck. When Arnold was very young, Deacon placed his hands on a club and said, "That's the way you hold it." Arnold held it that way.

"I wouldn't have dared change for fear he'd catch me," Arnold said. "What he said do, I did."

In winning the 1954 Amateur Championship, Palmer defeated Robert Sweeny, a New Yorker who had spent a considerable part of his life in England. Sweeny had won the 1937 British Amateur and had been the runner-up in 1946. A member of a family with investments in South African gold mines, Sweeny was an accomplished flier. He joined the Royal Air Force early in the Second World War and rose to squadron leader. He and his brothers helped organize and finance the Eagle Squadron, a group of Americans in the RAF copied after the romantic Lafayette Escadrille of the First World War. Their uncle, Colonel Charles Sweeny, had flown with the Escadrille.

Sweeny was no pushover. A member of Seminole, he spent the better part of a month helping Ben Hogan prepare for the Masters by playing him even—no strokes given—in money matches. He didn't always lose.

Palmer almost didn't reach the final. Semi-final matches were played over thirty-six holes in those days, and Palmer was having a terrible struggle with Ed Meister, an executive with a publishing company in Willoughby, Ohio. All square after thirty-five holes, Meister had played his approach within ten feet of the cup while Palmer had overshot the green. He looked like a certain loser because his ball sat in a gully more than fifty feet away; he had a bad lie and the green sloped away from him.

Palmer was never short of confidence; he just *knew* he would win. He played a perfect wedge that hit the near edge of the green and coasted down the slope within three feet of the cup. Meister missed his birdie, Palmer holed, and Arnold won the match on the thirty-ninth hole. If he hadn't played that exquisite wedge, he might not have become Arnold Palmer.

By then Arnold had met Winifred Walzer, the twenty-year-old daughter of a canned-goods company executive, at a tournament at Shawnee-on-

Delaware. It was love at first sight on his part. Arnold needed money to buy an engagement ring. Some friends took him to Pine Valley, a course he'd heard of certainly, but had never played. They made him a bet: they agreed to pay him $100 for every stroke he shot under 72, but at the same time he'd have to pay them $100 for every stroke over 80.

"I was young and in love, and nothing could scare me," Palmer recalled later. He started poorly and had to hole a twisting thirty-foot putt to salvage a bogey 5 on the first hole, but he recovered smartly, shot 68, and won $400.

Palmer won his first important tournament in 1958, possibly by knowing the rules better than an official on the course. Arnold came to the twelfth hole, the little par-3 across a creek at the Augusta National Golf Club, leading Ken Venturi by one stroke in the Masters. Putting too much into his 7-iron, he carried the ball over the creek and over the green as well. The ball embedded itself in the ground, softened by the heavy rains that had hit Augusta early in the week.

The embedded ball rule was in effect, which meant that Arnold would be allowed to lift his ball out of the mud, clean it, and drop it. The official wasn't sure about this, though. After a short debate he insisted that Arnold play the ball as it lay, without lifting it from the depression, then play another ball under the embedded ball rule. Palmer fumed, but he did as the official demanded. He made a double-bogey 5 with the original ball, but saved a par 3 with the alternate ball and headed off to the thirteenth not knowing how he stood.

Still fuming, Palmer lashed into his drive, then absolutely killed a 3-wood that everyone knew would reach the green as soon as it left the clubface. The ball rolled within about eight feet of the cup, and Arnold holed the putt for an eagle 3. Moments later he learned his par 3 on the twelfth would stand. He shot 284, beat Doug Ford and Fred Hawkins by one stroke and Venturi and Stan Leonard by two. He had won the first of his four Masters Tournaments.

Watching how Palmer charmed the Augusta galleries, and how they swarmed after him, Johnny Hendrix, an Augusta sportswriter, named Palmer's fans Arnie's Army. The name stuck.

Favored to win the 1960 Open after having won the Masters earlier in the year, Palmer had just shot 72 in the third round, dropped to fifteenth place, and trailed Mike Souchak, the leader, by eight strokes. To live up to his expectations, he'd not only have to make up those seven strokes, he'd also have to pass fourteen other golfers.

At lunch before the afternoon round, he ate a cheeseburger and drank iced tea while he sat at a table with Bob Rosburg, Ken Venturi, and Bob Drum, a reporter for the *Pittsburgh Press* who had followed Palmer since Arnold's days in junior golf. The talk was quick and forced.

Venturi said, "I wonder if Souchak can hold on."

"I don't see why he can't," Rosburg responded, "but it's a funny game."

Then Palmer brought the conversation to life. "I may shoot 65," he announced, "What would that do?" Drum scowled, "Nothing. You're too far back." Angered, his eyes blazing, Palmer snapped, "The hell I am. A 65 would give me 280, and 280 wins Opens."

Drum riposted, "Of course if you drive the first green and make a hole-in-one, that would help."

"I might just do that," Palmer said, and stalked from the room.

He had tried and failed to drive the first green in each of the first three rounds. He succeeded in the fourth. Throwing himself into the shot, Arnold sent his ball rocketing in a straight line toward the green. His ball bounded through a belt of rough, rolled on, and stopped about twenty feet from the cup. He barely missed the eagle putt, but saved the birdie.

Palmer birdied six of the first seven holes, shot 30 going out, came back in 35, giving him the 65 he wanted, and with 280 won the championship by two strokes.

Shortly after Palmer won the 1961 British Open, he returned home to Latrobe, Pennsylvania, and set out one bright and cheery Saturday for a round of golf with three friends at the Laurel Valley Country Club, in nearby Ligonier.

As Arnold and his friends teed off, they noticed two men were to play behind them. Within a few holes the men behind began pressing Palmer and his friends, and so Arnold invited them to play through. The two men declined, claiming they were enjoying Palmer's game more than they enjoyed their own.

The two groupings continued like this for eleven holes, until finally Palmer and his friends had had enough. After holing out on the eleventh they played their drives on the twelfth and then waited for the two men behind them to finish. Then they insisted the two men play through.

They agreed. Palmer and his friends had played from the extreme back tees, but the other two would have none of that and instead played from the regular men's tees, yards ahead. To the considerable surprise of one of them, all his gears meshed at the proper time and he hit a really fine drive. It split the fairway, and when it stopped rolling it had gone past Palmer's ball. It didn't matter that he'd had a substantial advantage, it only mattered to him that he had outdriven Palmer.

With his usually friendly grin, Palmer said, "Nice drive." The man answered, "My grandchildren will hear about those words." Actually, he didn't wait for grandchildren. For the next several weeks he told everybody who'd listen of how he had outdriven Arnold Palmer, and with each telling the distance between the tee Palmer had used and his own grew ever shorter. A week later he returned to Laurel Valley for a game with three friends who hadn't heard the story. As they stood on the twelfth tee the man once again ran through the tale, but as he reached the climax he noticed his caddie looking at him rather strangely. When he had finished the story, the caddie said, "Sir, could I speak to you privately?" The man agreed. "Sir," the caddie said, "I caddied for Arnie that day."

"Yes," the man said affably.

"Would you like to know what he said to me after he said, 'Nice drive' to you?"

"Yes."

"He said, 'With a swing like that I don't see how he hit the ball at all.'"

Jack Nicklaus played the last twelve holes at Oakmont, near Pittsburgh, in three under par and tied Palmer at 283 in the 1962 U.S. Open, setting up an eighteen-hole playoff the following day. Nicklaus won it handily. He led by two strokes coming to the eighteenth hole, and when Palmer stumbled badly, taking a double-bogey 6, Arnold picked up Jack's ball, conceding him a bogey 5 and the playoff.

Quickly, though, Joe Dey swept onto the green and stopped both men. Because this was a stroke-play competition, Palmer could not concede a hole. Nicklaus must replace his ball and play out the hole. He holed the putt and shot 71 against Arnold's 74.

At the height of his popularity, during the 1960s, Palmer often played exhibition matches nearly always followed by a dinner, where he would speak. With the coming of the jet age, though, and with demands on his time, he usually skipped the dinner and delivered his talk on the eighteenth green, just before holing his final putt of the day, then signing autographs before dashing home.

Arnold played at the top of his game in a match at the Westwood Country Club, near Cleveland, where he and Mike Souchak played with two local amateurs. He reached the eighteenth green needing a birdie to break the course record. Jack Cook, the chairman of the committee that put on the exhibition, had offered an additional $200 for the player who broke the record. Before putting, though, Arnold delivered an entertaining talk for about five minutes, and closed his address by saying, "And now, ladies and gentlemen, I'm going to take Jack Cook's money."

With that he stepped up to his twenty-footer and holed it. Naturally, the crowd cheered, but when the applause died down, Arnold noticed a small girl standing at the edge of the green sobbing her heart out. She was Sally Kennedy, the daughter of the club champion. Arnold had just broken the record her father had held. When it was explained to Palmer, he hugged the girl, gave her an autographed copy of his book *My Game and Yours*, and wrote her a letter when he returned home.

Palmer and Johnny Pott were locked in a tight playoff in the 1962 Colonial National Invitational tournament, at the Colonial Country Club, in Fort Worth, Texas. Just off the ninth green, Arnold was looking at a tough chip shot that he had to lay close to the hole in order to protect his one-stroke lead.

An enormous crowd had gathered around the green as Arnold prepared to play his shot, but just as he stepped up to the ball, a small boy began a conversation with his mother. Distracted, Arnold stepped away from the ball and turned toward the noise, but seeing the young boy, Palmer laughed. The crowd laughed along with him, and Arnold stepped up to his ball again.

Now the boy began crying because of his mother's scolding. Once more Arnold stopped and laughed, and again the crowd laughed with him. He stepped up to the ball for a third time, but now he heard the weird sound of a muffled scream. Turning, he saw a desperate mother with her hand over her child's mouth and the boy turning pink. Palmer broke up. Kneeling, he patted the boy and said to his mother, "Hey, don't choke him. This isn't all that important."

On the fourth attempt, he chipped close to the hole, saved his par, and won the playoff.

Flying to Atlanta in 1963 for his first appearance in the Ryder Cup, Johnny Pott sat next to Palmer, who had played his first match two years earlier. Confessing he felt nervous, Pott told Palmer he hoped he'd hold up under the pressure of playing on a team rather than for himself alone.

Trying to console Pott, Palmer said not to worry; since Arnold was the team captain, he'd pair the two of them. Not happy with that suggestion, Pott cried, "Hell, don't do me any favors. Put me with somebody who'll be just as nervous as I am."

True to his word, Palmer paired himself with Pott in the opening foursomes. As they stepped onto the tee, "Bands were playing the *Star Spangled Banner* and *God Save the Queen*; it was like the start of a football game," Pott claimed.

Assigned to drive on the odd-numbered holes, Pott drove from the first tee into the left rough. Palmer played their second shot onto the green about forty feet short of the hole, and Pott followed with a timid putt that pulled up about four feet short of the cup. As the ball died, Palmer, eyes blazing, stepped up to Pott, grabbed him by the shoulders, and snapped, "Don't ever leave me short again."

First Aid, with Limits

The rivalry between Palmer and Nicklaus reached its highest pitch during the 1964 season, after Palmer had won the last of his four Masters Tournaments. Within a period of two months they faced each other in six tournaments, and neither man placed worse than fifth. Palmer was second and Nicklaus third by two strokes in the Cleveland Open, in June. The following week Nicklaus closed with 67 in Philadelphia, passed

Palmer, and won, with Palmer third. Two weeks later, in Columbus, Ohio, Nicklaus shot 64 in the last round of the PGA and tied Palmer for second place behind Bobby Nichols. Palmer finished second, second, and third in the next three tournaments while Nicklaus placed fifth, third, and fourth.

The season reached its climax in the World Cup, played that year at Kapalua, Hawaii. Although Palmer and Nicklaus teamed together to represent the United States, they were rivals as well, because the competition included an individual title.

Opening with consecutive scores of 66, 67, and 67 Palmer led Nicklaus, who stood in second place, by six strokes, and the United States held a firm grip on first place in the team competition.

Arnold had injured his thumb earlier in the week and had asked for help from Jack's father, Charlie Nicklaus, a pharmacist, who had come to Hawaii to watch the tournament. Charlie had obliged and given Palmer something to ease the pain. Now, with one round to play and Arnold holding a solid lead over his son, Charlie approached Arnold and told him, "Well, now, Arnie, we've got to take you off that medicine." Flinching on nearly every shot from the pain in his thumb, Palmer shot 78 in the last round, Nicklaus 70, and Jack won by two strokes.

Palmer stood well in front of the 1967 Florida Citrus Open, at the Rio Pinar Country Club, in Orlando, when he reached the fifteenth hole, a 510-yard par 5 that sweeps around a pine thicket on the right to a small green protected by a wandering creek. A well-struck and well-placed drive should reach the go or no go line, a depression marking a buried television cable just beyond the fairway's bend. Short of the depression there's no chance; beyond it there is.

Palmer hit his drive a bit off-center. The ball ran down the right side, just missed the lower branches of the trees, cleared the corner, and rolled about fifteen yards short of the buried cable. Now Palmer faced a decision. Should he go for the green and risk falling into the creek, or should he play safe? Palmer hesitated. The group ahead still on the green, Palmer paced back and forth, puffed on his cigarette, and wrestled with the decision. Accustomed to Arnold's swashbuckling, daring tactics of old, his gallery urged him to go for it. Bob Blair, his caddie, wanted him to go for the green as well.

Instead of the 3-wood, Arnold drew an iron. As his hand closed over the head of the iron, Blair cringed. "No. No," he said. "You can get there with a good 3-wood." Arnold hesitated, stepped away, then made up his mind. The green was out of reach; he'd lay up.

As he drew the iron from his bag, Blair reeled away, throwing a hand at the sky in disgust. The General was chicken.

Blair's gesture stunned Arnold. He glared at Blair, then slammed the iron back into the bag and snapped, "Okay, you son of a bitch, I'll show you."

He snatched the 3-wood, and with the Army cheering him on, ripped into the shot with all his fury. The ball tore into the sky, still rising when everyone thought it must surely begin to descend, then began its fall.

The gallery held its breath when the ball began to drop. They could see it would be close, and they willed it on, some calling, "Go. Go." They couldn't help. The ball slammed into the far bank of the stream, hung for a moment, then tumbled into the stream bed.

Palmer bogeyed the hole and lost the tournament to Julius Boros by one stroke. He stormed into the locker room, slammed open the locker, yanked out his gear, and headed for the parking lot. Blair stowed Arnold's clubs and other belongings into the trunk while Palmer glared, then quickly slipped into the back seat, as far from Palmer as he could get, while two of Palmer's business associates climbed in. With Arnold at the wheel, the four of them drove off in silence.

Palmer hated air conditioning. As he eased his way out of the parking lot, he lowered his window. Just then an automatic sprinkler turned on. A stream of water shot through the window and drenched Arnold. Seeing it, Blair muttered, "Maybe that will cool that son of a bitch off."

The businessmen blanched and unconsciously lowered themselves in their seats. Water dripping from him, Palmer turned slowly and fixed Blair with a blazing scowl. Slowly, though, Arnold's eyes began to glitter, his lips turned up at the corners, and he couldn't hold back the smile. "Dammit," he screamed, "I told you I couldn't hit that 3-wood that far."

Palmer became a licensed pilot and often flew his own plane, sometimes cross-country, sometimes transatlantic. During a speaking engagement the Monday evening after the 1970 Bing Crosby Pro-Am, at Pebble Beach, Palmer decided suddenly to fly home after the dinner rather than follow the original plan and wait for Tuesday morning.

Palmer, Winnie, and the co-pilot arrived at the Latrobe, Pennsylvania, airport at 6:30 the next morning, half an hour before the control tower opened for business. With no one on duty to describe ground conditions, Palmer made a low pass over the runway. From the air it looked clear, but it actually lay under half an inch of ice.

With Arnold at the controls, the plane sat down softly. Then things began to happen. Immediately it began slipping and sliding sideways down the runway. Struggling to regain control, Arnold overloaded the nose-wheel circuit breaker. Useless now the wheel flopped around as if it might break off while the plane skidded along the ice. Palmer fought to right the plane with the rudder. It worked until it lost too much speed, then lurched out of control once again. At the end it rammed into a snowbank and stopped.

No visible damage had been done to the plane, but it was stuck. With Arnold on the ground shouting instructions, the co-pilot gunned the engines, the plane broke free, and thundered straight at Palmer. His eyes bulging, Arnold fell to the ground, slamming his head against the ice, and managed to roll clear of the landing gear. He survived with nothing more serious than a gash to his head.

Watching how female spectators reacted when Palmer strode past, a member of the gallery at Pittsburgh's Oakmont Country Club during the 1973 Open turned to a companion and asked, "Can you imagine being Arnold Palmer and single?"

Palmer shot 77 in the last round of the 1975 U.S. Open and finished three strokes behind Lou Graham, the eventual winner, and John Mahaffey, the loser in a playoff, at the Medinah Country Club near Chicago. Disgusted with his familiar blade putter, a model made popular by Tommy Armour years earlier, Palmer had switched to a "Zebra," a mallet-headed club with alternating black and white stripes from front to back across the top—hence the name.

Palmer had putted atrociously and lost the Open on the greens. He was in a sullen and surly mood as he stepped off the final green and into the scorer's tent. Nevertheless, the press always wanted an interview with Palmer, and Arnold was always willing to oblige. The route from the eighteenth green to the press tent took Palmer and his escort across a wide sweep of ground behind the practice area. Still morose after his pitiful last round, Palmer skulked toward the press tent with his head hanging low and with rage boiling within. Suddenly, with no prompting, Palmer snapped, "That's the end of that fucking Zebra."

Palmer had just shot 32 on the first nine of the Carrollwood Village Country Club, in Tampa, Florida, and rushed to the front of the 1981 Michelob Senior, but he bogeyed his way home, shot 42, and missed a playoff by one stroke. Don January beat Doug Ford in sudden death.

Reeling from his terrible finish, Palmer retreated to the clubhouse, where he sat beside Stan Musial, the great ballplayer, in a VIP lounge. Musial, who had had his bad days as well, did his best to give Palmer comfort. Then the telephone jingled. Musial grabbed it to save Palmer the agony. A spectator who had left early was calling to find out Palmer's final margin of victory. Musial's response that Palmer had butchered the home nine stunned the caller.

"No, no," Musial mumbled. "Arnold didn't' win. . . . January, in a playoff . . . What happened? Well, uh, he, uh . . . ," Musial stammered. His lips set in a tight line, Palmer glared at Musial, leaned closer to the phone, and snapped, "He got something caught in his throat and he choked on it."

After the tournament ended, Arnold and a gang of friends climbed aboard a helicopter for the trip to Palmer's home near Orlando. The mood was sullen following Palmer's collapse. Everyone spoke quietly until Joe Tito, an old friend, shattered the gloom with a short burst of laughter at something one of the other passengers had said. Still sullen, Palmer turned wide-eyed and glared at Tito. "Look, pal," Tito snorted. "We're not the ones who blew a seven-shot lead, so there's no reason for us to pout all the way to Orlando." Palmer's expression changed from anger, to remorse, and finally to a sheepish smile. Recovered, he threw a half-filled can of beer on Tito, everyone laughed, and the mood was broken.

In Stockholm for a match against Sven (Tumba) Johansson, who stood at the top of Sweden's list of athletic heroes, Palmer was assigned a caddie the gallery thought looked awfully familiar. He should have; he was, after all, Prince Bertil of the royal family. Word circulated that the prince had been assigned the job, but no one stated by whom. Nor did he actually sling that heavy bag across his shoulder à la Creamy Carolan, Palmer's regular caddie of the time. Instead, Arnold's bag was lashed to a pull cart and the prince towed it around the golf course in full view of an estimated 4000 loyal subjects.

Playing an exhibition match with Palmer would be enough to make nearly anyone nervous. John Gerring, a club professional from Marietta, Georgia, felt the pressure. After a good drive from the first tee, Gerring turned away, tripped over the tee marker, and fell on his backside.

Another man had entered a Port-A-John just as the advance guard of Arnie's Army began filtering past. After his mission had been completed he noticed a forbidding silence and assumed he was by now alone among the trees, where those things tend to be located—well away from the field of action.

The hinges screeched as he began opening the door, and the bright sunlight blinded him momentarily, but there, just outside the door, stood an enormous phalanx of people lining the hole all the way to the distant green, and not ten yards away stood Palmer himself, hunched over his ball, ready to play a shot.

At the sound of the door's opening, Palmer looked up squarely into the man's eyes. The man quickly stepped back inside and drew the door closed.

After a few seconds he heard this tapping on the door. Curious, he opened it a crack, and there was Palmer.

"Listen," Arnold said, "come on out; there's no hurry."

Seeing everyone in that huge gallery watching him and hearing them snicker, the man said, "No, I'm in no hurry either. You go right ahead," and once again softly closed the door.

A few more seconds passed, the crowd murmur rose, and there came this tapping on the door once again.

"Listen," Palmer pleaded through the closed door, "I'm finding it hard to concentrate thinking about you shut up in that box. I'd appreciate it if you'd come out."

Then the door creaked open once again, the man stepped out with everyone grinning at him, then, embarrassed, he turned away from the Palmer gallery and watched someone else play.

Part Six
The Age of Nicklaus

No one has shown such strong will-power as Jack Nicklaus. He began as a pudgy and obviously overweight crew-cut collegian, then changed his image, shed weight, let his hair grow longer, and became a hero.

Nicklaus grew up in a different atmosphere from Arnold Palmer. While Palmer wasn't allowed inside the clubhouse of the Latrobe Country Club, Nicklaus grew up as a member of the fashionable Scioto Country Club in Columbus, Ohio, where Bobby Jones had won the 1926 U.S. Open.

Palmer had his father as his teacher; Nicklaus was taught by Jack Grout, who as a young man in Texas had played golf with Ben Hogan and Byron Nelson. Grout occasionally used novel methods to make a point. As Nicklaus played practice shots, Grout would stand in front of him and grasp a handful of Jack's blond hair and hold it steady. In later years Nicklaus remarked that he learned early, "You don't move your head in this game if you want to keep your hair."

Jack's father, Charles Nicklaus, who owned a number of pharmacies, often played with Jack. When Jack was fourteen he shot 34 on Scioto's first nine and had a chance to break 70 for the first time. Charlie, though, insisted they had to go home because Jack's mother expected them for dinner. Jack pleaded for them to stay, but Charlie absolutely refused. Finally he said they would rush home and have a quick dinner—because that would please Jack's mother—and then go back to the club and see if they could finish. They made it. Jack holed a thirty-foot putt for an eagle 3 on the last hole and shot 69.

The 1956 Ohio Open field included a number of very good golfers with national and even international reputations: Frank Stranahan, a former

British Amateur champion who had tied for second place behind Ben Hogan in the 1953 British Open; Leo Biagetti, an off-and-on tour player; and Gordon Jones, well known in Ohio golf. It also included Jack Nicklaus, a sixteen-year-old high school boy who had set himself a busy schedule for the week.

The Ohio Open was played over seventy-two holes in three days—single rounds the first two days and then a double round the last day. Following the round on the second day, Nicklaus was to play an exhibition match with Sam Snead, then return for the double round the following morning. The Open was played at the Marietta Country Club in southeastern Ohio, just on the West Virginia border. The exhibition match was to be played at the Urbana Country Club, west of Columbus, more than halfway across the state.

Nicklaus opened with 76 at Marietta, a hilly course with lush fairways. Because they knew of the exhibition match, tournament officials had given him an early starting time the following day; he responded by shooting 70 and jumping into contention. The round over he hopped into a plane arranged by Warren Grimes, the promoter of the exhibition, flew to Urbana, where he hit every fairway and every green, and shot 72 against Snead's 68.

Back at Marietta the next day and evidently inspired by the fluid grace of Snead's legato swing, Nicklaus shot 64 in the morning and blew the Ohio Open apart. With 70 in the afternoon, he won the championship.

Happy Birthday

That Nicklaus is color blind is pretty well known. He must also have problems with his hearing. The telephone rang in his room in the Del Monte Lodge on January 21 one year during the Crosby tournament. Picking up the phone, Jack heard a rich baritone voice singing "Happy Birthday."

Telling the story, Jack said, "I'm wondering what's going on. Who is this singing in my ear, and why me? . . . When he finally finished, I give him a very stiff business-like 'To whom am I speaking?' "He said, 'Bing Crosby. Happy Birthday, Jack.'"

What Drives Jack

After he turned fifty and obviously had passed the time when he could be a consistent winner at the game's highest levels, Nicklaus was criticized for even trying. Gary Player commented that Jack and others who had passed fifty "should quit knocking their heads against the wall and realize that the senior tour is the place for them."

Player, however, overlooked the competitive urge. While admitting he was indeed knocking his head against the wall, Nicklaus said, "Until I'm sure my head is split open, I'm not going to quit. I want to be the best I can be. If I give up that goal, I have nothing else to play for. I've tried to be a ceremonial golfer, and I just can't do it."

Prophecy

Nicklaus played the last thirty-six holes of the 1960 U.S. Open with Ben Hogan, who hit thirty-four of those greens in regulation figures. Nicklaus was an inexperienced twenty-year-old college student at the time, but Hogan recognized the genius in him. Bitterly disappointed at playing the last two holes in four over par and throwing away his chance at a fifth Open, Ben remembered that Jack had lost three strokes over the last six holes and lost to Palmer by only two. Sitting in front of his locker, Hogan said, "I played thirty-six holes today with a kid who if he had a brain in his head should have won this thing by ten strokes."

While he attended Ohio State University, in his home town of Columbus, Nicklaus also sold life insurance, even though normally he would not have been allowed to under the law, which required a person to be twenty-one years of age to have a license. Nicklaus was given a special exemption at twenty, thus becoming the youngest person licensed to sell insurance in Ohio. He earned about $25,000 selling insurance, all the while attending school, "A pretty good living for a kid," he admitted.

Will Power

A few weeks after he had beaten Arnold Palmer in the eighteen-hole playoff for the 1962 Open championship, Jack saw something he didn't like in the highlights film. Seeing himself standing over a critical putt with a cigarette dangling from his lips created such a poor image and offended him so badly he determined he would never smoke on the golf course again. He held to it, even though he continued to smoke in private for years afterward.

Some time later he talked about it with Ken Bowden, a former editor of *Golf Digest* who wrote articles for him under the Nicklaus name. A smoker himself, Bowden couldn't understand how Nicklaus could go so long without a cigarette. Jack had the perfect answer. "Kenny," he said, "I just don't think about it."

Among the leaders going to the eighteenth hole at Pebble Beach during the 1966 Bing Crosby Pro-Amateur, Nicklaus hooked his drive into the Pacific Ocean, teed up another ball, and played only a slightly milder hook, one that landed on the beach. While Jack stared down at the beach from a sheer bank, George Walsh, a rules official, walked over to help. Suddenly the ground gave way under Walsh and he plunged to the beach.

As he wallowed in the ocean, swept gently seaward by the ebbing tide,

Nicklaus, not knowing anything better to do, waved and called "Bon Voyage."

Come Ride with Me

After becoming the first man to win successive Masters, in 1965 and 1966, Nicklaus missed the cut in 1967. Naturally he felt down and disappointed. Seeing his reaction, Cliff Roberts told him, "I want you to ride with me in my cart tomorrow. I want the people here at Augusta to see you and get to know what you're like the same way I've known you these years."

Recalling Roberts's kindness, Nicklaus said later, "I was deeply moved by that gesture, especially since Roberts had a tournament to run. He knew how keenly disappointed I felt, and that I was very dejected."

Doug Sanders was giving Nicklaus all he could ask for in the playoff for the 1970 British Open at St. Andrews. They came to the eighteenth tee with Nicklaus only one stroke ahead. Obviously Sanders might birdie, so Nicklaus had to make something happen. The last hole measures just over 350 yards. Looking it over, Jack felt that with a following wind he might reach the green, but he would need a little more freedom to swing his arms. Jack pulled off his yellow sweater, lashed into the ball and hit a mighty drive. His ball almost hit the pin. It ran over the green and nestled in the rough behind. Jack chipped within six feet and holed the putt.

As the ball tumbled into the hole, Nicklaus flung his putter high overhead. When it came down it barely missed hitting Sanders on the head. Jack shot 72, Sanders 73.

Touché

Jim Thorpe—the golfer, not the great Indian athlete of the early twentieth century—was about to win the 1985 Milwaukee Open, his first victory as a professional. He was paired with Nicklaus in the last round and led Jack by three strokes as they walked down the eighteenth fairway.

Realizing anything could happen, Jack tried to shake Thorpe's confidence and upset his composure. "How does it feel to be walking down the last fairway with a three-shot lead over the greatest player the game has ever known?" he needled. Without missing a step, Thorpe smirked, "It feels like you can't win."

Luck of the Draw

After a blind drawing to assign starting times for the 1988 British Open, Michael Bonallack, the secretary of the R and A, announced that Nicklaus would begin his first round at 7.48 a.m., his earliest starting time in his seventeen appearances. With a sheepish grin, Bonallack explained, "The draw was being televised, so it wasn't possible for us to say 'Oops' and drop his name back into the hat." Ironically, Nicklaus had complained bitterly in previous years about starting late in the afternoon, after earlier starters had left spike marks all over the greens.

Don't Press Your Luck

During a practice round leading up to the 1988 British Open, Nicklaus tossed three balls into a bunker beside the fifteenth green at Royal Lytham and St. Annes. He holed the first ball, and for an encore holed the second. He picked up the third. When someone called, Why didn't you try the third?" he grinned and answered, "You've got to be kidding."

Professor Nicklaus

Coming across Arnold Palmer working on his ailing game with Jim Flick, a teaching professional, after Palmer had opened the 1993 Tradition by shooting 76, Nicklaus soon took over as the instructor. Fifty-three at the time, Nicklaus told Palmer, sixty-three, to bend over from the waist before bending his knees. Within minutes Palmer was driving the ball thirty yards farther than he had been before Jack showed up.

Lee Trevino

Before he joined the pro tour, Lee Trevino made his living playing money matches around Texas. He worked for a time at a pitch-and- putt course and took on all comers using a quart-sized Dr. Pepper bottle wrapped with adhesive tape. He rarely shot over 30.

In the early 1960s, while Lee was working as an assistant pro at the Horizon Hills Golf Club, in El Paso, Martin Lettunich, a wealthy cotton farmer and enthusiastic golfer, set up money matches for him. Lee played one against Fred Hawkins, a consistent money winner during the 1950s; Trevino beat him. In the best-known of these matches, Lettunich set him up against Raymond Floyd, even then a well-known tour player. As Floyd drove up to the Horizon Hills clubhouse in his white Cadillac, Trevino rushed out to greet him and asked politely if Raymond wanted his clubs taken to the golf shop. Floyd said that indeed would be fine. Trevino led him to the locker room, unpacked his bag, and took his shoes and cleaned them. Floyd paid him, then asked, "Who 'm I playing?"

"You're looking at him," Trevino said.

"You?" Floyd exclaimed, looking over this short (5'-7"), pudgy (180-pound) scruffy-looking bag boy with nut-brown skin, coal black hair, and white even teeth. "You mean they bet on you?"

When someone asked Floyd if he would like to check out the course, Raymond yelled, "Hell, no. I'm playing this locker room guy. I don't need to look at no course."

The match was set up for fifty-four holes over three days. Floyd shot 66 the first day and Trevino 65. Floyd shot 66 again the second day and Trevino shot 64. The next day Floyd eagled the last hole and beat Trevino by one stroke. As he drove away, Floyd leaned out his car window and said, "I'll see you all later; they're easier games than this on tour."

Returning to the tour, Floyd told the players,"Boys, there's a little Mexican kid out in El Paso. When he comes out here, you'll have to make room for him."

Friendly Warning

Twice on his way to winning a U.S. Open, Trevino prevented an opponent from committing a rules violation. The first took place on the eleventh green of the Oak Hill Country Club during the third round in 1968. Bert Yancey had marked his ball, then moved it a putter-head's length to the side, off Trevino's line. Lee holed his birdie putt. In the excitement of the moment, Yancey replaced his ball without moving his coin back to the proper spot. Noticing Yancey's mistake, Trevino called out, "Hey; did you move your coin back?"

"Uh," Yancey stammered, "no."

Had he putted from the wrong spot, Yancey would at best have been penalized two strokes. If he didn't account for the penalty, he would have been disqualified. As it was, he went into the last round leading Trevino by one stroke. Trevino outplayed him and won the first of his two Opens.

Three years later Trevino was locked in a battle with Jim Simons, a twenty-year-old amateur, at Merion. Leading after fourteen holes of the third round, Simons teed his ball on the fifteenth while Trevino watched. A veteran by then, Trevino looked at Simons's ball and quietly said, "Son, you're ahead of the markers. This game's tough enough without a two-stroke penalty." Trevino beat Nicklaus in a playoff. Simons placed third.

Helpful Advice

Looking over an important putt, Trevino turned to his caddie for advice, asking, "What do you think?" Herman Mitchell, the overweight caddie, pondered the line for a moment, then said, "Keep it low."

Beman at Baltusrol

After winning two U.S. Amateurs and one British Amateur, Deane Beman turned professional in 1967. He was already twenty-nine, rather late to start. He made his professional debut in the U.S. Open, at Baltusrol, New Jersey.

One of the shortest hitters ever to play professional golf successfully, Beman made an ostentatious debut. At slightly under 500 yards, Baltusrol's first hole plays as a par 5 for members. It is shortened to about 475 yards for the Open and the par changed to 4. When Beman stepped onto the tee there was genuine concern that he might not be able to reach the green with his second shot. Those who wondered didn't take into account Beman's fierce fighting spirit.

As it developed, Deane Beman owned the first hole. On the first hole he ever played as a professional, Beman holed a full-blooded 4-wood for an eagle 2. The following day he birdied. The day after that he birdied again. In three rounds he had played that hole in eight strokes against a par of 12. Reality set in in the fourth round. He parred.

Stay Alert

George Haines, an amateur from New Jersey, had just shot 76 at Oak Hill and missed the cut in the 1968 U.S. Open. Sitting in the scorer's tent beside the eighteenth green, he was going over his card with P. J. Boatwright, then the assistant director of the USGA. Haines called out his score on each hole while Boatwright checked the card. Haines called out a 5 for the thirteenth hole. Boatwright stopped him. "Wait a minute," he said, "You have a 4 on the card." With a sheepish smile Haines said, "I just wanted to see if you were paying attention."

Roberto De Vicenzo

Roberto De Vicenzo celebrated his forty-fifth birthday on April 14, 1968, the last day of the Masters. From the beginning of the round, De Vicenzo was buoyant. The gallery cheered his eagle 2 on the first hole by singing "Happy Birthday, dear Roberto."

He could do nothing wrong. He played the first nine in 31, five under par, and his gallery grew as he built up momentum. He reached six under par when he birdied the twelfth, and he fell to seven under par with still another birdie at the fifteenth. A stunning approach to the seventeenth stopped just five feet from the hole, and he ran the putt home. Eight under for seventeen holes, he was on his way to a round of 64, but he bogeyed the eighteenth. Even so, he shot 65 for the day, 277 for the 72 holes, and tied Bob Goalby for first place.

Millions had watched the Masters on television and had seen De Vicenzo's 65. Or so they thought. Excited by the pressure of the competition and the crowd's reaction, De Vicenzo signed his scorecard without noticing that Tommy Aaron, who kept his score, had written 4 instead of 3 as his score for the seventeenth. Under the rules, Roberto was stuck with the 4 no matter how many fans in person or on television has seen him birdie the seventeenth; instead of 65 and 277, he had shot 66 and 278.

After Roberto left the scorer's table, Aaron noticed his mistake and told the official responsible for checking the scores. The official passed word to the rules committee, and one of them intercepted Roberto and brought him back to the table while the rules committee determined if the facts warranted an exception to the rules. No precedent covered this situation. They took the matter to Bob Jones, bedridden by then. Ike Grainger, John Winters, and Hord Hardin, all former presidents of the USGA and Augusta National members, explained what had happened. Jones listened, then asked, "What is the rule?"

Grainger read it to him: *If the competitor returns a score for any hole lower than actually played, he shall be disqualified. A score higher than actually played must stand as returned.* Jones looked at the three men and said, "Well, that must be the decision."

Grainger told De Vicenzo the bad news. Roberto refused to blame Aaron, who was quite upset, and he accepted full responsibility. He had dinner with Grainger that evening, and at the end Roberto told Grainger in his somewhat imperfect English, "I sorry I cause you so much trouble." Later he told the press, "I am a stupid."

Two years later, a week or so after De Vicenzo had received the Bob Jones Award for distinguished sportsmanship in golf, the Golf Writers Association of America presented De Vicenzo with the William D. Richardson Award for significant contributions to the game. As he accepted the Richardson award in the packed ballroom of the Americana Hotel, in New York, Roberto stood before the audience fondling the trophy and smiling in a strange way. Finally he turned to the microphone and in his slightly broken English said: "Golf writers make three mistakes spelling my name on trophy. Maybe I not the only stupid."

Success at First

Visiting Pine Valley for the first time, in the spring of 1968, Thomas S. Fothringham, a Scot who had recently played himself in as Captain of the Royal and Ancient Golf Club of St. Andrews, found the first nine crowded. Since he was to play with John Arthur Brown, the despotic president of the club, J. Ellis Knowles, a very well-known senior golfer, and Captain Pliney Holt, the four of them had no trouble in skipping over to the tenth, the devilish little par 3 with the bunker not only shaped like an inverted cone but so deep a player can disappear into the bottom.

Aware of the perils, Fothringham teed up his ball and stared long and hard at the job ahead of him. But he was not Captain of the R and A for nothing. His tee shot—his first at awesome Pine Valley—plunged into the cup, the first hole-in-one of his life.

Jim McKay, the television announcer, heard the story, and when he met Fothringham a few days later asked, "Do you still have the ball?" Alas, not a sentimentalist, Fothringham had to answer, "No. I lost it in the water on the fourteenth," an equally difficult par 3 over a pond.

Orville Moody

Orville Moody had driven into a fairway bunker on the eighteenth hole of the Lakewood Country Club during the second round of the 1968 New Orleans Open. Facing a difficult shot from a downhill lie in the sand, he turned to a couple of spectators and asked if they had been watching the scoreboard. They replied that they had indeed been watching carefully. "What's it going to take to make the cut?" he asked. W. K. Springer, one of the spectators, said, "My guess is 143." Moody grimaced and said, "Damn. I'm 142 right here. I'd better hole this one."

Orville set himself up carefully, swung into the ball with a fluid, measured tempo, and sent the ball soaring dead on line toward the flagstick. Gazing after the ball, the spectators felt Moody might indeed hole the shot. The ball hit the green, took one bounce toward the hole, hit the pin, and settled no more than six or eight inches away. With 144, he missed the cut by one stroke.

Part Choctaw Indian, with a round, moon face, thinning dark curly hair, a perpetual puzzled expression, and a roll of fat hanging over his belt, Moody had spent fourteen years in the army and risen to sergeant. Leaving the army, he had joined the tour in 1967 but showed nothing to lead anyone to believe he might win something of merit. He shocked everybody by winning the 1969 U.S. Open, played at the Champions Golf Club, in Houston.

An extraordinary ball-striker, Moody nevertheless ranked as the worst putter who ever won the Open. Using an awkward cross-handed grip, he stroked putts that never once looked as if they might fall. Toward the end of the last round in 1969, he floated still another superb iron shot into the sixteenth green of Champions that braked about six or eight feet from the cup. Another putt ran wide of the hole. Wearing a dejected expression as he walked off the green, Moody spotted his pretty wife. He said to her, "Honey, I'm hitting the putts real good but I can't buy one." Tears welling in her eyes, she followed behind until he won.

After drilling a big iron into the final green, Moody wanted to be absolutely certain of how he stood. He knew only one way to find out for sure. With his putter in his hand he stepped off the green and into the little tent set up for players to check and sign their scorecards after each round. Stepping up to Frank Hannigan, the USGA's assistant director, Moody asked, "What's the low score in?" Hannigan answered, "You have two putts for the championship." Moody used them both.

After checking his scorecard Moody retreated into the clubhouse to make himself ready for the presentation ceremony, which was to take place rather quickly. The gallery gathered around, committee members took their seats, and the USGA president stood at the microphone waiting to begin the ceremony. Everything was ready except for the champion.

Someone rushed back into the clubhouse and yelled, "Where the hell is he?" A group of marshals charged with escorting him to the prize ceremony pointed toward an office and answered, "He's on the phone." He was indeed, and saying things like, "Yes, sir," and "Thank you very much, sir." He was not about to cut this telephone conversation short; he had answered a call from Richard M. Nixon, the President of the United States.

Mr. Nixon commented that not many people serve so long in the army and then become the national golf champion. "No, sir," Moody answered, "I'm the only one."

Never brimming with confidence, Moody had almost failed to qualify. With nine holes to play he needed two birdies he didn't think he could make.

About to give up, he told Bobby Cole, his playing partner, "I don't mind not qualifying because I wouldn't have a chance in the Open."

"Don't be stupid," Cole snapped. "Keep trying."

Moody made the last birdie he needed by holing a bunker shot on the thirty-fifth hole.

Even in victory, Moody remained a humble man. Some weeks later, Frank Chirkinian, a television producer who arranged teams for a series of televised matches, recruited Orville and asked whom he'd like for a partner. Chirkinian suggested Gene Littler, the 1961 Open champion, who owned one of the finest swings ever seen. Surprised at Frank's choice, Moody refused, saying, "Oh, no, Frank. Gene would never want to play with me. None of those good golfers would."

"But you're the National Open champion," Chirkinian argued. "Believe me, they'll play with you."

"No, Frank, " Orville insisted. "Why there's a sergeant down on the base who can beat me five days out of seven."

Where's the Bar?

In contention after three rounds, Miller Barber played Champions' twelfth and thirteenth holes in 5 and 6 against a par of 3 and 5, finished the round in 78, and dropped to sixth place, at 284, three strokes behind Moody. Sprinting from the eighteenth green, he tore into the locker room, almost knocked over a policeman, and instead of apologizing snarled, "Get out of my way. I just lost the National Open." When a friend asked if he wanted a drink, he snapped, "Hell, yes. I might as well get drunk; I can't play golf."

Understatement

Deane Beman shot 68-69-73-72—282 at Champions and shared second place with Al Geiberger and Bob Rosburg, just one stroke behind Moody. During the last round Beman needed wooden clubs to reach the greens of eight par-4 holes and one par 3. Nevertheless, two of his three birdies followed wood shots to greens, and the other followed a 2-iron shot. He birdied the first with a 4-wood to fifteen feet, the fourteenth with another 4-wood to nine feet, and the fifteenth with a 2-iron to three and a half feet. Stepping out of the scorer's tent following the round, Beman made the understatement of the week. "I didn't waste too many shots today," he said.

No Consideration

Joe Carr, the Irish amateur, played a practice round with Julius Boros and Gene Littler early in the week of the 1969 Masters. They had played only a few holes when a blue Lear Jet buzzed the course. As it zoomed toward the airport, Boros looked up and said, "Well, Arnold's here."

A little later in the round, at about the thirteenth hole, a red jet buzzed the course. "Jack's here," Boros announced.

Then, as they putted on the eighteenth hole, a propeller plane flew lazily overhead and headed for the airport, not far away. Looking up again, Boros said, "You'd think that with all their money they'd get their caddies here ahead of time."

Inspiration?

Even par after three holes and two strokes ahead of Vinny Giles in the fourth round of the 1969 U.S. Amateur championship, Steve Melnyk played two strong shots and had reached a greenside bunker on the fourth hole at Oakmont, a par 5. Stepping into the bunker with his wedge in his hand, Melnyk had set himself up for the shot, but before he played it he took one last look at the flagstick. By accident he also saw Arnold Palmer watching him. Palmer had come to see who would join him as an Amateur champion.

Melnyk had met Palmer briefly at an All-American college dinner in New York a few weeks earlier, but he didn't know him at all well. Seeing him in the gallery gave Melnyk such a bad case of nerves he had to climb out of the bunker and settle himself. Calm again, he climbed back into the sand and holed the shot for an eagle 3, then played the next five holes in birdie, birdie, par, par, birdie—six holes in five under par. By then no one could catch him. He held an eight-stroke lead with nine holes to play, shot 70 for the round, 286 for the seventy-two holes, and beat Giles by five strokes.

Gary Player

Gary Player's experiences at Dayton, Ohio, during the 1969 PGA Championship, at the NCR Country Club, had to represent the most abuse a player has ever had to endure during a major golf championship. A small group of radical agitators tried to disrupt not only his game but also Jack Nicklaus's, who was paired with him in the third round. Even a force of 400 policemen stationed on the course couldn't control some of the more vicious demonstrators.

As Player took his stance on the fourth tee, someone tossed a program over the heads of the gallery standing in front of him. It fell at Player's feet. Without saying a word he stepped away. Someone removed it and he played his shot. As Nicklaus was about to putt on the ninth, someone yelled at him. He stepped away, then missed the putt.

As the two players left the ninth green and walked toward the tenth tee, a bearded man threw a cup of ice in Player's face.

Astonished, Player confronted the man and asked, "What have I ever done to you?" The man shouted, "You're a damned racist."

The atmosphere grew worse. As Nicklaus crouched over an eagle putt on the tenth green, a black man charged from the crowd directly toward him. Jack raised his putter to defend himself, and then the police moved in. They arrested eleven troublemakers.

Nicklaus never figured in the championship again, but Player maintained his concentration and finished second to Raymond Floyd, the winner.

The first time he ever played a round of golf, Gary Player parred the first three holes. Later, in a round with his wife, Vivienne, Gary shot a round of 71. Vivienne beat him.

"That's when I decided to have six kids and keep her from practicing," he claimed. Vivienne once shot two holes-in-one in the same round and had a third possible ace pull up an inch short. "Beat that," she told Gary. A year later Gary scored an ace and won $100,000 while Vivienne was watching. When the ball dived into the cup she ducked under the gallery ropes, ran to him, and gave him a big kiss. Never without a comeback, Player asked Vivienne, "Would you rather have your two or my one?"

A faddist who at various times dressed all in white to reflect the sun's heat, then changed to all black to absorb the sun's heat and keep the muscles supple, Player seemed uncertain which way to jump when he won the 1959 British Open. He stepped onto the hallowed turf of Muirfield's first tee wearing trousers with one white leg and one black leg. During his all black stage, he was asked truly why he wore black. He confessed, "It's just a trademark."

Player often followed bizarre diets—or at least ate foods said to have energy-inducing properties. He ate raisins at one time, and bananas at others, chewed candy bars and wheat germ, and carried a jar of honey in his golf bag, pausing occasionally to take a sip for a quick charge of energy.

In 1968, while he was going through his bananna stage, he had a caddie named Wally Armstrong, who later had a short career as a player. Player carried his bananas in his golf bag and Armstrong was given the responsibility for handing him one whenever he was asked. Once Wally stuffed a hand of bananas into the bag and forgot about them between tournaments. They remained there ripening for days.

Then came the fateful last round of the New Orleans Open. Player and Armstrong stood on the twelfth tee while stormclouds closed on on the golf course. Not wanting to get wet, Player told Armstrong to give him his rainsuit.

Armstrong opened the pocket of the bag where he had stuffed the rainsuit. This was also the pocket where he had stuffed the bananas so long ago. The aroma brought tears to the eyes.

Catching a whiff of the bouquet, Player frowned and demanded, "What's in the bag, Wally?" Armstrong reached down into the gook and recalled, "It was like black tar." Besides the stench, all the mush and goo from rotten bananas had congealed onto the rainsuit. "Oh, Wally, Wally," Player moaned. When the round ended, Wally found himself unemployed. Player had fired him.

Relaxed

Leading the 1969 British Open after three rounds, Tony Jacklin sat up late watching a television movie rather than going to bed and "playing around Royal Lytham and St. Annes 100 times in my imagination and getting no sleep." Jacklin had taken a sleeping pill, though, and fell asleep in an armchair, a cigarette dangling from his fingers. Bert Yancey, an American friend also playing in the championship, was staying in the same house. Finding Jacklin asleep, Yancey slung him over his shoulder, carried him upstairs, undressed him, and slid him into bed alongside his wife, Vivienne.

Mistaken Identity

A woman spectator collapsed in Pensacola, Florida, during the third round of the 1969 Women's Open at the Scenic Hills Country Club. Frantic calls for assistance flew across the airways. One call stated a doctor could be found near a television tower beside the sixteenth green, since he had been assigned to help as a spotter for the announcing crew.

Scrambling into a golf cart, the USGA official sped to the tower and seeing a man perched at the base called, "Doctor." The man on the tower responded and hopped into the cart. They raced down a service road toward the eighteenth green, where the woman had swooned. While the USGA man explained the situation, his passenger looked at him strangely and said, "Just a minute. You're making a mistake. I'm a Ph.D." Fortunately, the stricken woman had revived by then and was on her feet, apparently undamaged.

George Knudsen

George Knudsen, a free-spirited Canadian, had done rather well over the first three rounds of the 1969 Australian Masters tournament, played at the Victoria Golf Club, in Melbourne, and decided he deserved a night on the town. His celebration climbed to man-sized proportions.

The next day, well along in the fourth round, George suddenly collapsed on the course. With thoughts of cardiac arrest flashing through their minds, medics rushed to him and pumped him full of oxygen. Knudsen revived, claimed that all was well, continued his game, and won the tournament by two strokes.

The 1970 U.S. Open

No one could remember stronger winds than those that whipped across Hazeltine National Golf Club, near Minneapolis, during the first round of the 1970 U.S. Open. They roared out of the northwest at thirty-five miles an hour at their weakest, and gusted over forty. They blew with such strength they almost uprooted a massive scoreboard anchored in place by six-by-six-inch pilings driven four feet into the ground.

Drives covered ridiculous distances. Playing downwind, Jay Dolan, a club professional, drove his ball 318 yards and reached the second green with his second shot. The hole measured 585 yards. Scores were as ridiculous as driving distances. Jack Nicklaus hit only four greens on the first nine and shot 43.

The scene in the locker room was part tragedy, part comedy. Tommy Bolt stormed through the door, his face a raging crimson, his eyes glaring, and his teeth bared, like an attack dog that had cornered his prey. Watching his march toward his locker, everybody drew a breath and waited. Bolt looked straight ahead. He flung open the door to his locker, read a message, then slammed it shut.

Then, his face burning red, nostrils flaring, his eyes blazing, Tommy Bolt erupted. "All right, dammit," he shouted, "I shot 80. Now haven't you newspaper sons-of-bitches got something to do on the golf course? If it's too cold for you, I'll lend you my jacket."

With 69 in the second round, Dave Hill had climbed into second place, with 144, three strokes behind Tony Jacklin. Hill didn't like Hazeltine, a fairly new course designed by Robert Trent Jones, and he wasn't shy about his feelings. In a press interview following his 69, Hill was asked if he liked the course better now that he had moved into contention. "No, sir," he answered. "If I had to play this course every day for fun, I'd find me another game."

Someone else teased, "What does it lack?" His tongue firmly implanted in his cheek, Hill parried, "Eighty acres of corn and a few cows. They ruined a good farm when they built this course."

"What do you recommend they do with it?"

"Plow it up and start all over again. The man who designed this course held the blueprints upside down. My two boys could have done as well. If I didn't have some friends here with me, I probably would have left Tuesday."

"What do you think of your chances?"

"Oh, I'll win," Hill predicted, "but I won't enjoy winning on this course. It discourages great golfers like Palmer and Nicklaus."

"Why doesn't it discourage you?"

"I'm not smart enough. I just hit the ball, go find it, and hit it again."

Not only did Hill not win (he finished second, eight strokes behind Jacklin), he was fined $150 by the PGA Tour as well.

As first prize, Jacklin was awarded $30,000 at the presentation ceremony. Later in the evening he had dinner with some friends, then flew home to England the following morning. As one of his first acts once he reached home, Tony sent his clothes out to be cleaned. Jacklin had stuffed his $30,000 check in the pocket of his suit. That is exactly where the cleaner found it.

Even though he didn't like Hazeltine, Hill deposited the $15,000 check for second place in his bank. Some days later a letter arrived at USGA headquarters.

"Gentlemen:" it began."Enclosed is the check Dave Hill received for second place. We are returning it for signature."

Sure enough, the check hadn't been signed. Some say it was deliberate.

After shooting 79 in the first round, Sam Snead tried to sneak away without signing his scorecard, which would have caused him to be disqualified, the only way he could avoid coming back again the next day. Realizing what Snead was up to, Lee Trevino called, "Sam, come back; you forgot to sign your scorecard." Trevino then muttered to Frank Hannigan, the USGA official checking scorecards in the scorer's tent, "If I've gotta come back here tomorrow, he's coming back, too."

A Long Long Shot

Back in 1928, the year he won his third British Open, Walter Hagen boasted that he hit a ball from the roof of the Savoy Hotel, in London, across the River Thames. Early in 1970, the year after he won the British Open, Tony Jacklin climbed onto a platform on the Savoy roof, took a ball from a silver tureen, where it was kept warm for maximum resiliency, and flailed away. He, in fact, flailed several times.

From the vicinity of the Savoy, the Thames stands at least 400 yards wide—maybe a little more. Jacklin's best effort carried an estimated 365 yards. If Hagen had actually driven across the Thames, he remained the only one who had.

Merion in 1971

Merion Golf Club, on the outskirts of Philadelphia, ranked as the second shortest Open course in the years after the Second World War. Following a practice round leading up to the 1971 U.S. Open, Lee Trevino called it a nice little course, but "If you keep the ball down the middle, you've got to burn it up."

Two days later, after more exposure to Merion's subtleties, he confessed, "This is the hardest course I've ever seen. I don't see how anybody can break par."

Backing up Trevino, George Archer said, "Ninety-seven percent of the field, myself included, are not equipped to play this course. We just don't have the shots. I heard Byron Nelson say on television that 272 could win here. I couldn't shoot 272 if I got a mulligan on every hole."

Trevino had bought a toy snake to Merion. Early in the week, after one of Lee's rare excursions in the rough, a photographer shot a picture showing him standing in deep grass holding the snake draped over a club.

After he and Nicklaus had tied for the championship at 280, even par, they stood on the first tee waiting to begin their playoff. As Lee rooted through his bag, Nicklaus spotted the snake hanging out of one of the compartments. Laughing, he called across the tee asking Lee to let him have a look at the snake. While the gallery gasped, Lee flung it to him. The gallery broke up laughing.

Later, after Nicklaus played badly and lost the playoff, Trevino was accused of gamesmanship for tossing Nicklaus the snake.

Worthwhile Injury

From a bad lie on the final fairway as he battled Lee Trevino through the final stages of the 1971 British Open, at Royal Birkdale, Lu Liang Huan, the Taiwanese golfer who became affectionately known as Mr. Lu, hooked his approach into the gallery and felled Mrs. Elisabeth Tipping, a spectator. Lu needed a birdie there, but he parred the hole and finished second to Trevino.

Three years later, when the championship moved on to Royal Lytham and St. Annes, Lu came across the Tippings once again and invited them to an all-expenses-paid trip to Taiwan.

Extra Holes

The British and Irish celebrated long and hard following their 13-11 victory over the United States in the 1971 Walker Cup match. The players had hardly tucked themselves into bed when at about three in the morning, they were aroused and led onto the first tee of the Old Course at St. Andrews to replay their great moments.

Navigating at best unsteadily, they played many a badly struck ball down the wide fairway while barely conscious pursuers wove their zigzag patterns toward their target. David Marsh, one of the heroes of the day (his one-hole victory over Bill Hyndman had cemented the match for the British/Irish team), felt so full of himself he believed he could fly. Approaching the Swilcan Burn, he disdained detouring far enough right to cross by way of the footbridge and instead tried his own version of the Great Leap Forward.

A stream so narrow it can be stepped across by elderly ladies wearing tight skirts was simply too much under the circumstances, and Marsh plunged headlong into the stream bed.

Other farcical events followed. Rodney Foster became so disoriented that, facing collapse, he wandered off in search of the private home where he was being put up for the week. Failing to home in on its flickering beacon, he took the old seaman's policy of seeking anchorage in any port in a storm and climbed to the front door of a strange house. He found the door unlatched, crept inside, sought and found an unmanned bedroom, and quickly drifted into slumber.

The lady of the house found him zonked out the following morning and demanded to know what the hell he was doing there. When she learned he had played on the winning team, she fed him breakfast—most likely what they call a full Scottish breakfast.

The New Math

As customary, the organization televising the 1971 Lancomb Trophy tournament, an annual affair in Paris featuring an eight-man field, prepared a special scoreboard for the telecast. As technicians posted the names of the players and their standing with par, the man preparing the board asked a spotter, "How does Palmer stand?"

"Minus 12," he was told, and he placed -12 on the board.

Moving on he asked, Player?"

"Minus 10."

"Roberto De Vicenzo?"

"Zero," came the answer.

Pondering for an instant, the technician asked, "Is that a minus zero or a plus zero?"

Trevino at Muirfield

Going into the last round at Muirfield in the 1972 British Open, Jack Nicklaus trailed Lee Trevino by six strokes, a lot of strokes to make up on a player of Trevino's quality, but playing two holes ahead, Jack ran off a series of birdies and pulled even by the ninth and moved ahead after the tenth.

Standing on the tee of the ninth, a par-5 hole, Trevino turned to Jacklin, his playing partner, and said, "Look at that. Jack's gone crazy. We're beating each other to death and that son of a gun has caught and passed us." Then, speaking to his caddie, Trevino said, "We're behind, son. Give me that driver. We've got to make something happen." He ripped a solid drive and drilled a 5-iron onto the green.

Jacklin followed. Since Trevino's ball lay farther from the hole, he would putt first. Nicklaus, meantime, stood on the eleventh green, about to putt for still another birdie, his sixth of the round. Before Nicklaus could putt, though, Trevino rolled his ball into the cup: an eagle 3. The crowd roared, and Nicklaus stepped away. Jacklin next. Again the ball found the cup. Another eagle. Again the crowd roared. Nicklaus stepped away once more. Trevino quipped, "I guess that will give Jack something to think about."

Just then Nicklaus holed his birdie, and *his* gallery thundered its applause. Up to the challenge, Trevino noted, "I think the man has given *us* something to think about."

By the time Nicklaus finished his round, with 66, both Trevino and Jacklin had passed him and stood one stroke ahead. Muirfield's seventeenth is a par 5 of 542 yards. Jacklin played two solid shots within a few yards of the green, but Trevino played the hole badly. He drove into a fairway bunker, could do nothing more than play a safe shot out, then hooked his third into heavy rough short of the green. Discouraged, and thinking he had given the championship away, Lee hit a careless looking shot over the green.

After Jacklin chipped on, perhaps fifteen feet from the cup. Trevino stepped up to his ball, and without any thought or hesitation, played a hasty chip.

Earlier in the week Trevino had thinned a bunker shot on the sixteenth that hit squarely against the flagstick and dropped into the cup for one birdie, and had chipped into the hole from behind the eighteenth for another. Once again, Trevino's ball dived into the hole. He had his par 5. Possibly unnerved by watching Trevino hole three miracle shots, Jacklin three-putted, and Trevino won his second consecutive British Open.

A Record Double-Eagle

Double eagles, holes played in three under par, are rare enough, but Bill Graham's 2 on the sixteenth hole of the golf course at Whiting Field, an auxiliary naval air station located in Milton, Florida, near Pensacola, in 1972, was something special. A strapping lad of twenty-seven years who stood six feet five and weighed 240 pounds, Graham crushed his drive and sent it screaming 325 yards, then nailed a 3-wood the rest of the distance, into the cup. The hole measured 602 yards, the longest double eagle anyone knows of. Graham, of course, was helped along by a 25-mile-an-hour tailwind.

Aerial Hazard

Airports and golf courses have never been compatible neighbors; jet planes screaming overhead unsettle the concentration of even the greatest players. In Birmingham, England, the inconvenience worked two ways. The course shared a common boundary with the airport. Golfers being what they are hit the occasional shot over the fence; what to do but climb over and recover the ball? But that's where the other trouble came in. Whenever a golfer climbed this particular fence he was picked up as a foreign object on the base radar, shutting down the airport's entire instrument-landing system.

Her Best Shot

While it didn't look like it at the time, Dinah Oxley's wild drive may have been the best shot of her career. Playing a practice round leading up to the 1971 British Women's Amateur, at Alwoodley, near Leeds, England, Miss Oxley, a member of the British Curtis Cup teams of 1968 and 1970, slashed a shot that veered well off line, soared past the club boundaries onto a nearby road and onrushing cars. The ball shattered the windshield of a car driven by James Henson, a pilot with what was then the British Overseas Airways Company.

Understandably shaken, Henson pulled off to the side of the road. Miss Oxley rushed up, found Henson hadn't been injured, and invited him into the clubhouse to recover. They became friendly. Then they became more friendly. Then they married.

United We Stand

Midway through the last round of the 1972 U.S. Open, three bearded young men slipped under the gallery ropes and chained themselves to a huge Monterey pine in the eighteenth fairway at Pebble Beach to demonstrate against the war in Viet Nam. Feeling sure they would stand silently and not attempt to interfere with play, the Pebble Beach police force let them stay. If they expected their protest to be seen by anyone other than nearby spectators, however, they were disappointed. ABC television network policy forbade showing these situations unless they influenced the outcome.

Meantime, Arnold Palmer was back on the tee waiting to play. Not sure of what was going on, he asked a rules official for his binoculars. Television showed him with the glasses but didn't explain he was looking at the protesters. The following week the USGA was flooded with calls claiming that Palmer should have been penalized for using an artificial aid to help him judge distance. The USGA explained that normal binoculars have no range-finding feature; Arnold had broken no rule; he was looking at people chained to a tree.

Sarazen at Troon

Seventy-one years old in 1973, Gene Sarazen made one last trip to play in the British Open. He made it memorable. The championship was played that year at Royal Troon, a course than runs alongside the Firth of Clyde on the bleak west coast of Scotland.

Its eighth hole, a short par 3 of just 126 yards, ranks as the shortest hole in championship golf. Known as the Postage Stamp, it is Troon's best-known hole. It was called Ailsa at first—after the granite rock Ailsa Craig that rises from the sea bed some miles off the shore—but one year when Troon had been burned yellow by the drought, a visitor marveled at the lovely green color of the well-watered putting surface and said, "It looks like a postage stamp." The connection was made because Britain's most common postage stamp of the time was green. The name stuck.

Although he missed the thirty-six-hole cut, Sarazen played the Postage Stamp twice without using his putter. Playing with Fred Daly, an Irishman, and Max Faulkner, an Englishman, both former champions, Sarazen squirted a half-5-iron into a slight breeze in the first round. The ball landed dead on line, hopped a couple of times, and jumped into the hole for a hole-in-one.

The next day, as the crowd encouraged him to do it again. Gene misplayed his tee shot into a bunker, and then, miraculously, holed his bunker shot. In two rounds he had used only three strokes. The next morning a London newspaper proclaimed, "Sarazen Licks Postage Stamp."

Strangely enough, another player had needed just three strokes in two days on the eighth. David Russell, a young English amateur also holed in one the first day, although *his* ball bounced off a mound rising on the left and curled down and into the hole. The following day he played a remarkably fine tee shot and made an orthodox birdie 2.

Lost Crib Sheet

By the 1970s the modern professional golfer had become so dependent on knowing the exact yardage not only to greens but to the exact pin placement that they were nearly helpless without their notes. At the time of the 1973 U.S. Open, Johnny Miller had measured distances at the Oakmont Country Club by his own stride and noted them on a a card he carried in his pants pocket while he played, and then stored it overnight in a zippered pocket on his golf bag.

On the evening after the second round, after he had shot 69 and climbed into a tie for third place, Miller slipped up. He forgot to put his notes in his golf bag, took them with him to the house where he and his wife, Linda, were staying, and placed them on the nightstand when he went to bed.

Rushing to the golf course the following morning, he didn't notice he had left his notes behind until he reached the first tee. Realizing what he had done, he said, "I broke into a cold sweat."

One of the finest strikers of iron clubs the game has known, he didn't know what to do. He couldn't rely on his caddie because the USGA insisted players use caddies supplied by the club, and they weren't familiar enough with the individual golfers to offer reliable advice. With no other recourse, Miller had to ask his caddie for help nevertheless. Meantime, he sent Linda after his yardage card.

With no reliable reference, his whole game suffered. Nervous, unsure of what he was doing, and indeed choking a little because he was so close to the lead, Miller struggled through the early holes. He overshot the first green and left his approach to the second short, bogeyed both, double bogeyed the sixth, and stood five over par after seven holes, thinking he might shoot in the eighties.

After an hour's drive, Linda arrived back at Oakmont with Johnny still struggling through the first nine. Once he had his yardages, his game turned around. Helped by an eagle 3 on the ninth, he played the last ten holes in even par and staggered off the course with 76 and the certain knowledge he had blown himself out of the championship.

The next day he shot 63 and won by one stroke. He never took his yardage notes home with him again.

Shortly after Miller won the Open, two executives from Sears, Roebuck and Company met on an airplane. Seated together, one asked the other if he was aware that Sears had signed Miller to a contract as a company spokesman. The second executive pondered a moment and then said, "I'll bet he cost us plenty."

Faulty Horoscope

An astrology enthusiast, John Schlee had been playing the best golf of his life through the first three rounds of the 1973 Open. With scores of 73, 70, and 67, he went into the fourth round sharing first place with Arnold Palmer, Julius Boros, and Jerry Heard.

Trying to explain why he was playing so well, Schlee said, "My horoscope is just outstanding. Mars is in conjunction with my natal moon."

Something tilted overnight. Schlee began the fourth round by driving out of bounds. Still, he recovered and finished second, one stroke behind Miller, whose natal moon must have been in somewhat better alignment with Mars.

I'll Do It Myself

Hale Irwin played a wild shot during the 1973 Heritage Classic that sailed toward the gallery, struck a woman on the chest, slipped under the neck of her blouse, and lodged inside her bra.

Under the rules of golf Irwin was to remove the ball and drop it without penalty. Under the circumstances, the woman decided she'd rather do it herself.

Hot Round

Scoring twelve birdies and an eagle, Pat Fitzsimons shot 58 in a casual round at the Salem Country Club, in Salem, Oregon, in 1974. Finishing with five consecutive birdies, he shot nine-hole scores of 29-29 over the par-36-36—72 course. With the tees moved back from the winter settings only a day earlier, Fitzsimons played a course that measured between 6000 and 6100 yards.

A newcomer to the PGA Tour, Fitzsimons summed up his round by saying, "I don't suppose it will ever happen to me again. I actually played a round without missing a shot."

Through the Barroom Window

His adrenalin flowing as he played the eighteenth hole at Moortown Golf Club, near Leeds, in the 1974 English Amateur, Nigel Denham put a little too much into the shot. His ball not only soared over the green, it hit a paved path, cleared a stairway, and rocketed through an open door into the clubhouse. Occasionally, in quirky instances, some clubhouses are considered integral parts of the course; the ball is in play no matter where it wanders, and the rules offer no relief without a penalty. This was one of those occasions, and so there was no relief for Denham short of declaring the ball unplayable and trying the shot again. He would have to play the ball no matter where it lay.

First, though, he'd have to find it. That was easy. It had bounced off a wall, caromed into the club's bar and settled under a table. There stood Denham, lying two under a table in a bar filled with players who had finished for the day, facing a shot that while it might not be difficult could definitely qualify as unusual.

While the drinkers cheered him on he moved the table and chairs. Looking up through a window, he saw the flagstick of the eighteenth green, its pennant fluttering in the slight breeze, and suddenly realized he had a clear shot at the pin except for one minor problem—the window was closed.

Easy solution there. Denham stepped over and opened the window, then played a nice little flip that squirted through the opening and pulled up within twelve feet of the hole. His barmates cheered, and Denham, smiling all the way, stepped onto the green and holed the putt for an extraordinary par 4.

Some time later the English Golf Union had second thoughts about Denham's procedure, especially his opening the window, and asked the R and A for its opinion. The R and A first argued that by opening the window Denham had improved his line of play and should have been penalized. Later, in discussions with the USGA, the R and A ruled that windows are made to be moved. Denham had acted within the rules. His par stood.

At the Walsall Golf Club, in England, two players learned a new dimension of the term "lost ball." Walsall's fourth hole, a par 3, stretched just 182 yards, but because of high frontal walls of a greenside bunker the green cannot be seen from the tee. Both men hit shots that covered the flagstick all the way and came down obviously close to the hole. Reaching the green they saw only one ball, lying less than a foot from the cup. On the off chance, they looked inside the cup and there rested the other ball. Now the problem: Only then did they realize they were both playing the same brand of ball with the same identifying number. Since it was impossible to say for certain which ball belonged to whom, both balls had to be considered lost. The two men took themselves back to the tee and played the hole again.

The Size of It

While he had never been compared to a rocket scientist, Lou Graham, the 1975 U.S. Open champion, had a certain earthy wisdom, at least as it related to golf. He understood it better than one might expect. "Golf is a dumb game," he explained. "Hitting the ball is the fun part of it, but the fewer times you hit the ball the more fun you have. Does this make any sense?"

Lightning

Chicago had gone through a wet spring in 1975. From the first of June until the first round of the U.S. Open, the National Weather Bureau measured three inches of rainfall, much of it during violent thunderstorms. During evening hours giant shards of lightning streaked from low-flying thunderclouds and lighted the endless flat prairieland below.

Tom Watson was scheduled to tee off at 1.04 in the second round and had actually worked his way onto the tee. Suddenly he turned and fled back into the locker room of the Medinah Country Club. He had seen lightning in the distance and estimated the storm to be half a mile away. Nobody could persuade him to come out. Within minutes another lightning bolt flashed close by, and the USGA suspended play until the storm blew over.

The following week the players moved on to the Western Open at the Butler National Golf Club, so close to Medinah the players didn't have to change hotels. Still the threatening weather continued, with more serious consequences than the Open's simple rain delay,

The Pro-Amateur tournament that preceded the Western had to be cut back to nine holes. The first round was played in hot, steamy weather under a thick overcast, and the next day brought even worse conditions. With players on every hole and about half the field finished, lightning flashed, rain began falling, and tour officials suspended play. Lee Trevino and Jerry Heard huddled under an umbrella by the thirteenth green, Trevino lying back against his golf bag. Bobby Nichols, the third member of the group, was standing perhaps 100 yards away. As Heard held his hands in front of him, making a point to Trevino, a savage bolt of lightning struck. Trevino and Heard fell backwards and lay stretched out on the ground. Muscles contracted, Heard's hands spasmed into fists and his arms closed tight against his body. He thought, "I can't open my hands; I'll never play golf again." He was nearly right.

No one is quite sure what happened, but it is believed lightning struck a nearby lake and the electric charge shot through the wet grass to Trevino's bag. Lee suffered burns on his back, and Heard was burned in the groin.

Meantime, lightning struck close to Bobby Nichols. Because of injuries in an automobile accident just after he left high school, Nichols wore a steel plate in his head. He also had an 8-iron in his hand. When the bolt struck, the 8-iron flew twenty feet or so. Nichols has never been sure if he threw it or the power of the lightning knocked it away. Tony Jacklin turned around and saw Nichols somersaulting across the ground.

Describing what he remembered, Nichols said, "There was a tremendous pop, like a cannon exploding. The lightning knocked me to the ground. I got up and ran for about ten yards before I fell again. I got up again, and I'm not afraid to admit it, I was hysterical. I didn't know what I was doing. Then lightning started cracking again and again and again. I started running again in a half crouch, like a soldier running between foxholes. I was really scared. I though, 'Oh, God, just let me live. Let me live.'"

Speaking later to Red Harbor, an official at Butler National, Nichols said he felt strange. "I don't have any equilibrium," he said. "I knew there was something wrong when I smelled his breath," Harbour said, "I'm in the construction business; I know what burned wire smells like. That's just what I smelled. I called for an ambulance right away."

All three men were rushed to a hospital and held for thirty-six hours. Both Trevino and Nichols withdrew from the Western; only Heard continued. He shot 288 and placed third, four strokes behind Bobby Cole, the winner.

They all suffered from the incident. One of the most promising young players in the game, Heard was twenty-eight, had won four tournaments, and had finished among the leading ten money-winners three consecutive years, but he was never the same again. He eventually left the tour and became a club professional.

Nichols never played as well again until he joined the senior tour, and he, too, became a club professional.

The shock caused Trevino severe pain in his back and for a time threatened his career as well. He submitted to surgery, took a few months to recover, and ultimately returned to his old form. He in fact won his second PGA Championship nine years later, in 1984.

Sixteen years later lightning struck at two important championships. Two spectators died, one at the Open, the other at the PGA, the only known fatal accidents at major golf championships.

The 1991 Open was played at the Hazeltine National Golf Club, near Minneapolis, once again on the broad midwestern prairie, and the PGA at Crooked Stick Golf Club, near Indianapolis. Opening day at the Open had begun under unsettling conditions. Great shards of lightning had rent the dawn sky and drenching rain soaked the dry ground. The downpour had stopped about the time the first group left the tee, but the threat lingered.

The skies turned black shortly after noon, the wind came up, and more rain fell. An electrical storm moved in faster than anyone had expected, and play was suspended. Marshals warned spectators a storm was on its way and urged them to find shelter.

The USGA had suspended play at 12:49. Players dashed for buses stationed around the course for emergencies such as this, or else into the clubhouse. Some fled into a maintenance shed. Meantime, the galleries, estimated at 40,000, had nowhere to go. A marshal approached the sixteenth tee, heard a crackling over his hand-held radio, and told the growing group of spectators play had been called off. In spite of the warnings, only a few headed for their cars; most either sat where they were, huddled under the steel-framed grandstand, or under nearby trees.

As the storm grew, six men squeezed under a young willow. A few of them had dashed over from other trees, believing the shorter tree would be less likely to attract lightning. Crowding as close to the trunk as they could, they felt they had found the safest place; they were at the lowest spot on the course, and under a tree only about thirty feet high, the shortest of any around them.

In a moment of black humor, one man joked it would be just their luck for lightning to strike there. It was 1:07. Suddenly lightning flashed. Then again. Some said a third time. Somebody said it sounded like gunshots. The bolt flashed past the tall trees, missed the steel-framed grandstand, and ripped into the little willow. It burned the bark from a long, narrow patch a little over head-high. The air around it smelled of smoke. Six men dropped to the ground and lay around the trunk like spokes of a wheel.

Momentarily stunned, the crowd began to react. Seeing the men sprawled on the ground, none of them moving, they became frantic. They screamed, "Doctor. Doctor. Somebody get a doctor."

A doctor from St. Paul sped to what looked like the worst case and gave cardiopulmonary resuscitation to William Fadell. Two other doctors ran to the tree. Paramedics from a medical station within sight of the accident sprinted across the open ground and within minutes began treating those on the ground. One man cried he had no feeling in his legs. Another, his hands still in his pockets, sprawled across the legs of still another victim.

Scott Aune, one of the stricken men, yelled Fadell's name. No answer. A paramedic tried to treat him but he said, "I'm okay now. I just want you to take care of my friend."

Doctors inserted a needle into Fadell's arm and tried to revive him with intravenous injections. They didn't work. Two paramedics hooked up a defibrillator, a machine that generates mild electric shocks intended to restart the heart. They attached it to Fadell's chest. No response. They tried again. Still no response. Within minutes an ambulance bumped across the course. The crew lifted Fadell inside and sped to St. Francis Hospital close by. They tried the defibrillator again. Again no response.

Not even a golf fan, Fadell had been given tickets by his father, Mike Fadell, a gallery marshal. Word reached him that his son had been among those hit by lightning and taken to St. Francis. He rushed to his car and raced toward the hospital. Along the way his car phone buzzed. It was the hospital. His son had died. He pulled off the road, not able to drive any farther. A policeman took him to the hospital. A priest was waiting.

All the other victims recovered.

Two months later another vicious storm interrupted the first round of the PGA Championship. When the siren sounded at about 2.15, players and spectators fled to shelter from the driving rain. Some huddled inside hospitality tents, most fans rushed to their cars.

Hurrying to his car, parked in a field near the fifteenth fairway, Thomas Weaver, a thirty-nine-year-old town councilman from Fishers, Indiana, about six miles from Crooked Stick, sheltered under his umbrella. Notices printed on daily pairing sheets warned spectators to lower their umbrellas when the sirens sounded, but Weaver either ignored the warning or wasn't aware of the danger. He and a friend had closed within 100 yards of his car when lightning flashed. It caught the metal tip of Weaver's umbrella. He fell to the ground. A friend standing next to him had his umbrella knocked from his hand, but he wasn't hurt. A nurse and several doctors saw the bolt strike. They rushed to help. Taking turns, they gave him cardiopulmonary resuscitation. He didn't move. An ambulance reached him within minutes. The Rescue Squad worked to revive him before taking him to St. Vincent Carmel Hospital, within two miles of Crooked Stick. Nothing worked. He was pronounced dead at 3:42 p.m.

Quick Round

The fourteenth hole at Royal Portrush, on the northern coast of Northern Ireland, an unusually difficult par 3 of 211 yards across a deep chasm, bears the name Calamity Corner.

During the 1970s, when fighting in Northern Ireland reached its worst, an American drove up to the Royal Portrush Golf Club and approached the club secretary. Showing all the proper credentials, including a letter of introduction from his home club, he paid his green fee, changed his shoes, handed his bag to a caddie, and said, "Let's go."

With the American in the van, they walked past the first tee, along the sixteenth fairway, and out onto the far reaches of the course. They were gone no more than half an hour. When they came back, the American paid his caddie, changed his shoes, and stuffed the golf bag into the trunk of his car. Asked what had gone on, the American explained, "I read somewhere about the best eighteen holes in Europe, and I've set about playing them. Your fourteenth is one of them. I just played it twice and made a 3 and a 4. Now I'm off to the ninth at Royal County Down." Then he drove away.

Great Ruling

Out of contention after three rounds of the Greater Jacksonville Open, Tom Weiskopf spent the evening at an all-night piano bar. As one of the highest scorers, though, he had an early starting time the last day, even though he was in no condition to face the glaring sun.

He needed help, so he called on Ed Sneed to drive him to the golf course at A.J. Foyt speed, raced to the first tee, hopping occasionally to put on a shoe, bogeyed the first two holes, then drove into the woods on the third. Leaning against a tree, he sent word he wanted to speak to Clyde Mangum, the tour's leading rules official.

Mangum drove up in his cart, found Weiskopf still leaning against his tree and holding his head, then stared at the ball. Turning back to Weiskopf, Mangum said, "What's the matter, Tom? Looks like you have a pretty good lie." Tom groaned, then said, "Clyde, can you get me an egg sandwich and a carton of milk?"

Mangum smiled, drove off, and returned with the deli order. Weiskopf played the remaining fifteen holes in six under par and shot 69. Tom believes this is the best round of golf ever shot by a dead man.

What To Do?

Looking ahead to the day when he could retire, George Archer, who won the 1969 Masters complained, "Baseball players quit playing and take up golf. Basketball players quit and take up golf. Football players quit and take up golf. What are we supposed to take up when we quit?"

Watch Your Aim

With one round to play in the 1976 Masters, Raymond Floyd led Jack Nicklaus by eight strokes and Larry Ziegler by nine. Approaching Ziegler, a reporter asked, "What would you have to shoot tomorrow to win?" Ziegler thought for a moment and answered, "Raymond Floyd."

What Flyers?

All day long players coming off the Atlanta Athletic Club's course after the first round of the 1976 U.S. Open complained bitterly that the USGA had set about humiliating the greatest players in the game by not cutting the fairway grass, causing what is known as flyer lies. When too much grass gets between the clubface and the ball, the ball doesn't spin normally and flies something like a knuckleball.

Investigating what had happened, the USGA found that the company supplying new wheels for the club's mowing equipment had sent the wrong size, an inch in diameter bigger than the old wheels. Consequently the grass stood about three-quarters of an inch high rather than half an inch.

While most of the professionals looked daggers at the USGA and charged the association had been caught red-handed trying to embarrass them, Mike Reid, a short-hitting amateur, came off the course with 67, the best score of the day by three strokes. Asked if he had trouble with flyers, a puzzled Reid asked, "What flyers?"

Death Threat

Paired with Andy Bean, Hubert Green had just begun the fourth round of the 1977 U.S. Open when an extraordinary meeting took place inside the Southern Hills clubhouse. Harry Easterly, the president of the USGA, Sandy Tatum, chairman of the championship committee, and P. J. Boatwright, the executive director, were called from the course to talk with Charlie Jones, a Tulsa police lieutenant serving that week as the club's chief of security.

Jones told a chilling story. Someone had threatened to kill Green as he played the fifteenth hole that day. A woman had called the Oklahoma City office of the FBI with information that three men who had been in trouble with the law in the past were on their way to Tulsa and had told her they were going to shoot Green. "I know they're serious," she said, "because they showed me their guns."

The telephone call had been taken at about 3:30 on Saturday, but word hadn't reached Southern Hills until Green had completed the first nine holes on Sunday—about twenty-four hours later.

Everyone took the threat seriously. Suddenly a phalanx of uniformed policemen sprang up around Green. Fans were shuffled out of the clubhouse and the doors closed and guarded. Plainclothesmen patrolled the gallery, and a mysterious person materialized in the control center of ABC television, the *sanctum sanctorum* where no one is permitted, and quietly gave orders to Roone Arledge, then the president of ABC Sports, and

Chuck Howard, who was producing the telecast. He told them to have their cameras scan the gallery around the fifteenth hole and didn't tell them why.

Green had been leading the Open, but he had been playing shaky golf for the last few holes. Nevertheless, he had to be told. When he stepped off the fourteenth green, Easterly, Tatum, and Jones called him aside to an area where they couldn't be overheard, told him of the threat, and gave him three options:

1–He could withdraw from the Open.

2–He could ask for play to be suspended.

3–He could continue to play.

Green considered for a moment, said, "It was probably one of my old girlfriends," and chose to continue.

Taking precautions, like staying clear of Bean, in case a mis-directed shot might hit Andy instead, and telling his caddie to stay away from him, Hubert completed his round without a shot's being fired and won the Open. No one can be sure, but the threat probably was a hoax.

Watson and Nicklaus at Turnberry

The 1977 British Open ranks among the brightest moments in the history of competitive golf. If the game had ever seen a more gripping and dramatic episode it escapes memory. It drove scoring to new levels, it introduced Turnberry to the rota of British Open courses, and it ended as a two-man battle between Tom Watson and Jack Nicklaus, the two best players of their time playing a caliber of golf others could only have dreamed of. It was as much a display of spirit and determination as it was of their ball-striking skill. It drove the gallery to a frenzy as the spectators raced around the course straining for a glimpse of a magic moment they knew would never be repeated.

Through three rounds each man shot the identical scores of 70, 68, and 65. Only at the end did Watson push ahead, shooting a final 65 while Nicklaus shot 66. Appropriately, the struggle reached its climax at the final hole, where both men birdied, as champions should. Watson shot 268 for the seventy-two holes, and Nicklaus shot 269. Both men broke the fifteen-year-old record of 276 set by Arnold Palmer at Royal Troon in 1962. When it ended, Hubert Green, in third place, said, "I won the tournament I played in. They were playing in something else." Green, who had won the U.S. Open a month earlier, had shot 279, eleven strokes behind Watson.

With 138 for the opening thirty-six holes, Nicklaus and Watson stood one stroke behind Roger Maltbie, who led the field with 137, but twenty players were grouped within five strokes of first place. None of the others mattered after the third round. Paired together, both Nicklaus and Watson played Turnberry in 65s.With 203 for the fifty-four holes, they had opened a gap of three strokes over Ben Crenshaw, who had crept into third place by shooting 66. Maltbie was next, at 209, tied with two others, and Lee Trevino and Johnny Miller had 210s, too far behind to matter.

Such unprecedented scoring dazzled the galleries, but the golf course had been stripped of its defenses. Very little rain had fallen for weeks, the rough failed to grow except in usually remote patches, and the wind, which usually rips in from the southwest, withered to a light zephyr and even then blew from the wrong direction. As a result, Turnberry, which can be a magnificent test with some wonderful holes, especially its long par 4s, had lost most of its sting. While it wasn't a pitch-and-putt, too many holes were little more than a drive and pitch. For example, facing a stiff wind during the 1963 Walker Cup Match, the game's leading amateurs played wooden clubs for their second shots to the sixteenth, a par 4 of 410 yards. During the 1977 Open, the professionals seldom played more than a 7-iron, and frequently a wedge.

Robbed of its usual defenses, Turnberry lay helpless. Nevertheless, no one was quite prepared for the final act of this dramatic championship. Paired together once again, Watson and Nicklaus started at the end of the field, and before anyone was quite ready for it, Nicklaus had jumped two strokes ahead, the result of his birdie on the first and Watson's bogey from a nearly impossible position on the second.

With a birdie 2 at the fourth, Nicklaus moved three strokes in front, but Watson struck back with a birdie on the fifth, added another on the seventh, a par 5 he consistently reached with his second shots, and still another at the eighth. Now they were even with ten holes to play.

By then the huge crowd of close to 20,000, awed by this magnificent display of golf, had grown unruly. Spectators slipped under ropes and swarmed over the fairways. Unable to pick out their targets for the difficult drive on the ninth, both men told the marshals they wouldn't play until the fairways had been cleared.

The slight delay seemed to interrupt Watson's momentum, and he didn't seem as sharp as he had been. He dropped a stroke at the ninth and fell behind again. Out in 34 against Nicklaus's 33, he looked shaky. He left his approach short of the tenth green, and his tee shot to the little eleventh dropped into a bunker. Still, he saved his pars on both holes and held on, just one stroke behind.

Then Nicklaus opened his lead once more. After driving into the rough on the twelfth, he pitched on and holed from twenty-five feet. Two strokes ahead with six to play. It didn't last; just as he had on the twelfth, he drove into the rough on the thirteenth and couldn't pitch as close to the hole as Watson. Tom holed from twelve feet and climbed within one stroke once again.

Watson might have pulled even on the fourteenth but he missed from no more than six feet.

Now they played the key hole of the championship, the fifteenth, a par 3 of just over 200 yards. Watson mis-hit his 4-iron and missed the green to the left, at least sixty feet from the cup. Nicklaus's ball barely missed a greenside bunker but stayed on. Even though fluffy rough and the green's collar lay between his ball and the cup, Tom chose to putt. He gave his ball a firm rap, it streaked through the rough, bounced across the collar, rolled onto the green directly at the hole, slammed into the flagstick, and dropped right into the hole. A birdie 2. When Nicklaus could do no more than make 3, the two men were level once more with three holes to play.

His driving still off, Nicklaus missed the sixteenth fairway but he recovered to the heart of the green. Watson, though, nearly threw it away there. He tried a more dangerous shot, directly at the flagstick, cut in the front right. Played just a touch too fine, his ball barely cleared the chasm that knifes across the fairway and stopped on the hillside just short of the green, for one heartstopping moment looking as if it might tumble down the slope and into the little stream below. It didn't though; he chipped stiff and both men strode to the seventeenth tee still dead even.

They had played sixteen holes in three under par and had left the rest of the field far behind. No matter what happened over the next two holes, either they would finish first and second, or face a playoff the following day.

Turnberry's seventeenth is a par 5 of just under 490 yards played from a high tee into a valley below, and then to a green set atop a hill about level with the tee. Both men drove perfectly, long and straight into the valley. Away, Watson played a terrific long iron that carried over the crest of the hill, onto the green, and settled fifteen feet from the cup. When the ball stopped rolling, the gallery roared.

Perhaps the cheering prompted Nicklaus to try too hard to play his ball even closer to the hole, although he had no way of knowing just where Watson's ball had come to rest. Whatever the reason, Nicklaus, usually an impeccable striker of the ball, hit his shot fat. It stopped well short of the green.

Nevertheless, when he pitched his third within five feet of the cup and Watson missed his eagle opportunity, Jack looked safe enough. Instead, he borrowed too much on the putt; it slipped past the cup, and for the first time Watson moved ahead.

The eighteenth hole will be remembered as long as the game is played if for no other reason than for the memory of Nicklaus's fighting spirit. Up first, Watson drove with an iron to the center of the fairway, but sensing he needed a special effort, Nicklaus pulled out his driver. Once again he hit it badly. Normally when Nicklaus found himself in a tight spot and hit a bad shot he pulled it left. Here, though, he pushed it into the rough on the right, alongside a prickly stand of gorse bushes.

While Jack studied his problems, Watson played a stunning 7-iron a little more than two feet from the cup. The championship seemed over. Nicklaus, though, is never beaten while holes are left to be played. He ripped into an 8-iron, tore the ball from the heavy grass, and rifled it toward the green. It pulled up on the right edge about thirty feet from the cup.

Seeing Jack's ball roll onto the green, Watson told Alfie Fyles, his caddie, he assumed Nicklaus would hole the putt. Agreeing, Fyles said, "I expect him to, Sir."

Drawing on his uncommon strength of will, Nicklaus did indeed hole the putt and finished the round with 66.

With his own putt suddenly longer, Watson unhesitatingly rapped it into the hole. Two strokes behind after the twelfth, he had birdied three of the next six holes, shot 65, played the last thirty-six holes in 130, set a new British Open record, and won by one stroke.

Earlier in the year Watson and Nicklaus had fought another stirring battle for the Masters. Watson had led Nicklaus by three strokes going into the last round, and even though Nicklaus called on all his competitive fire and shot 66, Watson shot 67 and beat him by two strokes.

As they walked off Turnberry's final green together, with Nicklaus's arm slung over Watson's shoulder, Jack said, "I'm tired of giving it my best shot and coming up short."

Illegal Clubs

Looking into George Burns's golf bag during the 1977 Hartford Open, Jerry Heard said, "Nice clubs you have there, George. Too bad they're illegal."

For such a quick glance, Heard had no way of knowing if they were or weren't, but Burns thought he'd check to be sure. Clyde Mangum, deputy commissioner of the PGA Tour, examined the clubs with a measuring magnifier, a device resembling a jeweler's loupe, and found they did indeed violate the rules. The grooves on the clubhead's face were too wide and too close together.

Thereafter all the players' clubs were inspected before they played in the PGA Championship, at Pebble Beach. Gary Player had to replace half his irons, and both Raymond Floyd and Tom Watson had to replace entire sets. Under contract to play Ram clubs, Watson had an old set of MacGregors he had used in the past shipped in, but they failed the grooves test as well.

On the morning of the first round Watson borrowed an old set of Tommy Armour Silver Scots from Roger Maltbie, had time to play only eight shots from the practice tee, rushed to the first tee, drove the fairway, and then, using the 6-iron for the first time, laid his ball six feet from the cup. He shot 68 with clubs he had never seen before.

Valuable Lesson

Seve Ballesteros learned a valuable lesson in 1978 when he was paired with Gary Player in the final round of his first Masters. Seve was just twenty-one while Player was twice his age. Five strokes behind Hubert Green, the leader, well into the round, Player grew progressively more upset with the galleries because they weren't taking him seriously enough to suit him.

Even with the holes running out, Gary seethed. Turning to Ballesteros he said, "Seve, these people don't think I can win. You watch; I'll show them." He did. He played the last ten holes in seven under par, shot 30 on the home nine, 65 for the round, and beat Green by one stroke.

A Matter of Decorum

Leonard Crawley, who wrote for London's *Daily Telegraph*, and Henry Longhurst, of the *Sunday Times*, were close friends, but Henry had a rather bohemian attitude and didn't always conform to the proper British gentleman's idea of suitable conduct—particularly his table manners. Leonard, therefore, dedicated himself to making Henry conform—especially in front of Americans.

As the fame of the Masters grew, it attracted correspondents from foreign publications. They were welcomed by the Augusta National Golf Club, and were often provided accommodations and meals. At dinner one night, the club served lamb chops. Henry became frustrated trying to cut the meat from the bone with knife and fork, and so he picked up a chop and began gnawing as he held it in his fingers.

Seeing Henry slipping from grace, Leonard rapped on the table with his spoon and intoned, "Henry, that is not a chicken bone you're eating." Without stopping his munching, Henry growled, "I know it's not a bloody chicken bone, it's a lamb chop."

"In that case," Leonard persisted, "don't do it. Remember, Americans are watching us."

"Bollix the Americans," Henry snarled.

"If you do not desist, old boy, I shall have to leave the room."

Longhurst carried on. Crawley picked up his plate, his knife, his fork, and his spoon and headed up the stairs to his room, pronouncing, "Longhurst, you are a disgrace to the British party."

Watch Your Tongue

While he worked for the CBS television network, Jack Whitaker was assigned to the Masters. Those who do the telecasts have to be careful about what they say. For example, broadcasters must never refer to Masters galleries as crowds or mobs. They must be called patrons.

In a moment of excitement, Whitaker referred to the teeming mob as . . . well, a mob. For this offense he was barred from the Masters television booth and not allowed to foul the Augusta airwaves again. Not that Whitaker wasn't warned. Before the 1977 tournament, the Masters Committee had sent the CBS television network a memo laying down thirty-three rules for broadcasters. Jack had broken the first commandment: A sampling:

Never refer to the gallery or patrons as a mob or crowd.

Never estimate the size of the gallery.

Never refer to players' earnings.

Never refer to Masters prize money.

De-emphasize players' antics.

Do not compare any holes at Augusta National with those at another golf course.

The water in front of the thirteenth green is not to be called Rae's creek but a tributary of Rae's creek.

Make no reference to Masters tickets having been sold out.

Make frequent mention of the presentation ceremony to be conducted at the end of the final round.

Do not guess at where a ball might be.

Do not estimate the length of a putt.

Instead of identifying Lee Elder as the first black man to play in the Masters, say he is the first person of his race to play in the tournament.

Those involved in the telecast claimed they largely ignored the memo.

Nicklaus at Cherry Hills

Jack Nicklaus was closing in on the 1978 U.S. Open. He had begun the third round at 142, even par, which left him two strokes behind Andy North, the leader. He still held on to even par after ten holes, and then came to the eleventh, a downhill par 5 of 594 yards.

Jack had been using a driver he hadn't carried in three years, one with slightly more loft than the one he had been using. After an excellent drive, he tore into the driver again from the fairway, hit a tremendous shot six feet from the hole and eagled. Two under then, he parred the twelfth and moved on to the thirteenth, a 382-yard par 4 with a small creek cutting across the fairway perhaps thirty yards short of the green.

Passing up his driver, Nicklaus played an absolutely perfect 3-wood that stopped in the geometric center of the fairway. Pleased with himself, Jack began walking toward his ball. Along the way he spotted a Port-A-John and ducked in. Emerging to a cheering throng, he continued to his ball and the short pitch to the green.

Nicklaus had never been at his best with the wedge, but no one could remember so horrible a shot as he hit here. He took a divot the length of the hall rug. Forget the green; the ball didn't even reach the creek on the fly. It hit a few feet short and bounced in. He dropped out with a one-stroke penalty, mis-played another wedge over the green and into a bunker, slapped a scruffy shot out, and finally got down in two strokes. He had taken 7 on a par 4 hole, six of those strokes from no more than 100 yards.

Word spread through the gallery: Nicklaus had pissed the Open away.

Sands of Nakajima

Cut into the left side of the Old Course's seventeenth green, at St Andrews, the Road Bunker, a small, deep sand pit with a nearly vertical face, sits just beyond a series of slopes and swales that seem to swallow up timid shots.

Tsuneyuki (Tommy) Nakajima, a very fine Japanese professional, had opened with rounds of 70 and 71, and stood among the leaders of the 1978 British Open. He approached the seventeenth tee the next day needing two 4s for another 71 that would have kept him within range of the leaders. Two splendid shots put him on the front of the green. Then his troubles began.

With his ball lying about seventy-five feet from the hole. Tommy put a little too much into his stroke. The ball raced past the hole, curled left, and tumbled into the bunker. Barring a miracle he would make no 4 here, and he would have trouble salvaging a bogey 5. With the flagstick no more than twenty feet from his ball, Nakajima tried to play out gently, allowing his ball to roll close to the hole. He hit the shot too softly; it stayed in the bunker. He tried again. Again he left the ball in the bunker. Once more he tried, and once more the ball rolled back into the sand.

Nakajima finally made it back to the green with his seventh shot, took two putts, and played the hole in 9. He shot 76, finished with 288, and tied for seventeenth place. The Road Bunker has been known ever after as The Sands of Nakajima, playing on the old John Wayne movie *The Sands of Iwo Jima.*

It had been a bad year for Nakajima. Not only had he made a hash of the seventeenth hole of the Old Course, but before then he had scored 13 on the thirteenth hole at Augusta during the Masters.

After hitting his second shot into Rae's creek, the stream that knifes across the fairway close to the green, Nakajima climbed in. Before climbing out he had been assessed five penalty strokes for various violations, among them grounding his club in a hazard and for being hit by a ball that rebounded off the rocks and bounced against his foot. His 13 stands as the most ever on one hole at Augusta.

Memorial

As Doug Sanders stepped onto the eighteenth tee at St. Andrews leading the 1978 British Open by one stroke over Jack Nicklaus, a man dashed out of the gallery, handed him a white tee peg, and said, "Tony Lema used this tee when he won the 1964 Open. Would you use it in his honor?"

Not wanting to offend the man and figuring the tee wouldn't make a difference anyway, Sanders agreed, three-putted the hole, fell into a playoff, and lost the next day. Ever since there have been British professionals who wouldn't dare use a white tee.

Lost in Scotland

Frank Strafaci, a pretty good amateur from New York who finished his career as executive director of the Metropolitan Miami Golf Association, entered the 1978 British Amateur, which was played at Troon. He arrived at least a week early, spent three days learning the course, and since he had the time, decided to see more of Scotland before the championship began. He started by driving to Dornoch, about 250 miles farther north, and played Royal Dornoch, then drove to Shin Falls to watch the salmon run, then visited Nairn, another first-rate Scottish course a bit south of Dornoch.

Strafaci made a date for a round with the club professional at 9.15 the next morning, had dinner, then went to bed. He awoke suddenly about 4 a.m., not certain where he was, but knowing he was to tee off in the British Amateur at noon. His drive to Dornoch had taken nine hours, so he wasn't sure he could make it to the tee on time.

Rather than drive, Frank figured he could make his starting time if he flew from Inverness to Glasgow. He hurried to the Inverness airport and managed to catch the 7 o'clock flight. Once in Glasgow, he hailed a cab and raced down the road to Troon.

Strafaci changed into his golf clothes, trotted over to the golf course, and approached the starter. There Frank learned he indeed had the time of day right, but somewhere along the way he had miscalculated. He wasn't due until the next day.

Emergency Measures

After Fuzzy Zoeller, Ed Sneed, and Tom Watson tied for first place in the 1979 Masters, the sudden-death playoff began at the tenth hole to accommodate television. After the approach shots, Hord Hardin, chairman of the Masters committee, who acted as the referee, had to measure to determine which of the three lay farthest from the hole.

Involved in a conference when the regulation seventy-two holes ended, Hardin had rushed out to the tenth tee, where the playoff began, without picking up a measuring device, such as a tape measure or a long string. With nothing in his pockets, Hardin had an inspiration. He laid down the flagstick and used it as a yardstick. Not good enough; all three balls lay outside the flagstick's reach. With no other means of measuring, Hardin pulled an envelope from his pocket, laid it at the end of the flagstick, and with that combination determined that Zoeller was away. Fuzzy won on the second playoff hole.

The Sommelier

Out to dinner at the Clifton Arms, a hotel with an elegant dining room, in St. Annes on Sea during the 1979 British Open, Tom Weiskopf decided his table should have the finest wine in the hotel's cellar. Summoning the sommelier, he ordered him to bring "something ancient, red, dusty, and expensive."

The sommelier, whose name was Deiter, brought the wine and placed it before Weiskopf to taste. Taking a sip, Weiskopf made a face, crying, "This is terrible. Bring me a bucket of ice and some club soda." Aghast, Deiter complied nonetheless, Weiskopf dropped ice cubes into a water tumbler, filled it halfway with soda water and added the expensive wine, took a sip, and, smiling, gloated, "Now that's a good glass of wine."

Lost Stroke, Lost Hole

One hole up on Brian Marchbank and Peter McEvoy after thirteen holes of their foursomes on the second day of the 1979 Walker Cup, the Americans Jim Holtgrieve and Doug Fischesser gave away the fourteenth hole. With the honor, Holtgrieve hooked into the left rough while McEvoy pushed his drive into the right rough.

Since the Americans had the longer drive, Marchbank played first. The rough at Muirfield is unreal. Every blade of grass that ever grew on that ancient ground is still there, some of it dead, most of it vigorously alive. Finding a ball is hard enough; playing a decent shot is next to impossible. Marchbank had no option but to pitch back to the fairway, but in that thick and tangled grass he moved his side's ball only about twenty feet. Meantime, the Americans were so busy, first trying to find their ball and then preparing to play the next shot, they didn't notice Marchbank's futile slash. When McEvoy finally chopped the British ball back to the fairway and Marchbank pitched onto the green about twenty feet from the hole, they believed the British lay 3 rather than 4.

Holtgrieve did know, though, that he and Fischesser lay 4, a little more than three feet from the cup. McEvoy then played a lovely putt that rolled slowly across the slick green and toppled into the cup. When it dropped, Holtgrieve assumed the British had parred rather than taken a bogey 5. Believing he had no chance for a half, Holtgrieve picked up his side's ball, giving the hole away. Holtgrieve and Fischesser lost the next two holes and the match by 2 and 1. Nevertheless, the United States won the Walker Cup by 15½ to 8½.

Gladding at Sandwich

In 1979, Reg Gladding, at fifty-four, ranked as the oldest scratch golfer in England. Playing in the English Amateur, at Royal St. George's, Gladding and his quarter-final opponent had finished all square and gone on to extra holes. They halved the first three, then approached the tee of the fourth, a par 4 of 466 yards that called for a drive across a high hill scarred by a deep bunker, at least twenty-five feet from top to bottom, carved into its face. Failing to clear the hill, Gladding's ball splashed into the top of the bunker. Since he couldn't climb in from the top for fear he might cause a landslide and bury his ball, in the best tradition of Sir Edmund Hillary he slowly inched his way up from below. Club in hand, he gingerly took his stance, but as he reached the top of his backswing, the avalanche began. The sand shifted, he lost his balance, tumbled over backwards, did a reverse cartwheel and plunged to the bottom of the sand pit, where he lay in a heap. The ball and club followed.

On his inglorious journey to the bottom, Gladding had violated more rules than he could count: He had grounded his club, certainly, he had obviously caused his ball to move, and he could not deny that lying there as he was he had tested the texture of the sand. Nevertheless, neither he nor his opponent could figure out how many strokes he must penalize himself. Certainly, though, he learned the meaning of sudden death.

Double Trouble

George Forrest, a member of the Bathgate Golf Club, near Edinburgh, Scotland, was watching an exhibition match at his home club between Bernard Gallacher and Ronnie Shade on one side and Gordon Cosh and Charlie Green on the other. Gallacher, a professional, had played Ryder Cup golf, and the other three had been members of Walker Cup teams.

Green, a big man, hooked his drive on the second right into the ribs of Forrest. It was quite a blow, but Forrest continued to follow along. Then Cosh hooked his drive on the eleventh. The ball conked Forrest on the head. That was enough for him. He was taken to a hospital, where doctors stitched the wound together; then he went home.

Summation

Pondering over the attitude of one professional golfer to another, Mason Rudolph announced, "Out here you've got to realize that if you take an 8 on a hole, 90 percent of the other pros don't care, and the other 10 percent wish it had been a 9."

Wrong Trophy

When in 1979 Jay Sigel became the twenty-first American to win the British Amateur, defeating Scott Hoch, another American, by 3 and 2 in the final, he asked the Royal and Ancient Golf Club to ship the trophy to his home, near Philadelphia. Somewhere along the line the trophy was lost. It was a stunning loss to Sigel, who planned to display it in his house.

Someone suggested the trophy was no doubt insured and that it could be replaced, but Sigel would have none of it. "I want the same trophy Bobby Jones won," he said. The trophy turned up a few weeks later.

Your Resignation, Please

Robert Townsend, a member of the Augusta National Golf Club, had been chairman of the board of the Avis car rental company, and a director of Dun and Bradstreet, the organization that provides credit information on other organizations.

Townsend then wrote *Up the Organization*, a book critical of corporation policies and practices. It became such a best-seller Townsend was interviewed by *Time*, which did a cover story on him. The article mentioned he held an Augusta membership. The day the magazine hit the newsstands, Townsend received a telephone call from Cliff Roberts, the imperious chairman of the Masters Tournament and the man who actually ran the club. Roberts informed Townsend that he would accept his immediate resignation.

Ballesteros Disqualified

Seve Ballesteros, the most dynamic player in the game at the time, held both the British Open and Masters titles when he arrived at Baltusrol for the 1980 U.S. Open. Just twenty-three at the time, he had won seventeen tournaments and led the European Order of Merit three consecutive times. He was, therefore, one of golf's two or three greatest attractions.

As fine a player as he was, however, he was rather wild from the tee, a sin at a U.S. Open course, whose fairways are bordered by calf-high rough. Ballesteros had shot 75 in the first round, leaving him twelve strokes behind Jack Nicklaus and Tom Weiskopf, who had matched the Open record by shooting 63s in the first round.

Ballesteros was staying in a small suburban hotel, normally a drive of five or ten minutes from Baltusrol. The morning of the second round, however, with traffic clogging the two-lane road, the drive took longer. Furthermore, Ballesteros thought he was scheduled to tee off after 10 o'clock, instead of his actual time of 9:45. He left his hotel at 9.25, and when he arrived at the first tee to join Hale Irwin and Mark O'Meara, both men had reached the green. Ballesteros was disqualified.

Even though he was clearly to blame for his disqualification Ballesteros did not leave gracefully. He stormed into the locker room where he blamed the club's transportation system, then raved on that he would never play in the Open again.

Later, when his temperature dropped, he reconsidered. "I am a stupido," he said. "Nothing was right this week. For the first time in years I make no birdies in eighteen holes. Then I am disqualified. When you're hot you don't know what you're saying. I'll play in the Open again. Now I'm embarrassed. I come all this way for nothing."

The Human Factor

Just before the third round of the 1980 U.S. Open, Tom Watson was told that a Baltusrol member wanted very much to see him, an unusual request, coming as it did at a time when Watson was contending and understandably concentrating on the job at hand. The member, however, Dr. William A. Tansey, lay bedridden with terminal cancer.

Less than two hours before his starting time, Watson spent twenty minutes with Dr. Tansey. Asked why he would take the time when he might have been preparing himself for the last round of the Open, Watson said, "You can't take golf so seriously you rule out what is the human thing to do."

Ben Crenshaw

Ben Crenshaw is among the best-liked golfers who ever played the game. Recognizing something special in him, his father bought him his first set of clubs when Ben was ten. One day when the course was dry and fast running, Ben came home and told his father he had just shot 74.

His father said, "Son, that's a pretty good round. How did you do it?"

Ben took him through the round stroke for stroke. When he finished, his father said, "You didn't move the ball around, did you?" Ben said he hadn't. Then his father said, "Now you told your daddy, and I believe you. But only you know for sure."

Ben was talking to a member of the Honourable Company of Edinburgh Golfers in the changing room at Muirfield during the 1980 British Open, when the member suddenly stood and walked to his locker, pulled out a number of antique wooden-shafted clubs, including a valuable long-nosed play-club marked R. Forgan, handed them to Crenshaw and said, "I don't have anybody to give them to," and walked away without introducing himself.

The Baffing Spoon

The sun had fallen to within a few degrees of setting, casting a golden glow across the links of Muirfield the evening after Tom Watson won the 1980 British Open. Ben Crenshaw had left Greywalls, the elegant old hotel alongside the ninth fairway, and held an ancient club as he stepped onto the tenth tee. Over dinner he had announced he would play the tenth and eighteenth holes with the ancient clubs the club member had given him, and a gutta-percha ball dating, he believed, to 1890.

Crenshaw called for sheep dung—"Te tee ma ball"—he explained, as Watson and his wife, Linda, happened by. With no sheep dung in sight—Muirfield wouldn't stand for it—Watson trotted to the ninth green, scooped a handful of sand from a bunker, and molded a tee the old-fashioned way. As Ben waggled a wooden club with a short, hickory shaft and a long, unscored face that looked more like a hockey stick than a golf club, Tom Weiskopf asked the name of that evil looking instrument. In his best Scottish brogue, tainted by a strong Texas twang, Crenshaw answered, "She's ma baffing spoon, laddie."

Meantime, Linda Watson, who had disappeared into the mist, reappeared leading a piper, in full kilted dress, playing a rousing march. With the hypnotic sounds of the pipes echoing over the deserted links, Crenshaw, again using nineteenth-century methods, took a lusty swing. Its shaft whippy as a flay swatter, the old club had thrown Ben's timing off badly; the ball sailed about 160 or 170 yards far off course and into Muirfield's dense and impenetrable rough.

With the pipes wailing their mournful cry, a growing party set off into the dusk—the Crenshaws, the Watsons, Tom Weiskopf, Bill Rogers, Andy North, Bruce Lietzke, while Tony Jacklin ran ahead carrying those primordial clubs across his shoulder and playing the part of a madcap cockney caddie, cackling, "Hit it where ye will. Ye'll nowt lose it wit me 'ere."

A second whack with the baffing spoon cracked the ball apart, and so everyone agreed it should be retired from the game. Ben substituted a modern ball. Another shot, played with what looked like a primitive 5-iron, reached the green and drew back within six feet of the hole. He missed the putt and bogeyed, the same score he had made earlier in the day with the advantages of modern equipment. (Crenshaw had placed third in the Open.)

Inspired by the magic of an enchanted setting, the music of the piper, and the collegiality of the group, Watson, using as much Scottish burr as a midwesterner can muster, challenged Crenshaw to a one-hole match, claiming, "I'm from Gullane." the small town where Muirfield is located. Crenshaw answered, "And I'm a Musselburgh mon. Good luck to ye, Young Tom."

Using the same wobbly baffing spoon, both men drove well, Crenshaw's ball a little longer, perhaps because Watson, adopting the conventions of an earlier time, played in his coat. Watson's second rolled within eighty yards of the green, but Crenshaw, playing the longest shot of the match, hit his approach into a greenside bunker. Watson's third rolled off the green, and then Crenshaw faced a difficult choice. Which of this odd assortment of clubs could he use to play the shot?

He chose a rusted iron club with the approximate loft of a 6-iron. In his first attempt, the clubhead bounced off packed sand and into the back of the ball, cutting it badly. It still lay in the sand. Laying the blade wide open, Ben cleared the high wall of the bunker but left his ball on the lower level of the two-tiered green, at least seventy feet from the cup. He was still away. His putt, played with a creaking antique putter and a badly wounded ball, climbed the slope and hobbled within two feet of the cup. Shaking his head in wonder, North gasped, "Crenshaw could putt with a rake and an Easter egg."

With Crenshaw in with 6, Watson had to get down in two to win the hole. First he had to choose a club for his pitch, but his first choice, a rut iron, a lofted club with a small, rounded face designed to play shots from wagon ruts, looked too shankable, so he asked Jacklin for another iron and played a wonderful little flip that stopped within seven feet of the cup. One putt for the win.

After surveying the line, Tom took the creaky old putter, twisted himself into a crouched position hardly resembling his normal stance, and after two practice strokes, rolled the ball into the cup. Crenshaw offered his hand in congratulations and Watson, bowing from the waist, said, "A good match, sir."

"Aye," Crenshaw agreed, "a bonnie good match."

Not everyone was amused. Moments after the match ended, Paddy Hanmer, a former naval captain and the irascible and disagreeable secretary at Muirfield, stormed through the clubhouse door, confronted Crenshaw, and lectured him strongly on proper etiquette at his course. He told Crenshaw he had played without permission and warned him not to return.

Hanmer apologized later, saying he was only joking. No one believed him.

Misdirection Play

The small nation of Fiji, a group of islands in the South Pacific, entered a team in the 1980 World Amateur Team championship, in Pinehurst, North Carolina. Pinehurst has no airport, and so the Fiji team booked a flight to Fayetteville, a few miles north.

An hour before their scheduled landing time, the team members changed into their colorful native Fijian dress—bright flowered sarongs and thonged sandals—especially for the welcoming party waiting for them.

Unfortunately, when the plane landed, they stepped off not into the pleasantly mild autumn of Fayetteville, North Carolina, into the bitter-cold fall of Fayetteville, Arkansas.

Bobby Nichols

A tall, pleasant Kentuckian, Bobby Nichols went to Texas A&M on a football scholarship, injured himself, switched to golf, and won the 1964 PGA championship as well as ten other tournaments.

Through his career he carefully invested his money under the advice of Bill Boone, a prominent Louisville, Kentucky, attorney. Believing he had more than $500,000 in a bank account, Nichols tried to withdraw a small amount to invest in real estate. He found the account held only $1200. Understandably upset, Nichols asked Boone where his money had gone. Boone told him it had been invested in a company called JJ&B. Of course, the investment had been made without Nichols's permission, and Bobby wanted to know more about the company.

He asked several times without getting an answer. After pressing further and demanding to know just what people comprise this company, Boone finally confessed, "Me."

Boone had gambled most of the money away at Las Vegas and at the Churchill Downs race track.

Attempting to recover $531,332 in lost pension money, Nichols brought legal action against Boone, his two former law firms, and a bank. If newspaper reports were accurate Nichols recovered more than he lost. The Louisville *Courier Journal* reported Nichols realized more than $900,000 when the lawsuit was settled.

Ban the RAF

An RAF fighter pilot during the Battle Of Britain, Pat Ward-Thomas was shot down over the English Channel, captured, and spent most of the Second World War in a German prison camp. The author of a number of books and the regular golf correspondent for *The Guardian*, one of Britain's leading newspapers, Ward-Thomas was an enthusiastic golfer, but there were many better at the game.

Vain as a peacock, he stood on the first tee at St. Andrews one day and tried to impress a young American caddie by showing him his driver and boasting, "Personal gift from Arnold Palmer." Maybe Palmer could play that driver but Ward-Thomas couldn't. Faced with the widest fairways in golf, Pat fought to keep his ball on the course. He was losing and his passion rose to magnificent heights. In the quiet stillness of a clear autumn afternoon, Ward-Thomas dumped his second shot into the Strath Bunker before the eleventh green. The eleventh, of course, is a par 3.

A man of flaming temper, capable of inventive profanity and awesome rage, Ward-Thomas could not accept that he, himself and alone, had played that terrible shot. With gritted teeth he looked around for someone to blame. Finally, his face the color of blood, he gave up the battle with his rising passion. Spotting the contrail of an RAF warplane passing noiselessly miles overhead. He had found the reason for all his frustrations.

Arms waving and nearly foaming at the mouth, Ward-Thomas drew himself upright, looked heavenward, and screamed, "The fucking RAF ought to be disbanded."

Confidence

Playing the sixteenth at Cypress Point, a 233-yard par 3 across a bight of the Pacific Ocean, Mike Austin, a very long hitter, went for the green with a 3-wood. He didn't make it. Not giving up, he hit four more, and each ricocheted off the face of the cliff and tumbled into the sea. When his sixth shot made it, Austin cried, "I knew it was the right club." He was on the green in 12.

A Bellyful

Tony Green, who once played on the Scottish international team, often took his golden retriever Ben with him when he played. Returning home one day, Ben showed signs of being ill and collapsed. After an examination, the veterinarian decided on surgery. When he cut him open the doctor found Ben had swallowed eleven golf balls. Ben recovered, but he was banned from the golf course.

Cruel and Unusual Punishment

After the death of Mao Tse-tung, first chairman of China's Communist Party, Madam Mao, his wife, was convicted as one of the Gang of Four. She was confined to a guest house within a park near Shanghai. During her confinement she heard rumors that a golf course was under construction in the park. So furious that the vicious capitalist game would be played within view of her jail, Madame Mao spent the first night banging on the guest house walls. The golf course was never built.

Hobday's Lament

Simon Hobday, a South African who played the European Tour for a time during the 1970s, was known as an eccentric. On a dare he once took off all his clothes in a Swiss bar and stood drinking while other customers gawked. When the owner suggested he at least put on his shorts, Hobday agreed, and put them over his head.

Playing in a tournament at Wentworth, England, not far from London, he arrived late and found all the caddies had been taken except one, a hippie-like character with scraggly, waist-length hair. He was so disgusting looking, Hobday thought he was a wino, and asked him, "Are you pissed?"

Assured he was not, Hobday told him to come along, and off they went, the caddie with Hobday's heavy bag slung over his shoulder. Hobday and the others in his group had already driven, so they set off at a speedy pace. They had to climb over a mound along the way, and as soon as they reached the top and started downhill, the caddie fell and rolled all the way to the bottom.

Seeing him lying there, then struggling to stand up, Hobday reasoned he must be on drugs. Certainly all wasn't well with him.

"He'd fallen over thirteen times by the time we got to the seventeenth tee," Hobday recalled. "By now he's got sand on his face, his hair is thick with it, the bag must weigh eighty pounds it's so full of sand. You had to wipe off every club you took out. He was in terrible trouble, this boy.

"Now the back of the seventeenth green slopes steeply down to the next tee, so I took him over to the side of the green while the other guys were putting out, and I said, 'Do me a favor, please. Do not under any circumstances come off the back of the green. Go down the pathway where the slope is nice and gentle, and I'll see you on the tee."

When the others putted out, they all went over the back of the green. As they eased their way down the slope, they heard a terrible clattering at their backs. They swung around and, horrified, saw the caddie coming after them. He'd evidently decided he was stable enough to come down the hill.

"We look around and he's at the top of the slope. You had to see this guy's stride; he was doing sixty miles an hour. Every time he put a foot down, the bag would lurch forward and crack against his shoulder. It was giving him a terrible hiding. But the weird part was he still held the pin; he'd forgotten to put it back into the hole.

"Realizing now he's going to take a tumble, he figures the only way he can stop himself is by jamming the pin into the ground like a pole vaulter. Of course, the pin is only made of Fiberglas, so it snaps.

"How he didn't kill himself I've no idea," Hobday said. "He did a bellyflop and landed on his nose on the pathway," his legs splayed on either side of the bag.

"I never even picked him up," Hobday said. "I just thought, 'I told him.'"

He hobbled onto the tee, blood spurting over everything from a gash where Hobday's clubs had sliced into the back of his head. "Oh, God, he was a horror story. I gave him ten pounds and told him, 'Go away, and for God's sake get another bag. Don't come anywhere near me tomorrow."

Sitting in Wentworth's bar later in the day, Hobday heard everyone complaining about the seventeenth hole. "Bloody incredible," one said. "I thought it was an 8-iron, and I hit it over the back of the green." Hobday realized the reason. The broken flagstick stood only half the height of a normal pin, which distorted the judgment of distance. The green looked farther away than it actually was. Hobday ignored the remarks. "I kept quiet. It wasn't me."

Part Seven

Parity and the International Scene

The Second World War ended in 1945. By the middle 1990s the world had been free of wide-scale war for fifty years. For the first time since Napoleonic times generations of young Europeans and those from the outposts of the British Empire were allowed to live their full lives. They prospered, especially in athletics, and particularly in golf.

Whereas Americans had been dominant in the game since the early 1920s, they were at first forced to share their command of the game, and then, as the foreign golfers grew more confident, they gave way. Reasons other than the absence of war lay behind this shift in power.

First of all, the rise of the common man. Feudal society had broken down gradually over the centuries, but only during fairly recent generations had the former peasant class risen economically to a point where its members enjoyed leisure time. Second, golf had expanded enormously over those same generations, especially into continental Europe and particularly in Spain, whose Mediterranean climate attracted British tourists thirsting for golf during miserable winters. Third, with the improvement and explosion in communications, the former European peasant learned through television that scads of money could be earned by those good enough to compete at the game's highest levels.

Fourth, the success of the European Tour. Until the middle 1970s the British Open, for example, had ended on Fridays so the players could return to their clubs and take care of their members over the weekend. British professionals couldn't afford to do nothing but play the game for money; they earned their livings in their shops. American tour players had been free of that pressure since the end of the Second World War. With the European Tour prospering, the British and Continental players—at least those good enough to survive—raised their games to higher levels, and they found eventually they could compete with anyone.

Fifth, and this relates to the American player, there is the Eddie Arcaro Syndrome. The great jockey summed up the effects of success by saying, "It's tough to get out of bed when you're wearing silk pajamas." Translated, his message meant it's difficult to force yourself to work hard when even mediocrity earns a lot of money. The great players do work hard, though, but the American tour golfer of the late 1980s and early 1990s gives the impression he's content to live comfortably by finishing twentieth.

With all their success, though, there was no evidence that European, South African, Australian, or any other nation or group of nations had produced another Jones, or a Nelson, Snead, Hogan, Nicklaus, Palmer, or Watson. There was every evidence, however, that instead of the game's being ruled by great players, it had become the property of the very good.

Tony Jacklin had taken the first tentative step toward ending American domination by winning the 1970 U.S. Open, but Jacklin, like Gary Player five years earlier, had honed his game in the United States. Then, in 1980, Seve Ballesteros, a dashing young Spaniard, barely twenty-three, won the Masters, the first golfer developed exclusively in Europe to break the American grip at the game's highest levels. A fierce competitor who simply would not lose, especially to an American, he breathed new life into European golf and led others to the top—Nick Faldo, Jose Maria Olazabal, Greg Norman, Nick Price, Bernhard Langer, Ian Woosnam, Sandy Lyle. The game, then, had come full circle and had truly grown international.

Embarrassing Moment

A day or so before the 1981 U.S. Open began, Jack Nicklaus complained that a couple of portable toilets should be placed inside the ropes for the exclusive use of the players. The USGA agreed, but rather than order an additional pair, the association commandeered a couple already in place for spectators.

Mike Butz, the USGA's man responsible for such things, sought out the course superintendent, whom he knew had a fork-lift, the kind of machinery he needed for his mission. Finding one of the portable johns near Merion Golf Club's ninth tee, Butz ordered it moved to a position inside the ropes. The operator slid the two prongs of the fork-lift under the little house, engaged the gears, and up it went.

As it began its ascent, though, a woman's panicked voice rent the morning air demanding to know, "What's going on out there!?" The door flew open, and there she sat in some disarray. Seeing two men, mouths agape, staring at her, she quickly drew the door shut.

The operator set her down softly, he and Butz waited until her mission was accomplished, then, after she had walked away with as much dignity as she could summon, they continued with their assignment.

Butz apologized with as much sincerity as he could muster, then explained that the move was being made on Nicklaus's suggestion.

"It was for Jack Nicklaus?" the woman asked. Butz assured her it was indeed at Jack's request. "Well," the woman said, "If it was for Jack Nicklaus, then it's all right with me."

Yul Brynner

While he was arranging to appear in a 1981 production of *The King and I* in Philadelphia, the actor Yul Brynner asked his agent to rent him "a lovely, big house in an area where I may have peace and quiet during the day." The agent obliged by arranging for a place on Golf House Road, in the Philadelphia suburb of Ardmore.

On his first Monday morning in the house, Brynner was awakened by unusual noises. He climbed out of bed, looked out his window, and cringed. Masses of people swarmed over his lawn. His agent had rented him a house that sat across a narrow road from the Merion Golf Club. The U.S. Open was about to begin.

Up a Tree

Nearing the end of the third round of the 1981 Benson and Hedges Tournament, at the Fulford Golf Club, in York, England, Bernhard Langer's approach to the seventeenth green lodged in a tree overhanging the green. First he considered taking a penalty stroke and replaying the shot, but then he realized if he could reach his ball, a little tap would drop it onto the green. While the gallery gaped and laughed, Langer climbed the tree, found his ball caught in a notch between branches, and knocked it free, saving a bogey 5.

While everyone else thought the incident funny, Langer wasn't amused. "I was afraid I might fall out of the tree," he claimed. As he walked toward the eighteenth tee, Langer handed the ball to Mark Lunn, a young spectator, not realizing it was the boy's fourteenth birthday. Lunn had the ball mounted and displayed it in his house.

Acquisitive Germans

At the 1981 German Open, played in Stuttgart, not all the light touches were evidenced around the greens. Indeed, some sort of world record might have been set in the category of pilfering. By the time the players and galleries had left the scene, so had enough property to stock a golf supply house. Peter German, the appropriately named organizer of the tournament, gave this run-down of missing equipment: Fifteen tee markers, 2500 range balls, seventeen flagsticks with their flags attached, fifteen pairs of shoes from the locker room, and uncounted umbrellas. "Someone even took the parking sticker off my car," German moaned.

Ferret up the Leg

Ron Booth, a handicap player from Leek Golf Club, in England, stood on the club's sixteenth tee with driver in hand ready to give his ball a mighty whack. As he coiled his body to deliver the powerful blow, a ferret darted up his trouser leg.

Booth felt something had gone wrong, but he wasn't quite sure what. Then the belligerent ferret sank his savage teeth into Booth's calf, and he knew all was not well. Dropping his driver and falling to the ground, Booth screamed in pain and tried to shake the vicious animal free. No such luck; the more he shook, the deeper the varmint drove his teeth into Booth's leg. Eventually Booth's partner yanked the ferret free by his tail, receiving several bites for his efforts. Both men were rushed to a hospital close by for tetanus shots. The shots administered, they returned to the sixteenth tee to complete their rounds. The ferret turned out to be someone's pet.

Human Rights

Evidently players on the losing side of a match in India don't blame themselves for their poor showing: It's all their caddies' fault, so give them a taste of the whip. A sign posted on a bulletin board of the Chandigarh Golf Club warned against the seemingly common practice of caddie beating.

Sudhar Varma, the club secretary, warned, "This is an extremely bad habit, especially on the part of players who are losing. If there are any further reports of caddie-beating, strict disciplinary action will be taken against the measure."

Watson at Pebble Beach

After driving into a fairway bunker and bogeying the sixteenth hole, Tom Watson arrived at the seventeenth tee at Pebble Beach needing two pars to tie Jack Nicklaus for the 1982 U.S. Open championship, but the seventeenth and eighteenth holes are no pushovers. Clinging to a precipice hanging over the Pacific Ocean, the seventeenth played to 209 yards in the last round, and the eighteenth, at 545 yards, skirted the edge of the Pacific from tee to green.

Watson chose to play a 2-iron to the seventeenth. When the ball took off it looked as if it would be a fine shot, drawing slightly from right to left, but the farther it flew the more pronounced the draw. Watson had been a little too quick with his hands. His ball hit once on the left edge of the green, then hopped into weedy rough between two bunkers. His position looked grim, but he had two breaks. First, the ball had not snuggled too deeply into the grass. Second, Watson is one of the few players who had actually practiced this shot. He knew how to play it, and he had the confidence to believe he might still birdie. As Tom set himself to play the shot, his caddie, Bruce Edwards, said, "Knock it close." Watson responded by saying, "I'm not going to hit it close, I'm going to make it."

The ball popped out of the grass, hit the edge of the green, and began rolling toward the cup. As he saw how the ball was running, Watson squatted low and yelped, "That's in the hole." The ball hit the flagstick and dropped straight down.

The crowd roared, everyone leaped to their feet, and Watson flung his arms toward the heavens and ran around the green. He had the birdie that moved him into the lead. His birdie on the eighteenth increased his winning edge to two strokes over Nicklaus.

Nicklaus had seen the shot on a television monitor in the scorer's tent behind the eighteenth green. Jack was there waiting for him when he finished. They shook hands, then Nicklaus smiled and said, "You little son-of-a-bitch, you're something else. I'm proud of you." Reflecting on the incident later, Watson called it, "The best shot of my life. It had more meaning than any other shot in my career."

The following year the USGA wanted to capitalize on the shot for a television commercial. Watson was to replay the shot, this time at Sawgrass, in Ponte Vedra, Florida, the day before the Tournament Players Championship. With the television crew in position, Watson stepped to the edge of the green and said, "They say practice makes perfect. Well, in golf there might not be any such thing as perfect, but some shots are better than others."

Having made his pronouncement, Watson gripped his wedge, addressed the ball, and popped the ball into the hole. Perfect. A USGA person cried, "Terrific, Tom. Terrific," then turning to the cameraman he asked, "Get it on tape?" Casual as you please, the cameraman drawled, "No. I thought it was just a run-through."

Teeth grinding, the USGA type asked Watson if he would mind trying again, all the while knowing they were in for a long afternoon—those shots don't often fall. Unperturbed, Watson once again spoke his lines without a hitch, set up the shot, and, amazingly, holed it again.

The Missing Painting

Since 1954 a painting of a Dutch boy holding a golf club had hung in the museum of the United States Golf Association. It was a small painting, perhaps a foot square, done on wood. It was one of the museum's prized exhibits and thought to be valuable.

One Friday afternoon in 1982, as she showed a visitor around, Janet

Seagle, curator of the museum, noticed the painting had disappeared. Someone had taken it off the wall and sneaked out with it.

After filing a police report and advertising the theft in a publication of the Golf Collectors Society, the USGA was told the picture had been put up for sale by an art dealer in New York. Miss Seagle went to the dealer's shop, along with a lawyer, recognized the painting as the picture that had been stolen, had a blowup made of a color slide from the museum's files, and within a short time the Dutch boy hung once again where it had been on display for twenty-eight years.

At the same time, Miss Seagle learned quite a bit about the painting. Sadly, it probably wasn't as valuable as the USGA had believed. First, it was a copy, not an original, and, second, it hadn't been painted in 1820, as its donor claimed, but between 1830 and 1925. It had been painted on plywood; plywood wasn't invented until 1830.

Fourteen for Norman

Greg Norman is not one to give up easily. Un-nerved when a photographer accidentally tripped his shutter while he was in mid-stroke, Greg hooked his drive into a thicket on the seventeenth hole at Lindrick Golf Club, in England, as he defended his title in the 1982 Martini International.

Ever the optimist, Norman looked over his dismal lie, told himself, "Never surrender," then flailed at his ball, trying to chop it back to the fairway. Plan A didn't work.

Evidently a student of Shakespeare, Norman decided the better part of valor is discretion, declared his ball unplayable, took a penalty stroke and dropped clear of the bush. Then he hit it back into the bush. Another penalty shot, another shot back into the bush. Four penalty strokes in all, and Norman worked his way back to the fairway with his tenth shot, missed the green with the eleventh, made it with the twelfth, then two-putted for 14. After finishing the round in 82, Norman vowed, "I'll know that photographer if I ever see him again."

Soggy Fad

Like college kids swallowing goldfish in the 1930s, diving into ponds near the final green became a fad during the early 1980s. Exuberant over a tournament victory, the winner would dive into the pond and swim around to cool off. Jerry Pate, however, went a little further when he won the 1982 Tournament Players Championship, the first tournament played at the Tournament Players Club in Ponte Vedra, Florida, a course many of the players hated. Before going out for the final round, Pate swore that if he won he would throw Pete Dye, who had designed the course, into the broad pond alongside the eighteenth green, "Alligators or no alligators."

Sure enough, Pate cruised home with 67, shot 280 for the seventy-two holes, and won by two strokes. When the final putt fell, Pate ran not to Dye but to Deane Beman, the commissioner of the PGA Tour, and pushed him overboard first, then grabbed Dye and tossed him in, and ended the day by belly-whopping in himself. Not that they had anything against Pate, but quite a few of the contestants who had survived the week were hoping the water was ten feet deep and indeed teeming with hungry gators.

Special Guest

On the Saturday of the 1983 Masters, Paul Callahan, a twenty-five-year-old quadriplegic confined to a wheelchair since slipping and falling on a wet floor in 1979, decided he wanted to see the last round. He lived only 350 miles from Augusta, but he had no ticket, none were available, and he had no place to stay. Even if he had, Augusta had been saturated by rain, the ground was slippery and muddy, and no golf course is laid out to accommodate wheelchairs.

None of this mattered to Callahan; he had a history of overcoming the impossible. Doctors had told him he might never sit in a wheelchair, and he certainly would never walk again. After months of therapy, not only could he sit, he could move both his arms and legs; the previous September he took his first step.

The rain held the key, though. It had fallen so heavily the fourth round had to be held over until Monday, a day late. By the time Callahan reached Macon, Georgia, in his special van with his therapist and a friend, Hord Hardin, chairman of the Masters, had learned he was on his way. Some ticket-holders who couldn't stay for Monday had left theirs with Hardin. Callahan could buy the three he needed.

Then Hardin arranged for a place for them to stay, and Hardin and his secretary, Kathryn Murphy, figured out paths Callahan could follow, and pointed out spots where he could see the play. While he sat near the clubhouse, Arnold Palmer stopped to chat with him, and when Arnold left for the practice tee, Callahan thanked him for taking the trouble to visit. "No," Palmer said, "it was my pleasure."

Costly Whiff

Hale Irwin was locked in a tight battle with Tom Watson, Jack Nicklaus, Craig Stadler, Nick Faldo, and Lee Trevino as he approached the fourteenth green at Royal Birkdale during the third round of the British Open in 1983. On the green with his second shot, Irwin laid his approach putt within two or three inches of the hole, but when he stepped up to tap his ball into the cup, he tried to hit it with the sole of his putter. Instead, the putter hit the ground and bounced over the ball without touching it. He had whiffed; instead of a par 4, he had to take a bogey 5.

The next day, while Faldo, Stadler, Raymond Floyd, David Graham, and all the rest faded from the fight, only Irwin held on to challenge

Watson, who was going for his fifth British Open. Watson finished with 275. Irwin placed second, just one stroke behind him.

Combat Injury

Royal Birkdale's eighteenth hole can be the most vicious finishing hole on the British Open rota. A par 4, it measures 473 yards with a slight bend to the right, and it often plays directly into a stout wind, as it did on the final day in 1983. Coming to the eighteenth, Watson stood one stroke ahead of Irwin and Andy Bean, who had already finished. Watson hit a terrific drive, a fade that settled in mid-fairway just over 200 yards short of the green.

When the group ahead finally cleared the green, Watson tore into a 2-iron. His ball flew directly at the flag, but then began hooking. Alfie Fyles, Tom's caddie, stammered, "Don't hook t-t-too much."

"Don't worry," Watson said, "the wind will straighten it out." It did; the ball pulled up fifteen feet from the hole. Neither Watson nor Fyles saw it, though. As the ball rose, the gallery, as it did every year in those days, broke through the restraining ropes, scrambled for the green, and blocked the players' view.

Now came the hard part; players, caddies, and referees all had to work their way through the crowd, which didn't give ground easily. Evidently trodden on by the gallery, Fyles badly damaged his wrist. Watson won the championship, his fifth, but Fyles was never again able to play golf.

In the dark of night before the third round began, vandals conducting a terror campaign to release the convicted killer Dennis Kelly caused extensive damage to Birkdale's sixth hole, perhaps the most difficult test on the course. They dug holes in the green and painted slogans referring to their drive to have Kelly freed. The following night government officials sent a detachment of paratroopers to Birkdale to protect the course from further damage.

Wadkins at PGA National

Jack Nicklaus served as captain for the 1983 Ryder Cup match, at the PGA National Golf Club, in Palm Beach, Florida. Both the Europeans and the Americans had won thirteen points with only two matches left on the course. Then Tom Watson closed out Bernard Gallacher on the seventeenth, and in the last match still alive, Lanny Wadkins was coming to the eighteenth one hole down to Jose Maria Canizares. Unless Wadkins won the hole, Europe would halve the Ryder Cup.

Playing first, Canizares left his approach short, giving Wadkins an opening. Among the most reliable clutch players in the game, Wadkins played a stunning pitch that hit the green, danced around the cup, and braked within a foot of the hole. He won the hole, halved his match, and saved the Ryder Cup, After Wadkins nearly knocked the flag from the hole, Nicklaus felt such relief he dropped to his knees and kissed Lanny's divot.

What Happened?

Gary Player's ball lay beside a broad-leafed weed just off the green of the sixteenth hole at Desert Highlands, in Scottsdale, Arizona, during the 1983 Skins Game, the first in the series, matching Player against Arnold Palmer, Jack Nicklaus, and Tom Watson.

A total of $120,000 was at stake at the sixteenth, a long par 3 with a three-tiered green. To win the skin, Player would have to hole his chip shot.

He stood up to the ball, but what he did next became a matter of dispute. He bent over and seemed to do something with the ground, then set his club down behind the ball, ready to play. His chip missed, but the hole was halved, and the four men moved on to the seventeenth, which would be worth $150,000.

During the good-natured bantering on the tee, Watson seemed strangely removed. His lips set in a tight, grim line, he said nothing. Player birdied the seventeenth and won $150,000. With $20,000 from winning an earlier skin, Player collected $170,000 for the day. Watson had won just $10,000 by scoring a birdie on the first hole.

About an hour after the game had ended, Watson walked along a dirt path with Joe Dey, the former USGA executive director who had refereed the game. Watson asked Dey, "What would happen if a broad leaf of a weed overhung your ball and you moved it? Would that be the same as moving a tree branch?" Dey agreed it would, then said, "Why do you ask?" Watson explained then that Player had done exactly that.

Next, Watson confronted Player. They spoke for a moment, then stepped a few feet off the road, along with Nicklaus and Dey. Facing Player, Watson was heard saying, "I accuse you, Gary. You can't do that. I'm tired of this. I wasn't watching you, but I saw it." Answering the charge, Player maintained, "I was within the rules." Watson claimed Player had cheated by flattening a leaf of that weed, improving his lie. Player said he hadn't.

Speaking later, Watson said, "What I saw was a violation. I challenged Gary on it. I asked him if he was ignorant of the rule. As I saw it, he was moving a leaf of a weed right behind his ball so he could have a clear path to his ball for his clubface. I know the leaf was rooted, because it popped back up to its original position. When I took it up with Gary, he said he was touching something beside the ball, not behind it."

Player said, "I think there's been a misunderstanding. He [Watson] was under the impression I had moved a leaf, and I assured him, that was not the case." Dey could only say, "I wish I hadn't asked, 'Why do you ask?'"

Wrong Game

One of the leaders through the first thirty-six holes of the 1983 Canadian Open, Andy Bean wasn't having a good third round. He stood on the eighteenth green of the Glen Abbey Golf Club needing two putts to shoot 75. His first putt left him only inches from the hole. In a moment of levity he turned his putter upside down, gripped the clubhead, and tapped the ball into the cup with the handle.

Since he had violated the rules of golf. Bean was penalized two strokes. The next day he blistered Glen Abbey with a round of 62, five strokes better than anyone else, shot 277 for the seventy-two holes, and missed tying for first place by two strokes.

Remembrance of Things Past

During the 1954 Masters, Ben Hogan and Sam Snead were locked in a tight battle with the amateur Billy Joe Patton, a fast-swinging North Carolinian just beginning to build his reputation not only as a first-class golfer but as a witty and entertaining personality. Patton had scored a hole-in-one on the sixth, a shortish par 3, but he had fallen behind by the time he had played the little twelfth hole.

After a good drive on the thirteenth, a dog-legged par 5 measuring 485 yards, he felt he had to go for the green and set up a certain birdie and possible eagle. Patton mis-hit his ball; it fell short and tumbled into the creek that knifes across the fairway in front of the green. He fell behind and never caught up. Snead eventually beat Hogan in a playoff.

Thirty years later Ben Crenshaw came to the thirteenth hole leading the 1984 tournament and faced the same tantalizing choice of whether to go for the green and risk seeing his ball drop into the rock-strewn creek, or lay up and hope to pitch close enough for a birdie or perhaps a routine par. A student of golf history, Crenshaw was familiar with the story of Patton in 1954. As he stood in the fairway he had already made up his mind to go for the green. With his hand on his 3-wood, Crenshaw turned and saw Tom Kite hit his tee shot into the pond on the twelfth. A reprieve.

Then he turned again and began searching for his father in his gallery. Instead, he was attracted to a beam of sunlight that filtered through the tall Georgia pines. It spotlighted Billy Joe. Seeing Patton, Crenshaw shoved his wooden club back into his bag and laid up. He won the tournament.

Told later of how his being in Crenshaw's gallery influenced Ben's decision, Patton said, "He must had had visions. I wasn't there."

Chen's Double Hit

T. C. Chen, a Taiwanese, led the 1985 U.S. Open by four strokes with fifteen holes to play. Standing on the fourth tee of the Oakland Hills Country Club, near Detroit, he drove straight down the middle, but then pushed a 4-iron into a copse of trees right of the green. With a clear opening to the pin, he tried to play the shot too close to the hole, misjudged it, and left his ball in clumpy rough a yard or so short.

Drawing his wedge, Chen chopped down. The ball popped straight up and hung in mid-air. When the club, delayed an instant by the thick grass, broke loose it hit the ball once again, flipped it high and to the left, onto the collar of the green. Chen was stunned; his face a blank mask. By hitting the ball twice on the same stroke, he had incurred a penalty stroke; instead of lying 4, his ball lay 5. He chipped well past the hole, got down in two putts, and scored 8 on a par 4 hole. He fell to second place, a stroke behind Andy North. North eventually won, and Chen tied for second place, one stroke behind.

Flying Tackle

Four strokes behind Sandy Lyle after seventy-one holes of the 1985 British Open, Peter Jacobsen had missed the final green at Royal St. George's. He had no chance to win, but he would finish as well as he could.

Just as he was about to play a little chip, he spotted a male streaker breaking out of the crowd and running across the green in front of the thousands of spectators crowding the tall bleachers. Calling on his American background, Jacobsen reacted quickly and leveled the streaker with a first-class football tackle.

Denying any sense of outrage, Jacobsen explained he had only one reason for bringing the streaker down. "He was about to run across the line of my shot." He also said, "I put my shoulder in where it hurt the most."

Silence of the Lamb

During the 1985 U.S. Senior Open, a small boy in the gallery, upset about something that upsets small boys, was crying loud enough to disturb the golfers. Hearing his pitiful wail and realizing the boy was causing a problem, Miller Barber walked over to the gallery ropes, chatted for a moment, and handed him a ball. The caterwauling stopped, and Barber went on to win the championship. When the round ended, the little boy approached Barber, told him, "I have the ball, Mr. Barber," then planted a large wet kiss on his cheek.

A Gun in Hand?

As the final round of the 1985 Women's Moss Creek Invitational began, on Hilton Head Island, South Carolina, a female guard spotted a spectator sitting in his car cradling what looked like a pistol. The guard left to find help, but when she returned the man was gone. Searching the car, officials found a pairing sheet with circles drawn around the names of Beth Daniel, Betsy King, and Hollis Stacy.

Told of the incident, Suzanne Jackson, the LPGA's director of tournament operations, and Tim Moss, the club's tournament director, suspended play and called the players into the locker room.

Local police identified the car's owner through a check of his license plates and found him about two hours after he had been seen sitting in his car. They found nothing on him but a bottle of suntan lotion. Miss Jackson wasn't satisfied, however, claiming the guard who first saw the man was a nationally ranked sharpshooter who had identified the make and caliber of the pistol. She also reminded the police that Moss Creek had so many ponds and streams, the man could have disposed of the gun very easily.

After questioning, the man was escorted from the course and the tournament resumed.

Just to be safe, meanwhile, Sylvia Bertolaccini, who had been mistaken for Beth Daniel occasionally, took a magic marker and wrote on her visor, "I Am Not Beth Daniel."

A Sure Giveaway

A customs official at the English cross-Channel port of Newhaven approached forty-six-year-old Karl Melms. The official became suspicious when Melms told him he planned to stay at the Wentworth Golf Club Hotel and play the course, one of the best-known in the game. Feigning interest, Daryl Hickman, the customs official, asked Melms to show him his swing. Not only did Melms grip the club badly, he swung in the wrong direction as well.

Hickman ordered a search of Melms's car and turned up more than $400,000 worth of cannabis resin, which is used to make an intoxicating drug with the same effect as marijuana.

Lucky Loss

First-class persimmon, the wood of choice for the best clubs, became so scarce during the middle of the twentieth century, manufacturers were forced to use low-grade wood that hadn't aged enough and was consequently full of pores. The MacGregor Company developed what it called an oil-hardened process. Clubs were soaked in an oil bath to fill the pores.

As years passed and the quality of persimmon continued to decline, the pore became so large that one expert claimed that soaking the current supply of wood in oil would make the clubs so heavy a player wouldn't be able to lift them.

In this atmosphere, the Power Bilt company realized a benefit from some shoddy storage procedures. Someone had bought 5000 blocks of persimmon during the 1950s, consigned them to a warehouse in Chicago, and then forgot about them. Nearly thirty years later somebody else swept aside the cobwebs and found 5000 well-seasoned air-dried persimmon blocks ready to be shaped into golf clubs.

The Secret

During his long and colorful career, Tom Weiskopf placed second in four Masters, but he never won. Gradually he cut back on his tournament golf and began doing television commentary. He was in the television booth when Jack Nicklaus won the 1986 tournament in one of the game's major surprises. As Nicklaus lined up his putt on the sixteenth green, an announcer asked Weiskopf what Jack was thinking of at that moment. Without a pause, Weiskopf said over the air, "If I knew what was going through Jack Nicklaus's head, I would have won this golf tournament."

Two for One

Fran Wadsworth, a pro from Columbus, Georgia, hooked his drive into the woods bordering the third hole at Shinnecock Hills during the first round of the 1986 U.S. Open. He found his ball, which bore an identifying mark. He didn't realize it was sitting atop another ball of the same brand and number, embedded in the soft ground. Wadsworth took a mighty swing and dug out both balls. Standing well within the trees, with their vision blocked by overhanging branches, neither Wadsworth nor his caddie nor anyone noticed that two balls came flying out, Wadsworth's into the fairway, the other ball just a few yards forward of him. Wadsworth saw only the second ball, stepped up to it, and played it toward the green. At that point he saw the other ball, identified it, and realized what had happened. He sustained a two-stroke penalty for playing a wrong ball, scored 7 on a par 4 hole, shot 81, and missed the cut.

Out of Balls

July 11, 1986, was a very hot day in Williamsburg, Virginia, so hot that Bill Kratzert's caddie felt that lugging a dozen or so balls around the golf course bordered on the foolish. So he emptied Kratzert's bag of all but three balls as they set out in the Anheuser-Busch Classic.

Kratzert, though, was having a bad day. He lost one of those three balls with a very wild hook, lost another with an equally bad shot, and, strangely, lost still another with a terrible shot. When he asked for another ball, his caddie had to tell him there were no more. Kratzert was forced to withdraw.

Craig Stadler's Towel

Craig Stadler had yanked a drive so far under a low-growing pine tree during the third round of the 1987 San Diego Open he could play a shot only from his knees. With pine resin all over the ground, Stadler spread a towel to protect his trousers as he took his stance, so to speak. His round over, he turned in his scorecard properly signed and attested, then returned the next day to complete the tournament.

Early in the Sunday telecast the network showed a re-run of the previous day's incident. Then telephones began ringing. Viewers felt Stadler should have been penalized, because earlier in the year the USGA and R and A had issued an official Decision on the Rules of Golf stating that a player who kneels on a towel while he plays a shot has built a stance and has breached Rule 13-3.

When Stadler finished, the tour staff asked if he had indeed knelt on a towel to play the shot. Stadler admitted he had, without knowing he had violated a rule. Because his sin had been committed a day earlier and he hadn't added a two-stroke penalty, he had turned in a score lower than his actual score. He was disqualified.

A few months later Stadler lifted an embedded ball from the rough at Muirfield during the British Open, another violation. Told to add two strokes again, Stadler mumbled, "I better learn the rules."

A for Accuracy, F for Result

Raymond Floyd unleashed a perfect drive on the eleventh hole of the Stadium Course during the 1987 Players' Championship, but it turned out to be the most costly shot of the championship. His caddie had left his bag lying in the fairway. Floyd's ball hit the ground and rolled into his bag, costing him a two-stroke penalty for hitting his equipment.

Less Is Better

Just to see how well he could do, John Humm, a good amateur golfer from Long Island, played nine holes using only his 3-iron. He shot 34. The next day he played the same nine holes with his full complement of clubs and shot 40.

Nick Faldo

Nick Faldo, the great English golfer, began his career as a genuine prodigy. On the first hole of the first round he ever played, he hit a driver and a 3-wood onto the green of a 450-yard par 4. Men live their lives without ever doing that. At the time, Faldo was fifteen.

Even though he ranked as the best player in the game during the late 1980s and early 1990s, Faldo also ranked among the most disliked of all the great players. In his match against Graham Marsh during the 1988 World Match Play Championship, Nick pulled his drive into trees and had to play a sharp hook to reach the green. Nick put a little too much into the shot; his ball rolled over the green and among the spectators. One fan who had partied too much kicked Nick's ball back onto the green about 10 feet from the hole.

Not Faldo or Marsh or the referee saw it, but a marshal told the referee the ball had been deliberately kicked. The referee asked if the ball had been in motion when it was kicked. Informed that the ball had indeed been moving, the referee ruled the ball should be played as it lay.

The ruling would have been correct if the ball had been moved accidentally, but it wasn't. Instead, the referee should have had Faldo drop the ball where it might have rolled had it not been deflected. Fans gathered around the green felt that under the circumstances Faldo should have offered Marsh a half, but not realizing how his ball had stopped so close to the hole, he went ahead with the referee's instructions and won the hole.

As Nick walked to the next tee, an incensed marshal, furious that Faldo had taken advantage of a spectator's blatant mischief, leaped toward him, grabbed him by the throat, and called him a cheating bastard. Seeing the attack, someone remarked that this was a case of the fan hitting the shit.

A thick Scottish fog, or haar, settled over St. Andrews as Faldo approached the eighteenth green during the 1988 Dunhill Cup, a fog so thick he had trouble picking out the green and judging the distance to the flagstick. Des Smyth, playing for Ireland and leading Faldo by one stroke, played his approach through the mist, but Faldo refused. He delayed, walked around, and refused to play. As the delay wore on, the crowd gathered in the grandstand and along Links Road, which borders the right side of the home hole, grew restive. Through the mist a voice bellowed, "Get on with it, you plonker." After a half-hour delay, the referee, however, agreed with Faldo that conditions made the course unplayable and called it off for the day.

Blinded by the fierce tropical sun reflecting off his putter blade during the Kenyan Open, Sandy Lyle solved the problem on the second hole by placing a strip of tape along the polished top edge of his club, violating the rule that prohibits changing characteristics of a club in mid-round.

Everything went smoothly until the sixth hole, where Faldo, playing with him, noticed the tape. Nick asked about it, Lyle explained, and Faldo said nothing more. Not, that is, until the ninth hole, where he slipped away and reported the violation to tournament officials without telling Lyle.

Lyle was disqualified. To no one's surprise, he was also annoyed with Faldo. Since that incident, the rules have been changed to allow a player to protect himself against the sun's glare.

A Better View

Jack Nicklaus, Greg Norman, and Seve Ballesteros were playing together in a practice round leading up to the 1988 U.S. Open, at The Country Club, in Brookline, Massachusetts, followed by a huge gallery confined outside ropes that encircled each hole. One man walked with the players; indeed he stood at greenside as they putted out. Two members of the USGA staff patrolling the course noticed the man with the players and assumed at first he was part of the group, perhaps a relative, but he spoke to no one, nor did anyone speak to him. His role, at best, seemed obscure. Finally one of the USGA staff approached the man and said, "Pardon me, sir, but are you part of this group?"

Surprised, the man said, "No."

"What are you doing here?"

"I'm just a spectator," the man said.

"No," the USGA staff man said, "I mean what are you doing *out here*?"

"I can see better here," the man answered.

While they all agreed he had spoken the truth, they led him outside the ropes nonetheless, and left him to mingle with thousands of other spectators who lacked his sense of enterprise.

Lockerbie

Until the evening of December 21, 1988, Lockerbie, Scotland, a small village between Carlyle, England, and Glasgow, had one claim to lasting fame. It was there in 1867 that the Marquis of Queensbury wrote the rules for modern boxing. After December 21, Lockerbie took on a much more sinister association. Pan American Airways Flight 103 exploded over the village just after 7 o'clock that evening killing all 259 passengers and crew of the Boeing 747 and eleven residents of the village. Debris from the explosion fanned over the Lockerbie Golf Club, and one of the bodies landed alongside the first fairway. The accident caused the course to close for a month. The following year Lockerbie Golf Club observed its centennial.

A Call to Arms

Practicing for a Senior Tour Tournament at the Silverado Country Club, in Napa, California, Dave Hill pounded 4-irons into the wind from one position at the circular practice range while at the station opposite him, J. C. Snead hit 3-woods downwind. As Snead's shots flew closer and closer to Hill, Dave waved his towel overhead, signaling Snead that his shots were falling too close, endangering him and his wife, seated in a golf cart close by. Seeing Hill waving to him, Snead put his 3-wood aside, but then picked up his driver. His first drive slammed into the cart where Mrs. Hill sat.

Angry now, Hill grabbed his 5-iron and, eyes blazing, marched deliberately around the range. Approaching Snead, he gripped his iron in both hands, raised it overhead, and chopped it down toward Snead's head. Realizing what was about to happen, Snead raised his driver to parry the blow. The shaft snapped, and the two men flailed at one another with their fists. With both men bloodied, they were pulled apart by tournament officials. Both men were in their 50s.

Four Aces

The sixth hole at the Oak Hill Country Club measured 169 yards for the 1989 U.S. Open, a relatively simple par 3, played from a high tee across a narrow creek that swirls along the green's left side before it curls around the front, then sweeps across the approach to the fifth green. The stream crowds against a two-level green shaped like a fan, with the narrow handle end pointing toward the tee and the higher ground lying to the rear, where the green widens and spreads out. The hole was cut in a hollow toward the front for the second round of the Open, in a spot where balls landing within relatively close range would funnel toward the cup.

Doug Weaver, in the first group off the tee, had just double-bogeyed the third and bogeyed the fourth. Stepping onto the sixth tee, he drew his 7-iron, hit a lazy shot that carried past the flagstick, hit against the gentle slope that rises toward the upper level, took the backspin, then drew back down the hill and into the cup for a hole-in-one. The crowd cheered.

Six groups passed through before Mark Wiebe arrived. Like Weaver, Wiebe played a 7-iron that drifted a bit left of the hole, spun right, and dropped into the cup. Another hole-in-one. The gallery roared; two holes-in-one in one morning.

Another group passed, then Jerry Pate hit a 7-iron that hit the slope six or eight feet past the cup and rolled back in. Now three aces in one day.

In the next group, Nick Price was standing on the tee watching Pate. When his turn came, he too played a 7-iron that hit about ten feet right of the hole, jumped forward, then spun back into the cup. Another hole-in-one—the fourth in one morning. The gallery's thundering shrieks must have carried from Rochester to Syracuse.

Four years had passed since the last hole-in-one in the Open, and then, within an hour and a half, four men had aced the sixth at Oak Hill, three within four groups, and all by 10:30 in the morning. There were no more throughout the championship.

The USGA's P. J. Boatwright had set the pin positions for the day. Asked what degree of difficulty he had assigned to that position, he riposted, "Obviously very easy."

No Friend at All

The 1989 Million Dollar Challenge, a tournament of no significance except for the money it pays, ended in Bophuthatswana, in Africa, with Tim Simpson and Scott Hoch listed in a tie for second place. Simpson, therefore, was chagrined when he stepped up to the lectern and was handed a check for third place, about $25,000 less than he had expected.

Asking how come, Simpson found that Hoch had been refused a drop without penalty for line-of-sight relief from a television tower blocking his approach to a green. Instead, he had to declare his ball unplayable and accept a one-stroke penalty.

Later, though, Hoch learned that another player had indeed been given free relief from the identical spot by another referee. He raised the issue with the head of the rules committee, then led him to the spot and explained what he had been told. The chairman agreed that Hoch had been entitled to relief without penalty and duly subtracted one stroke from his score. Hoch took second place alone.

Even though he was among the least popular men on the PGA Tour, Hoch didn't deserve the indignity Simpson heaped on him. Almost blue, Simpson fumed unreasonably, "I was the only friend that asshole had, and now he's got nobody."

Whither the Walker Cup

Missing from Britain for eighteen years following nine successive losses to the United States, the Walker Cup went missing once more after the British and Irish side won it again in 1989, beating the United States by 12½ to 11½, in Atlanta. Later in the year, the Royal and Ancient Golf Club lent the trophy to John James, one of the men responsible for choosing the team's members. James had borrowed the trophy to show at a building society conference in a hotel in Gloucestershire, in England. James put the trophy on the bar after the conference, and was delivering a late-night talk on the trophy's sixty-eight-year history when suddenly it vanished.

Questioned about the trophy's disappearance, the hotel manager said, "One minute it was there and the next it was gone," then added in a masterpiece of understatement, "Mr. James was understandably anxious to find it again."

The cup had been taken as a prank; later in the night someone returned it to the hotel's front desk.

A Short Life

Turned down in their bid to buy the Augusta National Golf Club, a group of Japanese decided to build one of their own. Or one hole of it. Spending $12 million on it, they put up a replica of Augusta's clubhouse and the club's twelfth hole, a devilish par 3 calling for a pitch across the arm of a creek to a wide but shallow green. Part of the International Garden and Greenery Exposition, the project had Augusta's cooperation.

Each day 160 golfers had one shot at the green; if they missed, there was no second chance.

After the exhibition closed, the hole was torn up and the clubhouse torn down.

Giveaway

Probably no one has given away as many clubs as Larry Nelson. For various offenses he had given away drivers, 3-woods, and sand wedges he felt had misbehaved.

Taking a light-hearted view, Nelson explained, "I'm trying to pick up a gallery, but that might not be a good idea. They'd start hoping I'd hit everything into the water so I'd give away more clubs.

His putter came next. Larry had closed within reach of Scott Hoch during the second round of the 1990 Bay Hill Classic, in Orlando, Florida, dipping to eight under par by birdieing the fourth, fifth, and tenth holes. It was a short-lived streak; he two-putted from six feet on the twelfth, three putted the fifteenth, then missed from three and a half feet on the sixteenth. Stepping off the eighteenth green, he handed the putter to a young boy, saying, "If you had followed me around you'd know why."

Painful Misdirection

Just two strokes off the lead during the last round of the 1990 Australian Open, Bret Ogle found his ball behind a tree at the Royal Melbourne Golf Club. Trying to thread a shot past the tree, Ogle played a solid 2-iron, but he pushed the ball slightly off line. It ricocheted off the trunk, slammed into his knee, and broke his kneecap.

Foul Ball

Taking time out from the tour, Dan Pohl tried to teach the 8-year-olds on his son's baseball team how to drive through the ball. He knelt on one knee and lobbed balls from about five feet away. The kids were to hit the ball into a screen.

All went well until one youngster turned as he made his swing and drove the ball smack into Pohl's mouth. The blow split Dan's lip and loosened his front teeth. He needed 24 stitches to close the gash, and he withdrew from the Greensboro Open. No permanent damage, though.

Historic Stroke

Playing a wedge to the eighth hole at Nairn, Scotland, Bill Grant, a businessman and scratch golfer from Aberdeen, took a fairly deep divot and hit another object below ground. At first he thought he had hit a stone, but a closer look showed he had dug up another ball.

He put the ball in his pocket and later showed it to Ronnie MacAskill, the Royal Aberdeen Golf Club's manager, an expert on golf ephemera. MacAskill identified the ball as a gutta percha with mesh markings made between 1850 and 1860, worth between £300 and £400.

Nairn Golf Club wasn't founded until 1887, but the game had been played over the club's grounds for years before the club was organized. Rather than sell the ball, Grant donated it to Nairn for the club's collection of golf antiques.

Bombs Away

Harry Toan, a caddie on the European Tour who lives in Ulster, took a ferry to the island of Jersey, in the English Channel, and left his car parked by the terminal. After he had been gone for some time, the police, spotting the Ulster license plates, decided this was a suspicious vehicle, towed it away, and blew it up, afraid it might contain a bomb.

Irate at learning his car had been destroyed, Toan claimed the police had blown up nothing more than his golf clubs and two soccer balls.

Indian Uprising

The Mohawk tribe went on the warpath once again in 1990, when the Oka Golf Club attempted to add nine holes to its course on Reservation land in Canada. The Indians objected to expanding the course and tried to hold off the developers by force.

In the early days of the siege, a gun battle broke out and a Quebec provincial policeman was killed. As the dispute wore on, the Indians built a barricade on a bridge, seriously disrupting commuter traffic to and from Montreal. In response, the Canadian army blocked delivery of food, oil, and other supplies. The dispute was finally resolved when the Canadian government agreed to buy the disputed land and turn it over to the Mohawks.

Meantime, for seven weeks in July, after the barricades went up, no one swung a club at Oka.

Challenge . . . Sort of

Playing the eighth hole of the Hazeltine National Golf Club, near Minneapolis, in a practice round the day before the start of the 1991 U.S. Open, Peter Jacobsen bet Scott Simpson five dollars that Jacobsen's caddie could beat him on that hole. Simpson accepted the bet. While the gallery laughed, Peter's caddie, dressed in jeans and wearing sneakers, played his shot within eight feet of the cup. Simpson's ball settled fifteen feet away. Jacobsen bent over, lifted his bag, and lugged it to the green while the gallery cheered him on. He lost the bet. Simpson holed his putt for a birdie 2, and the caddie missed.

Adios

Given a special exemption form qualifying for the 1991 U.S. Open, Ronan Rafferty, from Northern Ireland, stood eleven over par after twenty-seven holes. Obviously going nowhere, he headed for the clubhouse, first telling Craig Parry, one of his playing partners, he had to visit the locker room, but then telling Corey Pavin and Ed Gowan, the referee, he was quitting. He caught a 6 o'clock flight back to Britain.

Once home, though, he ran into trouble from the European Tour. Since he had been given the special exemption, the tour felt his withdrawal, especially under those circumstances, reflected badly on the organization. Within a short time Rafferty was hauled into the tour's office where with fire spurting from his nostrils, Ken Schofield, the executive director, demanded to know if Rafferty would have withdrawn if he had been eleven strokes under par rather than eleven over. Rafferty admitted he would not. He was fined.

Reprise of a Sort

Johnny Miller and nineteen-year-old Seve Ballesteros were paired in the last round of the 1976 British Open, at Royal Birkdale. Miller shot 66 that day and won the championship while the young Ballesteros shot 74.

As a sentimental gesture, the R and A paired them again for the first two rounds when the championship returned to Birkdale in 1991. In the reprise, Miller shot the 74 and Ballesteros the 66. Neither man won. Miller missed the cut, and the Australian Ian Baker-Finch won the championship.

When Mark Calcavecchia, the 1989 champion, headed home after the 1991 championship, his baggage weighed considerably less than when he arrived. Piqued after shooting 79 in the second round and missing the thirty-six-hole cut, Mark gave away his clubs, growling, "I hate them."

Calcavecchia gave them to Jim Paton, a greenkeeper from a nearby course who was on duty raking Birkdale's bunkers. Paton's own clubs had been stolen. "Jim offered me his rake in exchange," Calcavecchia said. "Maybe I'd have done better playing with that."

Out in 34, Back in an Ambulance

Richard Boxall stood just three strokes behind Ian Baker-Finch when he stepped onto the ninth tee at Royal Birkdale during the British Open's third round in 1991. As he stood up to his ball and swung into the shot, spectators heard a loud "crack." To some it sounded like a gunshot. Boxall dropped to the ground, his face twisted in pain. One spectator thought he had snapped the shaft of his driver and embedded himself with the jagged end.

Nothing of the sort. The shin bone of his left leg had snapped. He was rushed to a hospital, where doctors diagnosed the injury as a stress fracture. Boxall had suffered pains in his left leg for a couple of days and had shanked two shots because he had flinched. He refused to flinch on the ninth.

Just a Naked Woman

After just teeing off at Royal Birkdale, Jose Maria Olazabal was walking toward his ball when out of the bushes streaked sixteen-year-old Sherrie Beavon, a modestly endowed girl from Yorkshire. Wearing nothing more than a few layers of purple eye-shadow, Miss Beavon raced up to Olazabal, threw her arms around him, gave him a warm kiss, then with gallery marshals in pursuit and the crowd cheering her on, sped off toward the green. There she was captured by two marshals, one of whom ungallantly refused her request to borrow his sweater, adorned with the crest of the Royal and Ancient Golf Club of St. Andrews. A more warm-spirited marshal lent her his coat.

All the while Miss Beavon's mother looked on and later tried to explain what had happened. While it was evident the exercise had been planned—Sherrie had worn nothing but a cotton dress to the course—Mrs. Beavon maintained, "We came to watch the golf, and the next thing I knew her dress was on the ground." Even though at least one photographer had known the episode was about to take place, Sherrie insisted, "I can't think what came over me."

The prank, meanwhile, had no evident effect on Olazabal. "It was just a naked woman, that was all," he said.

A Matter of Standards

Temperatures reeled into the 100s during the 1991 Women's Open, at the Colonial Country Club, in Fort Worth. The weather was so unpleasant spectators spent more time inside the air-conditioned clubhouse than outside watching the golfers.

The blazing sun threatened the greens as well. In the best of times the grounds crew must keep close watch to see that the grass doesn't wilt. Some greens have fans permanently installed close by to keep the air flowing, but even those weren't offering enough insurance against the heat.

The grounds crew found a solution. They covered selected greens with layers of ice overnight, then cleared what was left the following morning. The greens thrived.

A native of Fort Worth, Ben Hogan had stopped by to watch the women play. Stepping outside and into the searing heat, he sniffed the air and said, "It's a little cooler than it should be, isn't it?"

With rounds taking more than five hours in the oppressive heat, Lori Garbacz somehow contacted the local Domino's and had a pizza delivered to her at the seventeenth tee. When it arrived, she and her pairing mates had a very late lunch.

Turnabout

Jose Maria Olazabal had played a shot into a bunker during the 1991 Mediterranean Open, at the Saint-Raphael Golf Club, a course on the Mediterranean coast near Cannes designed by Robert Trent Jones. Olazabal's ball had embedded in loose and fluffy sand, leaving him a particularly difficult shot. Annoyed and feeling abused, Olazabal claimed the bunkers had been filled with too much sand—or else the sand was too soft. A shot from a fluffy lie is never much fun, but golf isn't bridge: you can't pass; eventually you must hit the ball. Olazabal felt otherwise. Evidently a closet martyr, he assumed a comfortable seat on his bag close by the bunker and waited. No one is clear what he waited for.

Eventually John Paramor, a rules official with the European Tour, arrived and asked, in effect, "What's up?" Olazabal pointed to his ball. Paramor suggested he should play it. Still enraged, Olazabal handed Paramor the club and suggested Paramor play it himself. More than a little annoyed at Olazabal's behavior, Paramor declined, told Olazabal to play the ball *now*, and went on his way.

Olazabal did indeed play the shot, moved the ball two feet with his first attempt, and eventually bogeyed the hole. He placed third and won £16,000. Because of the trouble he caused, he was fined £250, a small price to pay.

Light-fingered Golfers

Practice facilities in Britain are often below the standard of those in the United States. On the other hand, Ganton Golf Club, near the seaside resort of Scarborough, has an excellent facility—a big, level, wide open field with plenty of room for the longest shots and the wildest hooks and slices. There was a problem, however, during the 1991 British Amateur. First, since most of the American Walker Cup Team had entered and obviously had not carried practice balls abroad with their luggage, and second, since adequate practice ranges are so scarce in Britain that few British golfers lug practice balls to tournaments, the R and A, in a fit of good will, bought 1000 sparkling new Titleists and made them available on the range at no cost.

Within three days every ball had disappeared into players' bags. Vowing it would fight future seizures of generosity, the R and A said, "Never again."

Nature Calls

Heeding the call of nature during the final round of the 1991 World Cup, Ian Woosnam picked up his ball from the fifteenth green and ducked into the Port-A-John nearby. The problem was he forget to mark his ball and had to take a penalty stroke. The mistake cost Wales a share of first place. Sweden finished one stroke ahead.

War on the Shore

In the 1991 Ryder Cup match at Kiawah Island, South Carolina, nicknamed The War on the Shore, Mark Calcavecchia stood four holes up on Colin Montgomerie with four holes to play. He needed only to halve one of the four remaining holes to assure a victory for the United States. He wasn't up to it. Unable to handle the intense pressure, Calcavecchia lost both the fifteenth and sixteenth to Montgomerie's standard pars, then stepped onto the tee of the seventeenth, a demanding par 3 calling for a frightening 200-yard carry across a broad pond. You either hit the green or your ball plunged into the water.

Holding the honor, Montgomerie played first and hit a terrible shot that never once looked as if it might clear the pond. Wincing as if in physical pain, knowing he had handed the Ryder Cup to the Americans, Colin stepped aside to give Calcavecchia his chance to sew up the American victory.

With a tense and hushed gallery waiting to break out in a thundering cheer when his ball settled safely on the green, Calcavecchia hit an even worse shot than Montgomerie's. He dumped his tee shot into the water as well. As the ball splashed into the pond, Charlie Jones, doing television commentary, blurted, "Are you kidding me?"

Teeing up a second ball, Calcavecchia played a second ball just as badly, once again into the water. He lost the seventeenth, and lost the eighteenth, allowing Montgomerie to halve the match, win half a point, and almost save the European cause. The United States won when Bernhard Langer missed a six-foot putt on the last green and lost to Hale Irwin's bogey 5. The Americans won by one point.

Looking over Kiawah Island's Ocean Course before the match, the Irishman David Feherty described it as "Not like something from Scotland or Ireland; it's like something from Mars."

Different Twist

Fired as executive director of the Women's European Tour, Joe Flanagan, an Irishman, reflected, "I suppose this is a case of *women's* inhumanity to man."

Not So Happy Birthday

When David Senior, a thirteen-handicapper, holed his tee shot to the 160-yard fifteenth at Royal Lytham and St. Annes, in England, it seemed a perfect acknowledgment of his fortieth birthday. His celebration lasted only a few seconds. Bill Lloyd, his opponent, hit his shot into the cup on the fly, flipping both Senior's ball and his own out of the hole.

It didn't matter, though. Even though the rules state that no matter what happened because of Lloyd's shot, Senior had his hole-in-one, but his side still lost the match, 2 and 1.

Impregnable Defense

Living next to a golf course has its perils. For years golfers recovering mis-played shots from the Sunset Grove Golf Course, in Forest Grove, Oregon, knocked down a fence separating the course from the farm of Charles and Vicki Hertel, trampling on the Hertels' wheat, oats, and corn.

Covering a field with manure one day, Charlie Hertel found a new hole in the fence. Angry, he dumped his load of manure across the opening in the fence. Days went by; the sun baked the manure until it looked like a pile of dirt. A day or two later Charlie found a trail of deep footprints in his manure pile. They went only halfway. Pondering the victory, Vicki Hertel grinned and said, "Whatever works."

Million Dollar Hole-in-One

Jason Bohn, a chemistry student at the University of Alabama, made a hole-in-one and lost his place on the college golf team. He made his ace in a hole-in-one contest and didn't mind a bit losing his amateur status. One of twelve finalists in a charity tournament in Tuscaloosa, by holing his tee shot on the 135-yard second hole at the University's golf course, Bohn won $1 million. He was to receive $5000 a month for twenty years.

Harvey Penick

With more than 400,000 copies sold, Harvey Penick's *Little Red Book* is the best-selling golf book ever. Penick collaborated with Bud Shrake, a sportswriter and novelist from Fort Worth. Penick, in his eighties, lived in Austin.

Accustomed to dealing with publishers, Shrake and his agent negotiated the publishing deal with Simon & Schuster. In producing a book, it is customary for the publisher to advance money to the author in anticipation of a certain amount of book sales and resulting royalties. A kind, gentle man naive in the ways of the business, Penick didn't understand this. When Shrake told him, "Harvey, your share of the advance will be $85,000," he gasped, "Bud, I don't think I can raise that kind of money."

Penick had tutored Tom Kite as a young man and helped him develop the game that made him the leading career money winner and the 1992 U.S. Open champion. Because he had another commitment, Kite couldn't return to his home, in Austin, immediately after he won the Open at Pebble Beach. Instead, his wife, Christy, did the deed for him. Christy took the Open Championship Trophy to the Austin Country Club, found Penick and said, "I have something here you own a part of," and set it in his lap.

A Taste of the Vine

After two holes of a practice round on the eve of the 1992 U.S. Open, Fuzzy Zoeller and John Daly had worked up a thirst when they spotted a waiter drive past in an electric cart on his way to the Pebble Beach Lodge. "Bet you don't have any beer in there," Daly called out.

The waiter sped on, but five minutes later he showed up with a chilled bottle of white wine, still corked. A spectator offered a Swiss Army knife, the kind that has a corkscrew among its uncountable tools, they wrenched the cork out of the bottle, and each man sipped of the grape.

Faldo's Near Whiff

In his first practice day leading up to the British 1992 Open, Nick Faldo had reached Muirfield's sixth tee. Before he had a chance to play his shot, a young man slipped under the gallery ropes, ran up to him clutching a driver, and asked Nick if he would christen the club for good luck.

Smiling, Faldo teed up a ball, declared, "I christen this club 'straight down the middle,'" took a hefty swing, and while the gallery laughed, topped the shot. He explained later that compared with his own driver, this was so light it felt like a feather on the end of a string.

Straight, or with Soda?

Not exempt in 1992, Ben Crenshaw had to qualify for the British Open, but he shot 79 in the first round and lost his chance of winning a place in the field.

Later that afternoon he slunk onto a stool in the small bar at Greywalls, the hotel alongside Muirfield. The bartender approached and asked, "What can I bring you, sir?" Glancing up, the sad looking Crenshaw muttered, "Arsenic."

Capital Punishment

One of the longest hitters in the game, Sandy Lyle often drove with his 1-iron, but it was giving him fits during the 1992 European PGA Championship. He began hooking in the first round, and continued into the second. He hit a particularly bad tee shot on the ninth hole at Wentworth in the third round, played a worse shot on the eleventh, and when he duck-hooked his shot on the twelfth, he snapped.

So did the 1-iron. Usually the most even-tempered of men, Sandy grabbed the club with both hands and snapped it across his knee, crying, "It won't go left any more."

Adventures of Mark Brooks

Playing the seventeenth hole in the third round of the Las Vegas International, a five-round event, Mark Brooks hit a shot that evidently lodged in the upper branches of a palm tree. Since he had to identify his ball to avoid a penalty for a lost ball, Brooks had tournament officials commandeer a nearby cherry-picker to hoist him high enough to look for the ball.

Rooting through the palm branches, Brooks found eight balls—none of them his. He had to take the stroke-and-distance penalty. He shot 74 that day after opening rounds of 67 and 65.

After reaching the eighteenth green the following day, Brooks marked his ball and tossed it to his caddie. Error on the throw. When his caddie couldn't field his toss, the ball splashed into a pond alongside the green. Now he faced another lost-ball penalty. Still game, Brooks shed his shoes, socks, and shirt and dived in after the ball. He found eighteen balls, but, again, not his own. He shot 78.

The result—twenty-four balls found and four penalty strokes suffered. Spurred on by the certain knowledge he wouldn't finish among the leaders, Brooks raced around the course in an hour and thirty-nine minutes the next day and shot 71.

Scores To Remember

Chick Evans (not the original) played the last round of the 1992 Vantage championship, a part of the senior tour, in 126 strokes, a score that would embarrass nearly anyone. He did it because the rules of the senior tour require a player to finish a round to win prize money.

Evans had injured his shoulder playing from deep rough in the second round of the Vantage championship. He finished the round with 76 after opening with 71. Like most senior tournaments the Vantage was played over 54 holes. By the morning of the last round Evans's shoulder throbbed so badly he could draw back his club no more than hip-high. He wanted to withdraw, but if he did he would forfeit the $1000 prize that goes to everyone who finishes all three rounds.

Using only a 7-iron and a putter and moving the ball along with only half swings, he played the first nine in 70 and improved to 56 on the second nine. "A thousand dollars is a thousand dollars," he said. The previous high score had been shot by Mike Reasor in the 1974 Tallahassee Open. Also injured, Reasor played one-handed and shot 123.

Record Drive

Looking over the Oak Hills Country Club, in San Antonio, before the 1992 Texas Open, tour officials Glenn Tate and Vaughan Moise noticed a twelve-foot wide spillway with a twenty-four-inch curb. Designed to direct rainwater away from the golf course, it ran downhill off to the right of the third hole. For member play the spillway was designated out of bounds, but the tour doesn't like to declare ground within property lines out of bounds. Reasoning that it seemed unlikely a competent tour player would hit a ball so far off line, Tate and Moise moved the boundary to a fence on the spillway's far side. Unhappily, they reasoned without considering Carl Cooper.

From tee to green, the hole measured 456 yards, all downhill, with a gentle left-to-right bend. Cooper figured he'd play a little cut shot, letting the ball drift slightly right. Not one of the tour's great players, Cooper let his fade get away from him. It sailed off to the right, hit a cart path, bounced along it for a time, then ran onto the dry and empty spillway. Trapped inside the spillway's two-foot high curbing, the ball began picking up speed and running ever faster, as if it were on the ramp of a ski jump. Past the third green it raced, then past the fifth and sixth, the twelfth and thirteenth. Eventually the spillway petered out and the ball stopped at the end of the concrete.

The drive had covered 787 yards, not quite half a mile. No one could remember so long a drive in a professional tournament. Now his ball lay well past the third green, and he had to play 311 yards back. Given a drop without penalty off the spillway, Cooper slashed a 4-wood back toward the target and followed with an 8-iron that flew over the green. By then his partners had holed out, and another group stood on the green lining up their putts. They stood aside while Cooper chipped on and took two putts for his double-bogey 6. To play a 456-yard hole, Cooper had covered 1,098 yards.

Boots and Saddles

For some years the LPGA promoted an annual magazine that used its players for models and posed them as something they weren't. Danielle Ammaccapane posed as a cowgirl by wearing a white blouse, tan suede vest, dark brown leather chaps, and cowboy boots. The outfit wasn't inappropriate; Danielle was an enthusiastic horsewoman. Until.

After winning three tournaments in 1992 and collecting more than $500,000 in prize money she went out for a ride, fell off her horse, and suffered a severe concussion. She spent two weeks in a hospital, and once she was released she couldn't hit balls for a month. Returning to the tour in February, she was still bothered by dizziness and fatigue. Looking back at what her love of riding cost, Danielle said, "I won't do anything stupid like that again."

Vanity, Vanity

Throughout most of his life Tom Shaw hadn't looked his age; he always seemed to be younger. So he took advantage of it. Since records of his birthdate had been destroyed in a courthouse fire, Shaw decided to shave his age from 27 to 23.

Life was much simpler then, at least until he notified tour officials he planned to play the senior tour. Then he ran into trouble. Since the tour had a false record of Shaw's birthdate, he was told he'd have to wait four more years, until 1993. Shaw tried to explain, but the tour would have none of it; he would have to prove his age. Eventually he found a copy of his birth certificate on microfilm and satisfied the tour. "If I hadn't found that birth certificate," Shaw laughed, "they'd have cut off my legs and counted the rings."

Dull Life

Comparing his early days on the tour with the conditions in the early 1990s, carefree Fuzzy Zoeller complained, "The younger guys don't drink. They eat their bananas and drink their fruit drinks, then go to bed. It's a miserable way to live."

Watch the Backswing

Tom Heyward, a club professional from Myrtle Beach, South Carolina, was giving a demonstration of the swing during a convention. He climbed atop a high platform, drew his club back to the highest point of the backswing, and shattered an expensive crystal chandelier.

Trail's End

Nothing stops the golfer, not rain, not sleet, perhaps dark of night, but not a lost companion. Playing in a twice-a-week retirement league with forty neighbors from Swiss Village, Donald DeGreve collapsed as he putted on the sixteenth green of the municipal Willowbrook Golf Course, in Winter Haven, Florida, in October of 1991. DeGreve was stricken by an apparent heart attack at 10:30. A nurse playing on the course rushed to DeGreve and administered cardio-pulmonary resuscitation. It was too late.

Paramedics called to the scene covered DeGreve with a sheet, but through a county regulation they weren't allowed to move him to a funeral home. Meantime, Willowbrook officials closed the sixteenth hole. "We told everybody they had to skip the sixteenth," said Bob Sheffield, director of leisure services for Winter Haven. "If they didn't understand why at the tee, they understood by the time they reached the green."

While DeGreve's three playing partners abandoned their game and searched for DeGreve's widow, who was shopping nearby, the rest of the Swiss Village group played through.

Explaining why, Robert Alexander, the group's organizer, said, "We're all in our 70s. Knowing we're going to die, we're probably a little more conditioned to death. It was a real shock to all of us, but there was nothing we could do. None of us was in good shape to play. We just tried to finish up. At least he didn't suffer. We all thought to ourselves, 'That's a good way to go.'"

Narrow Escape

Angry because he had hit a bad tee shot on the par 3 fourteenth hole at King's Forest Golf Club, Rick Pearce, a Canadian golfer, slammed his club against a tree, but the shaft snapped and the sharp, ragged end embedded in his skull, entering at the temple. Pearce was rushed to a hospital where the club was removed and the damage to his head repaired.

Double Loss

An insurance company and a bank co-sponsored an amateur tournament in Britain. The insurance company wanted to reward the winner with a year's premium on a policy, but to be sure he wouldn't jeopardize his amateur status by accepting the policy, the company asked for a ruling from the R and A. Before informing the company of its decision, the R and A consulted with the USGA, which also deals with such matters.

The insurance company was told the winner could accept the policy, but in a whimsical P.S. it was also told that should the winner die before the year ended and his heirs collect the benefits, he would indeed forfeit amateur status.

Carr Drives In

After a night celebrating his election as Captain of the Royal and Ancient Golf Club of St. Andrews, Joe Carr, the great Irish amateur, along with his daughter and two of his sons, went for a swim in frigid St. Andrews Bay in September of 1991, the morning he was to drive himself in. Asked why in the world sane people would do that, Joe's son Roddy explained, "We had to get the old man sober somehow."

Treatment and Cure

Developing a sore back while he was playing in the 1992 Thailand Open, Seve Ballesteros visited a local doctor for treatment. After his examination, the doctor filled a drinking glass with alcohol, set it alight, and laid it against Seve's back. The pain from the ten burn marks the size of beer cans overcame the pain of the sore back, so the treatment succeeded on that score. After Seve's screams died down, the doctor suggested a second visit. Ballesteros declined.

No Sense of Humor

After birdieing three consecutive holes at the Mount Juliet Golf Club in Kilkenny, and closing in on the lead in the Irish Open, Ballesteros stood on the tee of the fifteenth hole, a shortish, uphill par 4, and asked his caddie, "Can I drive this green?" Without missing a beat, the caddie shot back, "Who do you think you are, God Almighty?"

Ballesteros was not amused. Summoning all his strength, he lunged at his ball and sent it screaming in the general direction of a distant farmhouse. Out of bounds. Now he was even less amused. The round safely over, Ballesteros fired the caddie.

Price of Fame

Greg Norman and Mark Wiebe both stood on the practice tee of the Houston Open working on their irons. Both men play Cobras. Noticing Norman using his old clubs rather then the new line, Wiebe sauntered down and asked why. "They didn't send them," Norman answered.

Wiebe handed him a club, Norman swung and rifled a shot far and straight downrange. Marveling at how well they felt, Norman looked at them, then frowned. "These have my name on them," he complained. "Where'd you get them?"

"At the factory," Wiebe said. Then, wearing a tight grin, he said, "When I talk to them, I'll tell them to send you a set."

Norman was part owner of the company.

Pushing his cart through the aisles of a supermarket during 1991, a rather dismal year for him, Norman was stopped by a woman shopper. "Aren't you the actor I see in television commercials?" she asked. Norman said he was, then pushed on toward the dairy foods.

Pampered Grass

At 155 yards, Augusta National's twelfth hole—a little par 3 that calls for a 7- or 8-iron for Masters caliber players—doesn't seem too imposing, but anyone who has seen the Masters knows this is probably the most difficult hole on the course.

The condition of the ground had been part of its difficulty until the Augusta grounds crew began tackling the problem early in the 1980s. Even a properly gauged shot often bounced off the hard surface and into the bushes on the hill rising behind the green. Putting was often chancey as well. The grounds crew dug up the green in 1981 and laid underground piping to carry either warm or cold water that regulates soil temperature and promotes growth.

Then, in the period leading up to the 1993 tournament, the crew strung sixteen 1000-watt bulbs above the green to simulate sunlight because the towering Georgia pines block the morning sun.

In the Bush

Billy Ray Brown overshot the thirteenth green with his third shot during the second round of the 1993 Masters and embedded his ball in a bank beside an azalea bush. Under the local rule for an embedded ball, Brown was permitted to lift and clean his ball and drop it without a penalty.

Billy Ray dropped once, but when his ball rolled down the bank closer to the hole he had to drop once again. He did, but as it fell the ball nicked a branch of the azalea bush and once again rolled closer to the hole. Since he had dropped twice, he could place the ball. Now the issue became, where?

The rule before 1993 had stated he should place the ball where it first touched the ground, but the wording had been changed. The latest version stated it should be placed where it first touched the *course*. The azalea bush clearly was part of the course.

Two groups played through Brown and Ben Crenshaw, his playing partner, while rules officials huddled as if they were interpreting the incantations of the Delphic Oracle and debating where Brown might place his ball. Finally the interlocution ended. Brown must place his ball several feet off the ground and in the azalea bush, for the bush is the first part of the course his ball had touched. Brown placed his ball about four feet off the ground, and then, trying to hit it as fast as he could before it moved, he gripped his putter like a baseball bat and nearly popped the ball into the cup. As it was, he saved a par 5.

Fish Story

Playing the thirteenth hole of the University of Georgia golf course during the 1993 women's collegiate championship, Holly Reynolds drove into a pond. Since her ball lay just on the edge of a bank, she didn't hesitate to play a shot rather than take a penalty stroke and drop outside the hazard.

She took a mighty swing and barely moved the ball out of the water, but as she hit her ball, her club swept under a little fish and flipped it onto the bank as well. Both ball and fish lay there on the grass, still within the boundary of the hazard.

The poor fish, only about four inches long, flopped about on the grass desperately trying to work its way back to the pond; if something wasn't done soon, it would die. Without thinking of the rules of the game, Holly quickly flipped it back into the pond. The fish was saved. But was Holly? It all depended on whether or not the fish could be considered a loose impediment.

Officials at the site invoked decision 18/4 in the *Decisions on The Rules of Golf*, which dictated that a live snake could be removed as an outside agency, but a dead snake is considered a loose impediment. Had Holly killed the fish she would have been saddled with a two-stroke penalty. Since she hadn't, she invoked no penalty. She still double-bogeyed the hole, shot 77, and placed fifteenth in the final reckoning.

Dick Metz

One of the best players of the 1930s, although not considered to be in the same upper tier as Nelson, Snead, and Guldahl, Dick Metz won fourteen tournaments, four of them in 1939, his best year, when he reached the semi-final round of the PGA Championship, where he lost by one hole to Henry Picard—the eventual winner—and placed among the leading ten scorers in six other events.

Like Snead, though, Metz is remembered best for the Open he didn't win. After fifty-four holes of the 1938 championship, Metz led the field by four strokes, but he closed with 79, and Ralph Guldahl beat him by six.

With dark, wavy hair, Metz was considered the handsomest player on the tour then. After he left competitive golf he retired to Arkansas City, Kansas, and turned to ranching, buying land considered too poor for farming and turning it into pastureland for his cattle. He still loved to play golf, but as he aged he developed arthritis that grew steadily worse and eventually forced him to give up ranching. Still he played golf, but his time was running out. He had a joint of the middle finger of his left hand replaced early in the 1980s, had a knee replaced in 1986, and in 1991, the day after he was inducted into the Kansas Golf Hall of Fame, he and his wife, Jeane, drove to Oklahoma City where doctors fused his left ankle. Eventually the condition began to affect his spine. On May 5, 1993, Metz drove his pick-up truck onto the parking lot of a funeral home, climbed into the back and stretched out, then fired a pistol shot into his head. He was eighty-four.

Take It All Off

Ian Baker-Finch hit his tee shot a little fat on the 178-yard thirteenth hole during the first round of the 1993 Colonial Invitational. His ball fell short of the green, then trickled down an embankment and into a pond. Approaching the green, Baker-Finch saw his ball near the edge in water shallow enough to play a shot, so he followed the customary routine of taking off a shoe and sock before climbing into the water to play his recovery.

Next, though, he went a step further. Not wanting to muddy his pants, he unbuckled his belt and took them off, and stood there in his boxer shorts and golf shirt. Stepping into the water he played a nice shot out, about twelve feet past the cup but missed the putt. On the next hole, the evidently aroused women in his gallery called for him to hit his ball into the water again.

Wishful Thinking

A long-driving contest arranged at the Woodlands Country Club before the 1993 Houston Open pitted John Daly, perhaps the longest driver professional golf has known, against Jim Dent, the biggest hitter on the senior tour. It wasn't much of a contest.

As it was set up, each player was allowed one practice shot with the option of counting it in the competition if he felt he hit it well enough. Up first, Dent smoked his first drive 318 yards and counted it. Up next, Daly sent his first drive 321 yards. Asked if he wanted it to count, Daly said, No. I just didn't hit that one well."

Officials estimated 10,000 people had gathered to watch the contest, all of them crowded around the Woodlands' tenth fairway. Daly launched one drive 340 yards. When it ended, Dent said, "I just enjoyed watching John hit. It was awesome to see." Daly said of Dent, "I just hope I'll be hitting it that far when I'm fifty-four." Then, "I just hope I live to be fifty."

Rental Clubs

In an untypical response after a shot from sand got away during the 1993 Las Vegas Senior Classic, Chi Chi Rodriguez flung his cherished R-90 sand wedge aside. Seeing the club on the ground, Robb Robinson, a young fan from Fort Smith, Arkansas, picked it up and offered to buy it from Rodriguez for $50. Surprising the gallery, Rodriguez agreed to the sale, saying, "Sure. I'll give the money to my kids."

Chi Chi faced a dilemma later when his approach to the eighteenth settled hole-high in a greenside bunker. As he studied his shot and tried to figure a way to get out without his wedge. Spotting Robinson walking with the rest of the gallery, Rodriguez called, "Hey, kid. Bring that wedge over here. I want to rent it." Robinson handed the wedge to Rodriguez, and Chi Chi handed him a dollar. Rodriguez stepped back into the bunker, wiggled his feet into a solid stance, then holed a forty-foot explosion shot, saving a birdie. As he stepped out of the sand he handed Robinson the wedge along with another dollar bill, and said, "Thanks for letting me rent it."

It Can Be Done

Sam Torrance was standing between the eleventh green and twelfth tee of the Old Course at St. Andrews during the third round of the 1993 Dunhill Cup talking to Alan Fraser, a columnist for London's *Daily Mail*. Fraser mentioned that Joakim Haeggman, of Sweden, had just scored 8 on the twelfth hole, at just 316 yards, the shortest par 4 on the course. Since it played downwind that day, a number of players had driven the green.

Hearing Fraser, Torrance said, "How in the world can you make 8 on that hole?"

Then Sam hooked his drive into a clutch of thorny, impenetrable gorse. Assuming it might be lost, he teed up another and hit it only marginally better. He couldn't find the first drive, which cost him two strokes, and he found his second in an unplayable lie, which cost him another penalty stroke. He dropped from the unplayable lie, pitched on with his fifth shot, and three-putted. Shaken, Sam still clung to his sense of humor. Walking toward the thirteenth tee he passed Fraser and mumbled, "I guess that's how you make 8."

Incredible Journey

John Daly's winning the 1991 PGA Championship stands among the most unlikely happenings in PGA lore, outranked only by his having played in it in the first place. Failing to win a place through the qualifying rounds, Daly stood no higher than ninth alternate, with as good a chance of playing as he had of winning the lottery.

The PGA was scheduled for the Crooked Stick Golf Club, in Indianapolis. After missing the cut in the Buick Open, in Grand Blanc, Michigan, the previous week, Daly drove home to Memphis. Driving through Indianapolis he spotted a van with a big PGA sign on its side and thought, "It would be nice to play." He didn't dream it would be possible.

Then this happened:

1–All former PGA champions are eligible to play, even Gene Sarazen, who was 89 at the time. Most didn't, leaving room for others.

2–Masahi (Jumbo) Ozaki, a first-class Japanese golfer had been invited to play, but he declined, claiming he had been injured.

3–The Australian Rodger Davis turned down his invitation, preferring not to interrupt his European schedule to spend one week in the United States.

4–The Irishman Ronan Rafferty declined his invitation for two good reasons: his wife was expecting and he had hurt his shoulder.

5–The head professional at the host club is entitled to a place, but Jim Ferriel, of Crooked Stick, declined, also for good reason. Responsible for the merchandising effort during the championship, Ferriel said, "I had other priorities that week. I had six [merchandising] tents going, and 400 volunteers while I usually have two assistants and a couple of kids in the bag room."

6–Trying to win a place on the European Ryder Cup team, which required him to move up on the money-winning list, Mark James, of England, withdrew.

7–Saying he'd talked to a few PGA officials who told him Crooked Stick was a long and difficult course, a monster, Lee Trevino withdrew, telling them "You can keep your monster."

8–With one round to play, Marco Dawson, a graduate of the most recent Tour School, held a two-stroke lead over Nick Faldo and Chip Beck after three rounds of the Buick Open. Had he held on he would have won a place and Daly would have been out. Instead he shot 74 and fell to tenth place.

9–By then three of the nine alternates had moved into the PGA field. Then Gibby Gilbert withdrew, giving a place to the fourth alternate.

10–Recovering from surgery on his shoulder, Paul Azinger played nine holes of practice on Tuesday and Wednesday, realized he couldn't compete, and withdrew, giving his place to the fifth alternate.

11–Nick Price and his wife were expecting a baby at any moment. He withdrew, giving his place to Bill Sander, the sixth alternate.

12–It was then Wednesday afternoon. Sander declined; his place went to Mark Lye, the seventh alternate.

13–Lye declined, not wanting to play Crooked Stick without a practice round. His place went to Brad Bryant, the eighth alternate.

14–Bryant's mother-in-law had fallen the previous week, and he and his wife, Sue, planned to visit her the week of the PGA. Besides, he hadn't been playing well, hadn't played in six days, and told Sue, "I don't want to spend another week in the rough." He declined.

15–Hoping for the miracles that swirled around him even as he drove, Daly had packed his car at about 5 o'clock Wednesday afternoon, when he stood no better than fourth alternate, and headed for Indianapolis. He reached his hotel around midnight and found a message light blinking. Ken Anderson, an administrator with the PGA, had left word that Daly had a place in the championship.

Without a practice round, Daly found a way to control his enormous drives, hit thirty-five of fifty-six fairways, fifty-four of seventy-two greens, putted like a dream, shot 69–67–69–71 for 276, beat Bruce Lietzke by three strokes, and won the first PGA Championship he ever played in.

Part Eight

The King Abdicates

Arnold Palmer played his last round in the United States Open on June 17, 1994. Thirteen months later he ended his career in the British Open as well. Both wrung tears and stirred memories of great moments in his long and glorious career.

Without question the most universally popular golfer of his time, Palmer played in the 1994 U.S. Open on a special exemption from qualifying. Arnold had reached the age of 64, and everyone knew he would play no more than two rounds over brutally hard Oakmont Country Club, outside Pittsburgh, not far from where Palmer had grown up and lived throughout his life.

Arnold had opened with 77; he would have needed 70 in the second round to survive the 36-hole cut. Instead he shot 37–44—81. With 158, he missed by 11 strokes.

None of this mattered. The closer he moved toward the final hole, the larger his gallery grew. No matter how badly he played, his fans applauded every shot. When he reached the final hole, the crowd stood five to six deep along the fairway ropes, and when he walked those final long yards to the home green, they cheered and applauded every step.

His eyes glistening with tears, Palmer took off his straw hat and waved to the crowd, and as he stepped onto the green, he turned and bowed.

He holed out, waved once more, and disappeared into the scorer's sanctum.

A few minutes later, choking with emotion, his voice wavered as he tried to speak over national television. He struggled to say, "It's been 40 years . . ." but his voice broke and he couldn't finish. Gathering himself, he went on, "When you walk up the 18th and get an ovation like that . . . I guess that says it all." Then he excused himself.

In the big press tent he sat before the same crowd who had seen him win so many times, and he struggled once more to speak.

"I think you all know pretty much how I feel," he said. "It's been 40 years of work, fun, and enjoyment."

Asked to name his greatest thrill, he answered, "The whole experience."

He paused to collect himself, then, his voice quaking, he muttered, "I haven't won all that much. I won some majors. I suppose the most important thing is the game has been so good to me."

His emotions welling, Palmer lowered his head and covered his face with a towel, hiding the tears. Then he said, "I think that's all I have to say."

He stood, walked toward an open seam in the tent, turned for a moment as the press corps rose and applauded, waved once again, stepped out, and walked away.

The following year, at the age of 65, Palmer filed his entry to the British Open and put the Royal and Ancient Golf Club on the spot.

Arnold had misunderstood the restrictions. As a former champion, he had been eligible to play *until* he reached 65. Arnold thought it meant he could play *at* 65.

No great problem. For Arnold Palmer, the R and A changed the conditions of the championship and allowed former champions to play until they *passed* the age of 65.

It seemed only fitting: the 1995 British Open would be played at St. Andrews, where 35 years earlier Palmer had breathed life into a moribund championship, hardly more than a provincial affair for golfers from the British Isles, a few Europeans, and some others from the British Commonwealth. The Americans played the best golf in those days, and they weren't interested.

With Arnold in the field, the British Open's importance grew and, when he returned in 1961 and 1962, it had climbed so in stature that the game's leading golfers felt *compelled* to play.

Arnold lost the 1960 British Open to the Australian Kel Nagle by one stroke, but he won the next two. While he continued to play, he couldn't win another.

Strangely enough, he shot a 36-hole score of 158 at St. Andrews, the same as at Oakmont a year earlier, and missed the cut once again. Just as at Oakmont, Arnold's score didn't matter. Galleries everywhere simply adored him. Fans jammed Links Road, which borders the right side of the 18th fairway, some wedged so tightly together they couldn't move at all. Others leaned from windows or stood on balconies of adjoining buildings. Grandstands on either side of and behind the green overflowed, and spectators in the grandstand beside the 17th green stood and applauded as Palmer struck his final drive.

As he crossed the Swilcan Bridge, a small stone structure that has spanned a narrow creek since the time of the Crusades, Arnold paused, turned to the gallery, and with a tight smile, he waved.

As a measure of Palmer's standing in the game, Lee Trevino, who had won twice himself, had said nothing of his own plans to make this his last British Open out of respect for Palmer. About to tee off for his last round, Trevino stepped off the first tee and waited for Arnold to finish.

Once Palmer putted out, Nick Faldo walked up to him and shook his hand. Nick had finished 25 minutes earlier, but he and a few others had stood by to see Arnold finish his British Open career.

"He's done everything for the Open here," Faldo said. "But for Palmer in 1960, who knows where we'd be; probably in a shed down on the beach."

Once again, as he had at Oakmont, Palmer reflected on the final walk up the 18th fairway.

"When I came up the 18th, I kept thinking about 1960 and what that led to—a lot of great years and a lot of happy times for me, both in golf and socially. . . . It was a warming and happy time."

For that final round, Palmer wore a blue cashmere sweater bearing the insignia of the Royal and Ancient Golf Club, as he was entitled to do as an honorary member. He said later that he planned to return in September and play in the club's Autumn Medal.

"Since I played like a member," he reasoned, "I might as well act like one."

Perils of the Press

Carnoustie Golf Club, a brutish course north of Dundee, Scotland, bordering the Firth of Tay, had fallen on hard times. Once part of the British Open rota, its maintenance had been neglected following the 1975 championship and it had fallen from favor. It was over its forbidding links that Ben Hogan won the 1953 British Open and where in 1975 Tom Watson won the first of his five British Opens.

Hoping to lure the championship back to Carnoustie, a public course, the town worked to improve its condition. To test it, Carnoustie staged the Scottish Open for 1995 and drew a fine field that included most of the European Tour's leading players.

As the third round began, gale-force winds roared in off the North Sea, threatening to shake tents apart and bending flagsticks horizontal. Soon a light drizzle added to the misery. Playing conditions couldn't have been more appalling.

Amid the worst of the weather, Colin Montgomerie staggered in with 75, not even close to his opening 64.

Dripping rain, his reddish blond hair a windblown riot, and looking as if he had just walked through a carwash, Monty stomped into the press arena to joust with the reporters who had tormented him throughout his career. He was not a happy man.

He took his place at the interview table, blood boiling, head bowed, and eyes staring at nothing, waiting for the first question.

As the writing corps held its breath, someone asked: "Colin, was the wind a factor?"

A man who lives on the brink of volcanic eruption, Montgomerie couldn't believe what he'd heard.

His jaw dropped, he looked up and repeated:

"Was the wind a factor?"

The other reporters giggled.

Then, more loudly, as Monty's normally ruddy complexion turned crimson:

"Was the wind a factor?"

The laughing grew louder.

Finally, with the press corps in full hysteria and Monty's complexion scarlet, he bellowed one last time:

"WAS THE WIND A FACTOR?"

End of interview.

Playing the first round of the Lancome Trophy tournament, in Paris, the English golfer Mark Roe yelped as an insect stung his left hand. It hurt, but Roe completed the round and left the course.

When he awoke the next morning, his hand had swollen so badly he wasn't sure he could play. Game to the end, he saw a doctor, who treated the sting and wrapped his hand in a bandage so huge that the hand looked twice its size.

Roe soldiered on, finished his round, then visited the press tent for the usual inquisition. As he fielded questions, he propped his bandaged hand on his left elbow in front of him, his hand pointing toward the ceiling.

The questions flowed, more concerning his hand than his golf, which, in truth, really wasn't worth mentioning. As the questions dried out and Roe made ready to leave, an inquisitor had a question.

Evidently blind to the massive bandage on Roe's left hand, he asked, "Now what hand was it, Mark?"

Rocca's Miracles, Daly's Open

While an international gallery of millions watched on television, Constantino Rocca lay prone in the Valley of Sin, his fists pounding the ground and his face buried in this depression in the 18th green at the Old Course.

Rocca had just holed what may be remembered as the most bizarre birdie putt in all the 124 British Open championships. It had covered 60 feet, followed a miserable pitch, and set up a playoff with John Daly.

The birdie followed an equally bizarre par on the 17th, the infamous Road Hole.

A 38-year-old former factory worker and boxer from Italy, Rocca had begun the round in second place at 209, two strokes behind Michael Campbell of New Zealand, who led at 207, and two ahead of Daly, tied for fourth at 211. Playing two holes ahead of Rocca, Daly shot 71 in the last round, one of the best rounds of the day, and took the lead at 282. Only Rocca could catch him.

When Rocca's approach to the 17th green caromed off a stone wall into a shallow hole in the road that borders the green's right flank and gives that holes its name, Daly's wife threw her arms around John, certain he had won.

With his putter, Rocca punched the ball. It hopped from the hole, shot across the road, climbed a bank, jumped onto the green and pulled up within four feet of the hole. He saved his par 4 of course, but he still needed a birdie to catch Daly.

Rocca played a big drive left of the 18th green, but feeling the tension, he played a pitch that would embarrass a 20-handicapper. He stubbed it; the ball ran no more than 15 yards and settled in the Valley of Sin.

Then Rocca holed the putt and fell to the ground.

Apparently spent by his wild finish, Rocca was no match for Daly in the four-hole playoff. John played the four holes in one under par and Rocca played them in three over, taking a 7 on the 17th.

Still, the 1995 British Open has been remembered more for Rocca's birdie than for Daly's victory.

True Grit

In 1976 the Royal and Ancient Golf Club accepted an entry for the British Open championship from Maurice Flitcroft, of whom no one at St. Andrews had ever heard. Flitcroft would play his qualifying round at Formby, on England's east coast, a few miles from Royal Birkdale, in Southport.

Flitcroft played his first 18 holes in 121 strokes and was led from the course. It was understandable. He'd taken up the game just 18 months earlier and made blessed little progress,

He entered again the next two years under phony names and played only a few holes before being shown the door. He managed to play nine holes in 1983, but with 63 strokes already, his cover was blown. Once again he was ridden out of town on a rail—figuratively speaking, of course. He made his last attempt in 1991 and didn't last long enough to add up his strokes.

As late as 1995 Flitcroft wrote to the R and A requesting entry forms for both the British Open, which was played at St. Andrews, and the

British Seniors Open. The response, phrased in more pleasant terms, told him to go away.

Asked why he felt the R and A should have sent him the entry forms, Flitcroft answered, "It is called the Open Championship, so it should be open to all. Besides, for all they know I may have improved enough to play."

Sweet Revenge

During the third round of the 1987 San Diego Open, Craig Stadler's drive on Torrey Pines North's 14th hole scooted off line and nestled close to a Leyland Cypress tree, whose low branches interfered with a normal swing. Only one option, Stadler decided. He'd play from his knees.

First, though, to protect his pants from soggy ground caused by overnight rain, he spread a towel and knelt on it to play his shot. At the end of the fourth round he believed he had tied for second place, at 270, four strokes behind George Burns. He soon found out he hadn't finished at all.

During the telecast of the fourth round, NBC showed footage of Stadler's imaginative solution to keeping his pants dry. Now, though, Stadler landed in trouble.

Soon telephone calls poured in pointing out that by kneeling on his towel, Stadler had built a stance, a violation of rule 13-3 of the rules of golf. More seriously, because he had not added a two-stroke penalty to his score for the third round, Craig had reported a score lower than he had actually shot, which meant disqualification.

Eight years later, word reached Stadler that the Cypress tree, dying from a fungus infection, had to be cut down. Who better to do the job, asked Tom Wilson, than Craig Stadler?

A member of the San Diego Open's organizing committee, Wilson made the arrangements, and Stadler showed up ready to do the deed. After donning a hard hat and sturdy gloves, Stadler picked up the chain saw. Before he placed saw against tree, someone asked if he'd practiced his technique.

"No. It doesn't look as if it takes much," he answered. "Just a little anger and an attitude."

Down came the tree, and Stadler walked away a happy man.

Agony at Augusta

Standing 13 under par after three rounds, Greg Norman looked like the runaway winner of the 1996 Masters. Nick Faldo lurked in second place, six strokes behind with just 18 holes left to catch one of the game's most dangerous players.

Hopeless? Not at all. In one of championship golf's more astonishing finishes, Faldo beat Norman by 11 strokes. Faldo shot 67, Norman shot 78, and Faldo won the Masters by five strokes.

On that heartbreaking day, Norman's game collapsed totally and completely.

The agony began quietly enough when Norman drove into the trees and bogeyed the first hole. Hardly noticeable damage since he still led by five. He still looked good when both he and Faldo birdied the second, but after dropping a shot in the fourth, Norman still led by five after Faldo bogeyed the fifth hole.

The unraveling began on the little par 3 sixth. Norman made 3, but Faldo made 2 and cut Norman's lead to four strokes. They parred the seventh, but Faldo birdied the eighth against Norman's par, and Norman played such a timid pitch to the ninth that his ball barely caught the front of the green and rolled back down a steep incline. He bogeyed, Faldo made his par 4, and with nine holes to play, Norman led by two.

Over the next two holes the lead that seemed so safe evaporated. Norman missed the 10th green and bogeyed, then three-putted the 11th. Faldo had played 11 holes in one under par and yet made up six strokes.

Still it wasn't over; before Norman knew what had struck, Faldo had moved ahead by two strokes. Greg underplayed his pitch to the 12th, the devilish par 3 over a creek, and his ball sank into the water. He made 5, Faldo parred, and now it was Nick's Masters to lose.

Of course Norman arranged it so Nick would have no worries by hooking his tee shot into another water hazard on the 16th. A second 5 on a par 3 hole. Four behind now, Norman parred the last two, but Faldo closed with a birdie. His drive caught a fairway bunker, but his 9-iron settled within 15 feet and he holed the putt for a closing 3.

When his putt fell, Faldo walked to Norman, threw his arms around him, and said, "I don't know what to say. I just want to give you a hug."

Junior Champion

At 11 years of age, Tiger Woods shot 71 and beat his father, a scratch golfer, for the first time. Later in the summer, according to David Owen in his book *The Chosen One,* he entered 33 junior tournaments and won every one.

In 1991, at 15, he entered his second U.S. Junior Amateur, shot 140, won the qualifying medal by four strokes, then went on to win the championship, the second 15-year-old winner in the Junior Amateur's 44-year history. Mike Brannan had won at 15 in 1971, but he had been about a month older than Woods.

Playing at the Bay Hill Club, in Orlando, Florida, Woods won his first five matches, fell behind Brad Zwetschke, of Kankakee, Illinois, in the early holes of the final, but, beginning at the eighth, he took five consecutive holes and won the match on the 19th.

At the Wollaston Golf Club, in Milton, Massachusetts, the following year, he shot 143 and won the qualifying medal by four strokes over Mark Wilson, of Menomonie Falls, Wisconsin, then beat him by one hole in the final for his second consecutive Junior Amateur. He was the first to win twice.

A year later he won his third, although Ryan Armour took him 19 holes at Waverley Country Club, in Portland, Oregon. With the match all square after 14 holes, Armour holed a 40-foot putt and won the 15th and took the 16th as well when Woods three-putted from 40 feet.

Woods struck back with a birdie at the 17th, let a 3-iron second drift into a bunker 40 yards from the 18th green, then played a demoralizing shot that settled 10 feet from the hole. Armour missed his putt for the birdie 4, but Woods holed for another birdie that sent the match to extra holes. Woods won the 19th with a par 4.

In the past, both Mike Brannan and Tim Straub had reached the final twice, and although both boys won a championship, neither had gone to a third final.

Woods's Junior Amateur career had ended. He had played in four, won 22 matches, the last 18 in succession, and lost only to Dennis Hillman, of Rye, New York, in a 1990 semi-final match when he was 14.

Daly vs. Woods

Even at 13, Tiger Woods followed a tournament schedule. Before high school he entered the Insurance Golf Classic, a moderately overvalued name for a tournament primarily for juniors that was played at the Texarkana Country Club, practically sitting on the Texas border in Arkansas's southwest corner.

First the juniors played 36 holes to whittle down the field, then those who were left joined professional players for the finish.

Surprising no one, Woods survived the cut and went into the third round paired with John Daly, who at that stage was known by hardly anyone outside Arkansas. Nevertheless Arkansas golfers knew him for his sub-orbital drives and all-around game. He had been named the state's Player Of The Year twice.

After five holes, Daly sensed he might be in trouble and told some friends, "I can't let this 13-year-old beat me."

After nine holes Woods had moved four strokes ahead. After 14 Daly had cut 3 strokes off the kid's lead, but he still lay a stroke behind and the holes were running out.

Now Daly rallied, birdied three of the last four, and beat Woods by two strokes.

It was 1989. Woods stood 5-foot-6 and weighed 105 pounds, and Daly was two years away from winning the 1991 PGA Championship and in six years would win the 1996 British Open, a pair of championships Woods was destined to win as well.

Three Straight Amateurs

Jack Nicklaus had won the 1959 U.S. Amateur at 19 by beating Charlie Coe, an established player, on the 36th hole at The Broadmoor, in Colorado Springs. In 1902, Louis James, another 19-year-old, beat the great Walter Travis in the final.

Until Tiger Woods came along, they remained the youngest Amateur winners. Woods won the 1994 Amateur at the age of 18 years, seven months, and 29 days.

Furthermore, just as he had done with the Junior Amateur, he won three successive Amateurs. Once again it wasn't easy; his opponents took him to at least the 36th hole in each final match.

First he beat Trip Kuehne by two holes at the Tournament Players Club in Ponte Vedra Beach, Florida, in 1994, then Buddy Marucci by two holes at Newport (Rhode Island) Country Club in 1995, and finally went 38 holes to beat Steve Scott at Pumpkin Ridge, in North Ridge, Oregon, in 1996.

Neither the Marucci nor the Scott finals matched the tension of the Kuehne battle. Six holes down after the 13th, Woods fought back, won two of the next five holes and went to lunch four down. Still trailing after eight holes in the afternoon, Woods rallied, pulled even at the 16th, then played a shot his mother will never forget.

The 17th at The Players Club ranks among the most unforgiving holes in big-time golf. At first the architect Pete Dye didn't know what to do with it, but Alice Dye, his wife, who happened to be a pretty good golfer herself, suggested he isolate the green in a body of water. He did, and created an all-or-nothing par 3; you either hit the green or keep trying until you do. Nobody knows how many golf balls it has buried at sea.

The hole measured 139 yards for the Amateur; Woods went for it with a pitching wedge.

His ball carried to the green, hit right of the hole, which had been set dangerously close to the green's right edge, took one hop toward oblivion, then spun back to safety.

Watching on television back home in California, Kultida Woods fell from her bed, afraid Tiger might have lost the match.

Of course he didn't. He won the 17th and the 18th, completing a string of six winning holes of the last ten, and took the championship. Still, his mother had the last word.

"That boy almost gave me a heart attack," she said later. "All I kept saying was, 'God, don't let that ball go into the water.' That boy tried to kill me."

No one, not even the great Bobby Jones, had won three consecutive U.S. Amateur championships, although Jones won five in six years. Jones had won the Amateur in 1924 and 1925 but he lost the final match of the 1926 Amateur to George Von Elm by 2 and 1 after beating Chick Evans in the quarter finals and Francis Ouimet in the semifinals. Jones then won both the 1927 and 1928 Amateurs, plus another in 1930, the year of his Grand Slam.

Jones aficionados like to point out that he had faced considerably stronger competition than Woods. Evans and Ouimet, both former

Amateur champions, had won the U.S. Open as well, and five years after beating Jones in the Amateur, Von Elm had lost the 1931 Open to Billy Burke over 144 holes, 72 in the Open proper and 72 more in an extended playoff. Both men tied after the first 36 playoff holes and had to play 36 more before Burke won by one stroke.

Orthodontist not Needed

As a student at Stanford University, Woods was mugged one night on his way back to his dorm from a charity dinner in San Francisco. A man with a knife demanded his wallet, his watch, and a gold chain with a Buddhist symbol that had been a gift from his mother. His loot in hand, the mugger punched Woods on the jaw with his knife handle, knocked him to the ground, then ran off.

Safely back in his dormitory, he called his father and told him, "Pops, you know that overbite I had? Its gone; my teeth are perfectly aligned."

Steve Jones's Grip

Steve Jones, the 1996 U.S. Open champion, began playing golf with the same old Vardon grip nearly everyone uses. When he won the Open, he gripped all his clubs the way most others grip nothing but their putters.

It wasn't by choice. Crashing his dirt-bike in the Arizona desert in November of 1991, he had torn a ligament and damaged a joint of his left ring finger, then hurt it again repairing his backyard putting green. Unable to grip a club, he dropped out of golf for two-and-a-half years. The accident might have ruined a promising career. Jones had won one tournament in 1988 and three in 1989, and had tied for eighth place in the 1990 U.S. Open at Medinah Country Club in Chicago.

Jones started his comeback late in 1994 by chipping balls using the reverse overlap grip. Instead of overlapping the little finger of the right hand over the index finger of the left, Jones overlapped the index finger of the left hand over the little finger of the right.

With his new method, Jones developed into one of the longest drivers in the game, averaging 279 yards in 1996, a yard and a half behind John Daly.

Over four days in June of 1996, he also played better than anyone else, shooting 74-66-69-69—279 and beating Tom Lehman and Davis Love III by one stroke.

Players with funny grips usually don't last. After his greatest moment, Jones gradually dropped from sight.

Pain Rewarded

A judge ruled that a player whose ball caromed off a railroad track and whacked her in the nose could collect damages from the offending club—not the weapon she used to play the shot but the course where she did the deed.

Playing the Fort Kent Golf Club, in northern Maine, Jeannine Pelletier hit a shot close to a railroad track. A local rule permitted dropping a ball on the other side of the tracks, but Mrs. Pelletier chose to play the shot instead.

Unfortunately, she didn't get quite enough loft. The ball ricocheted off the rail and hit her flush on the nose.

She sued the club, and the jury awarded her $40,000.

At the same time the jury turned down her husband's claim for damages, though, ruling he had not "suffered a loss of consortium" with his wife.

Santa Irwin

Mary Johnson, who lived in Brunswick, Georgia, had followed Hale Irwin for 20 years. By the time she arrived in Detroit for the 1996 U.S. Open, her two sons had enlisted in Irwin's camp. Irwin had seen them so often he recognized them in his gallery and had even learned their names.

A budding golfer, nine-year-old Chris Johnson had wanted a set of King Cobra irons for Christmas the previous December but didn't get them. Still hoping for them, and believing Irwin resembled Santa Clause, Chris wrote to him.

Irwin read the letter and remembered the young man. A short time later a Cobra representative called Chris, told him Irwin wanted him to have his wish, and asked Chris's height so the shafts could be cut to the right size.

On Christmas morning, Chris found a complete set of irons under the tree.

Mrs. Johnson said, "We were amazed. I mean we're not poor or sick—nothing like that—and still he did this. Chris will be a fan for life."

Brown's Dilemma

Henry J. Brown had a dream: He wanted to play in the U.S. Open.

But Mr. Brown also had a problem. He was in jail and, consequently, wouldn't be available for the qualifying rounds. His was not a violent crime; he'd been chucked in the slammer because he'd skipped his alimony payments.

Nevertheless, he had hopes, and so he wrote to the United States Golf Association asking for special consideration. If someone from the USGA would come down to Augusta, Georgia, get him out of jail, arrange a special 36-hole qualifier, and serve as his marker, he'd dazzle him with his superb game.

One more suggestion.

"If you set this up for me, you can even handcuff me between shots."

Brown never made it to the Open, but he came close, which was surprising since he played with a cross-handed grip and practiced in an automobile junkyard.

Out of jail finally, he filed an entry and tied for the qualifying medal in the local qualifying test at South Bend, Indiana. Moving on to the sectional rounds, his last barrier, Brown missed making it into the Open itself by one stroke.

A Lesson in Humility

A year after winning the 1996 British Open, Tom Lehman and his wife, Melissa, entered a mixed foursomes tournament at their club. Driving from the first tee, Tom ripped one about 275 yards down the middle and Melissa followed by shanking the second about 100 yards.

Next, Tom pitched onto the green within about 15 feet of the hole. Using a little too much power, Melissa ran the putt 20 feet past. Tom calmly holed the putt for the 5, one over par.

As they headed for the second tee, Tom suggested that Melissa needed to start playing a little better. To which she replied:

"Well, you took three shots. I only took two."

Tiger's Professional Debut

With his first shot as a professional, Tiger Woods flushed a drive that measured 326 yards splat down the middle of the first fairway of the Brown Deer Park municipal course, in Milwaukee, Wisconsin. He made his debut on August 29, 1996, in the Milwaukee Open with a 67 that featured one eagle, three birdies, and one bogey. Pretty good for a rookie, yet he finished the day five strokes out of first place. With hardly anyone watching, Nolan Henke shot 62.

The next day Henke shot 66 and kept a two-stroke lead, once again before a sparse group of fans. All but a small portion of the gallery followed behind Woods and gasped at his sheer power. Brown Deer's 18th hole, for example, measured 557 yards. Woods reached the green with a drive and 4-iron. In those days, distance like that seemed obscene.

Woods shot 69 in the second round and said later, "It's weird to be eight strokes behind after a 69."

Woods shot 67-69-73-68—277, seven under par but 12 strokes behind Jerry Kelly and Loren Roberts. Roberts birdied the first playoff hole and won the tournament. Woods tied for 60th place.

In a television interview before Woods played a shot in Milwaukee, Curtis Strange had asked what he would consider a successful week.

"If I can play four solid rounds," he answered. Then he added, "And a victory would be nice."

Strange was surprised that a young man playing his first professional round would even think of winning. Woods said he understood Strange's reasoning, but, he added, "I've always figured, why go to a tournament if you're not going there to try to win? That's the attitude I've had my entire life. As I've explained to my dad, second sucks, and third is worse."

With a sly chuckle, Strange told Woods, "You'll learn."

First Blood

As it developed, it was Strange who learned.

Woods won his first tournament in his fifth start as a professional. In the first week of October, he shot 332 in the 90-hole Las Vegas Invitational, tied the veteran Davis Love III, then beat him with a par on the first playoff hole. Both men bettered par by 27 strokes.

Blessed with intimidating power, Woods played the ninth hole of the Tournament Players Club at Summerlin, one of tournament's three courses, with a 3-wood and 6-iron. Keith Fergus, his playing companion, couldn't reach the green with two drivers.

Woods won $297,000. More significantly, by winning he earned a place in the 1997 Masters Tournament, where he would write some history.

What's in a Name?

Two days after the birth of his son, Earl Woods looked at him in the hospital nursery and decided he'd nickname him "Tiger."

His had been an easy choice.

As a lieutenant colonel in the Green Berets, Earl Woods had served two tours in Vietnam, his second, from August of 1970 until August of 1971, as an advisor to the deputy chief of Bien Thuan Province, who he remembered as Colonel Nguyen Phong.

Playing a more active role than advisor, he and Nguyen often fought together, more than once saving each other's lives. For acts of bravery, Woods was awarded the Vietnamese Silver Star. Woods called Phong "Tiger." He determined that if he had another son he'd call him "Tiger" (he already had two sons from a previous marriage).

Captivated by the story, Tom Callahan, a skilled and resourceful reporter and author, flew to Vietnam to track down Nguyen. In his book *In Search of Tiger,* Callahan wrote of how he avoided contact with the official escort assigned to him by the Vietnamese government and scouted the back country. He made some contacts, and after relentless digging learned that his name had not been Nguyen, the Vietnamese equivalent of Smith (in Vietnamese, as in most Oriental countries, the family name comes first), he was named Vuong Dang Phong.

Captured by the Vietcong, Vuong had been imprisoned and later sent to a forced-labor camp, where he died in 1976 not knowing his nickname would be made immortal one day by a young American golfer.

Tiger's given name is Eldrick, made up by his mother, Kultida Punsawad Woods, an office receptionist whom Earl Woods met in Thailand. She and Earl were married in 1969, and Tiger arrived on December 30, 1975. Creating her son's name, Kultida took the E from Earl and the K from Kultida, then filled in the letters between.

The Natural

Tiger Woods had evidently been born to play golf.

Working to improve his own game, Earl Woods set up a driving net in his garage and often sat his infant son in a high-chair close by, explaining this was one way of spending time together. Developments suggest young Tiger not only enjoyed it, he absorbed it.

In his book *Tiger Woods, the Making of a Champion,* Tim Rosaforte tells of how Earl brought home a little plastic club one day, obviously hoping to prod his son into taking a few swings. Bored watching his father having all the fun, Tiger climbed down from his perch one evening, picked up his baby club, and made a graceful golf swing—left handed.

Within a week, he turned around and swung right-handed, even changing his grip by setting his right hand lower on the shaft than his left.

Years later, the high-chair badly rusted, Kultida planned to toss it out with the trash. Earl stopped her. "No," he said. "That's going to the Hall of Fame some day."

Matinee Idol

Word of young Tiger's precocity spread after he won a pitch, putt, and drive contest against 10- and 11-year-olds. He was three years old at the time, which prompted a Los Angeles sports announcer to broadcast a story on him.

Within a short time *The Mike Douglas Show* put him on the air alongside Bob Hope and James Stewart. Tiger putted a few and played full shots from a mat.

When he reached five, another show called *That's Incredible* invited him to appear. He hit some shots, and when the show's host, Fran Tarkenton, the former NFL quarterback, asked what he planned to do when he grew up, Tiger forecast the future.

"I want to win all the big tournaments," he said. "I want to play well when I get older and beat all the pros."

What Price Glory?

Earl Woods financed his son's expanding golf schedule with a handful of credit cards and two mortgages on his house. Surprisingly, even at the age of 10, young Tiger realized the heavy cost to his family. One day he approached Earl and asked, "Daddy, do you think when I turn pro you could live on $100,000 a year?"

A Family Affair

By the time Greg Norman reached his mid-forties, he'd lost his standing among the more feared players and devoted much of his time not to playing golf courses but to building them. He designed Pelican Waters, just north of his boyhood home near Brisbane, on the northeast coast of Australia.

During May of 2003, Toini Norman, Greg's mother, scored a hole-in-one on the club's 14th hole, and immediately shot off an e-mail message to her son.

This was no easy golf course. The 14th at Pelican Waters calls for a 152-yard tee shot to an elevated green protected by bunkers both right and left and a slippery green canted decidedly right to left.

Nor was Mrs. Norman a neophyte playing a lucky shot. Seventy-two years old at the time, she had been addicted to the game most of her life, played to a 12 handicap, and had won the club's women's championship a year earlier. In more than 50 years of golf, this was her second ace.

As a much younger woman, Mrs. Norman had still been playing when she was seven months pregnant with Greg, and a few years later had introduced him to the game.

Listen Up

After three rounds of the 1996 British Open, Norman had played Royal Lytham and St. Annes in 210 strokes, even par but not good enough. Tom Lehman had shot 198 and led Norman by 12 strokes.

Greg's putting had not been at its best.

Between rounds he was approached by Jack Newton, a fellow Australian who had lost a playoff to Tom Watson for the 1975 British Open. Newton had turned to television commentary after losing an arm by walking into an airplane propeller some years earlier.

Newton told Norman that earlier in the week he'd spotted a flaw in Norman's putting stroke; Norman had been taking the putter back too far, then decelerating coming into the ball.

Instead of thanking Newton, Norman asked, "Why didn't you tell me earlier?"

"Why, mate?," Newton challenged. "You've never listened to me in your life."

Norman shot 67 in the last round, picked up six strokes on Lehman, who stumbled in with 73, and climbed into a tie for seventh place, at 277. Lehman won with 271, two strokes ahead of Mark McCumber and Ernie Els.

The Tiger and the Shark

One day after his 16th birthday, Woods played a round of golf with Greg Norman at the Old Marsh Golf Club, Norman's home club in Palm Beach Gardens, Florida. Woods outdrove him.

It was the last day of 1991, a time when Norman stood at the peak of his career, the most dangerous player in the game and among the longest drivers on the PGA Tour. He had, in fact, placed second in driving distance in 1990 with an average of 282.3 yards

They began at the second hole, a reachable par 5. Knowing the kid's reputation as a long hitter, Norman went at his drive with vengeance. Then Woods stepped onto the tee and lashed one of his best as well. From a distance, no one could guess whose ball had gone farther.

They piled into their golf cart, and what they saw shocked Norman. Slender, 140-pound Tiger Woods had outdriven 180-pound veteran professional Greg Norman by five yards.

Nor did it end there, according to Norman, who admitted: "That little shit was driving it by me all day."

Fee for Service

As David Frost stepped onto the first tee during the 1996 British Open, his spanking new golf shoes suddenly split. Spotting a marshal standing nearby, Frost asked if he'd dash to the merchandise tent and buy him a pair of shoes, size nine and a half.

With the speed of light, the marshal raced to the tent, bought the shoes and dashed back, to Frost's everlasting thanks.

He was given a reward as well. Frost told him to keep those split-open shoes.

Not a Meteor Shower

Lutz Richter lined up his putt carefully on a course in Amsterdam and stuck it just right. The ball dropped, and he and Erik Kemp walked off the green toward the next tee. Just in time.

As they stepped off the green, a huge chunk of ice struck exactly on the spot where Kemp had been standing, gouging a two-foot crater. It would have killed Kemp.

While no one can be sure, those who saw it believe the block of ice probably fell from an airplane.

Bold Words

When Woods visited Thailand in 1998, Ernie Els had been playing superb golf. A year earlier he had won his second U.S. Open, then beat Woods, who had won the Masters; Justin Leonard, the British Open champion; and Davis Love III, the PGA champion, in an affair the tour grandly named the PGA Grand Slam. Early in 1998, at Riviera Country Club, Els beat both Duval and Woods, who were playing behind him, in the Los Angeles Open, and both Woods and Love at Bay Hill, in Orlando, Florida.

When Els went into the last round of the Johnnie Walker Classic leading Woods and the rest of the field by eight strokes, a reporter asked Tiger if anybody could beat Ernie.

Woods answered, "I can."

Told of Tiger's comment, Els asked, "What's he on?"

Turned out, nothing more than conviction.

Woods caught Els over the closing 18 holes with 65 against Ernie's 73, then beat him on the second hole of a sudden-death playoff.

Slow, Really Slow

If Shinji Minagawa's excursion around his club in Kobe, Japan, wasn't the slowest round of golf known to man, it must have been the most expensive. From his tee shot on the first hole until his final putt on the 18th, Minagawa ate up 12 hours.

There were, however, mitigating circumstances. He must have spent substantial time rushing back and forth to the golf shop replenishing supplies. After all, he lost 197 golf balls.

In Memoriam
Ben Hogan

On July 25, 1997, just 19 days before his 85th birthday, Ben Hogan died in Fort Worth, Texas, where he had lived since 1921. Through dedication and endless practice, he had become the best player of his time. The great Bob Jones always insisted no one could do better than that.

Hogan was born on August 13, 1912, the youngest of three men born that year who would form the Great American Triumvirate—Byron Nelson in February, San Snead in May, and then Hogan. Each man achieved greatness, but none left a legacy to match Hogan's. He had a monumental year in 1953, winning the Masters Tournament, the U.S. Open, and the British Open. In recognition, New York staged a ticker-tape parade down Broadway, just as the city had given Jones after his Grand Slam of 1930.

Hogan had never fully recovered from colon cancer surgery in 1995 combined with an attack of bronchitis shortly afterward. After falling at his home, he was admitted to All Saints Hospital and died the following day.

Dave Marr, a former PGA champion and long-time Hogan friend, said of his death:

"We've lost the unicorn."

A myth. Had he really existed?

Herbert Warren Wind, whose essays on golf graced *The New Yorker* for many years, wrote of Hogan in 1955:

"In years to come, I am sure, the sports public, looking back at his record, will be struck by awe and disbelief that any one man could have played so well so regularly. Ben Hogan, the outstanding sports personality of the post-war decade, has, to be sure, secured a place among the very great athletes of all time."

From 1946 through 1953, Hogan played in 16 major tournaments and won 9 of them—four U.S. Opens, two Masters Tournaments, two PGA championships, and the 1953 British Open, his only attempt at the British Open.

The author Dan Jenkins liked to exaggerate and say that if 20 writers were covering a tournament, they'd send two out to watch what was going on while the others sat around telling Hogan stories. This tale from the British Open surfaced after Hogan died:

Hogan arrived in Scotland two weeks before the first round and gave in to a request from a Scottish newspaper to allow a reporter access to his practice sessions. The editor sent Harry Andrew, its golf correspondent, who happened to play a reasonable game of golf.

The championship would be played at Carnoustie, a severely demanding public course on Scotland's east coast, but since Carnoustie had no practice range, Hogan was given permission to use an especially nice range at Panmure, a private club a mile or so away. Andrew shadowed him every day and always liked to tell of a session when Hogan worked on his 3-iron.

Hogan dropped 30 balls to the ground and sent his caddie, Cecil Timms, down-range to field them. He began by playing one perfect shot after another. The ball would bounce once, Timms would catch it and drop it into Ben's practice bag.

Shot after shot flew either directly at Timms or else within easy reach, left or right, while Timms barely shifted his feet.

Suddenly, though, he took a step and lunged to catch one.

Hogan scowled, then dropped 30 more balls. Again Timms simply reached out to catch them—until he lunged again. Down went 30 more balls.

After about the fourth lurch, Hogan slammed his 3-iron to the ground, turned to Andrew and said, "Dammit, Harry, I used to be able to play that shot!"

There has never been anyone like him.

Sam Snead

Five years after Hogan's death, Sam Snead died on May 23, 2002, four days before his 90th birthday.

After Byron Nelson went into semi-retirement following the 1946 U.S. Open, Snead and Hogan became the tour's greatest attractions, the best in the game. Over his lifetime, Sam won 103 tournaments, although the PGA Tour record book recognizes only the 81 it cosponsored. No one won nearly as many. Jack Nicklaus ranked second in the tour's book, with 70, and Hogan third with 63. Nelson, who retired at the age of 34, is given credit for 50.

Snead won 11 tournaments in 1950. Through 2003, no one had won as many as 10 in one year since. Unfortunately for Sam, Hogan had won the U.S. Open. Even though he won nothing more, Ben was named Player of the Year. Sam never understood why.

Known for his fluid, rhythmic swing, Snead arrived on the tour with, according to George Fazio, "A three-quarter swing with a closed face." One of the two men Hogan beat in a playoff for the 1950 Open, Fazio witnessed Sam's swing close up when the two played a practice round before the 1936 Hershey (Pennsylvania) Open.

If that is true, Snead made a quick change in style, because the following year he turned into a consistent money-winner. After he won the Oakland (California) Open in January and the Bing Crosby Pro-Am in February, and placed second in the Florida West Coast Open and third in both the Houston and Metropolitan (New York) Opens, bookmakers named him the 8-1 favorite in the 1937 U.S. Open. He placed second, behind Ralph Guldahl, who set a new Open record in beating him by two strokes. Guldahl shot 281 and broke Tony Manero's year-old record by a stroke.

Always sure of who he was, Sam could be vain.

Anyone familiar with Snead knows he wore a straw hat with a colorful hatband. Spotting two men he knew standing on the first tee of the Homestead course, close to his home in Hot Springs, Virginia, Sam stopped his golf cart and walked over to chat. One of the men wore a hat with a floppy brim.

As they talked, the man wearing the floppy hat saw a young women holding a scorecard and pencil heading their way, obviously to ask Snead for his autograph. Arriving, she brushed past Snead, shoved the scorecard and pencil at the man in the floppy hat and said, "Pardon me, sir. Are you Sam Snead?"

Shocked and somewhat embarrassed, the man gestured toward Sam, and said, "No, Ma'am. This is Sam Snead." She explained that the people in the golf shop had told her Mr. Snead was the man with the hat.

Meanwhile, Snead had turned purple. He signed the scorecard without smiling, and as the young woman walked off, glared at the man in the floppy hat and snarled with all the contempt in his soul, "Are you Sam Snead!"

They talked for perhaps five more minutes, but twice Snead interrupted the discussion, glowered, and, oozing sarcasm, growled, "Are you Sam Snead!"

He could also tell a self-mocking story of mistaken identity. In the book *Slammin Sam,* done with the author George Mendoza, Snead told of a man who approached him and said, "I know you. You're the greatest golfer in the world."

Doing the "Aw, shucks," routine, Snead said, "Gee, thanks. That's real nice."

Then the man called to his wife and said, "Hey, Honey, meet Ben Hogan."

Snead took great pride in his record in head-to-head games against Hogan. In two of the more memorable, Sam beat Ben in a playoff for the 1950 Los Angeles Open, Hogan's first tournament following his automobile accident, and in the 1954 Masters.

So proud was he that when he attended Hogan's funeral, he spent much of his time telling anyone who would listen that Hogan never beat him head-to-head.

Like Hogan, there has never been another like Snead.

White Fang

Practicing on Baltusrol's putting green a few days before the 1967 U.S. Open's first round, Jack Nicklaus borrowed a putter from Fred Mueller, a salesman for Jansen sportswear and friend of Deane Beman. Mueller had painted his Bullseye putterhead white and named it White Fang. Jack liked its feel and asked if he could use it for the championship.

Mueller evidently agreed, and it worked so well that Jack set a U.S. Open record of 275 and beat Arnold Palmer in a duel that established Nicklaus as the game's leading player.

It has never been clear that Mueller actually gave Nicklaus the putter, but Jack kept it, eventually retired it, and added White Fang to his collection of favorite clubs. Sometime in the early 1980s, though, the club simply wasn't there. A series of searches turned up nothing, and so Nicklaus assumed he'd never find it again.

Then, in April of 2003, it turned up, returned by Joe Wessel, who played football for Florida State University with Jack's son Steve. Wessel had found it in his garage, the head a little worn but still showing flecks of white.

"I knew it was White Fang immediately," Nicklaus said, and put it on display in the Jack Nicklaus Museum, in Columbus, Ohio, his home town.

At last word he was still looking for the big-headed putter he used to win his sixth Masters Tournament, in 1986, which he claims is the only significant club missing from his collection.

And so far as anyone knows, Fred Mueller may still be looking for White Fang.

The Masters

In the first week of April in 1997, Tiger Woods won the Masters and established himself as the game's foremost player and successor to Jack Nicklaus.

A number of golfers had won the first tournaments they'd played as professionals, but no one in modern times had won his first major competition so convincingly after starting so poorly. He stumbled around the first nine in 40 strokes, yet ended the four days by setting the Masters' 72-hole record at 270 and its winning margin at 12 strokes. Nicklaus had set both previous records in 1965. Jack had shot 271 and beaten Arnold Palmer, Gary Player, and Dave Marr by nine strokes. Woods beat Tom Kite.

Woods began his first Masters as a professional by driving into the woods on the first hole and losing a stroke. He lost another at the demanding fourth, a par 3, then bogeyed both the eight, a par 5, and the ninth, a par 4. That rocky beginning behind him, he played the next 63 holes in 22 under par.

After his outward 40, Woods stormed in with four birdies and an eagle on the homeward nine, came back in 30, shot 70 for the first round, followed with 66 in the second, 65 in the third, and 69 in the fourth.

At 21 he replaced Seve Ballesteros as the youngest Masters winner. When he won in 1980, Ballesteros had been 23.

Back in 1953, Ben Hogan had set the 72-hole record at 274 and Nicklaus had lowered it to 271 in 1965. It had stood for 12 years. Nicklaus's 271 had stood for 32 years. Woods also became the first to play three rounds in the 60s.

Paired with Woods, the British golfer Colin Montgomerie went into the third round three strokes behind him, shot 74, and wound up 12 strokes back, in a tie for sixth place. With 201, Woods led Constantino Rocca by nine strokes.

Invited to a press interview nonetheless, Montgomerie waited for his introduction, but before anyone asked a question, he took the microphone and spoke.

"All I have to say is one brief comment," he began.

"There is no chance. There is no way Tiger Woods does not win tomorrow."

Someone called out, "What makes you say that?"

Perpetually annoyed by the press to begin with, Montgomerie grimaced in painful disbelief and asked, "Did you just arrive? Have you been on holiday?"

Reminded that Greg Norman had taken a six-stroke lead into the last round a year earlier, shot 78, and lost to Nick Faldo's 67 by five strokes, Monty scoffed.

"This is different—very different. Constantino Rocca is not Nick Faldo and Tiger Woods is not Greg Norman.

Woods's power overwhelmed Augusta National. At the second hole, a par 5 of 555 yards, he played a driver and a 9-iron that carried over the green. On the 15th, another par 5 of just 500 yards, he played a driver and pitching wedge.

Michael Williams

Seated among the press corps in the posh building that replaced the old Quonset hut at Augusta National, Michael Williams, the distinguished golf correspondent of London's *Daily Telegraph,* ground out his copy praising Tiger Woods and his smothering performance in the Masters.

Returning home to England, he cornered everyone who'd stand still and told all who would listen of the great things he'd seen. Then he'd sigh, "I'm glad I lived to see this day."

Two days later he took his son, Roddy, for a round of golf at the Chelmsford Golf Club, where he'd served as captain twice and was due to become club president. Reaching the short fourth hole, the two played their tee shots and started toward the green.

Suddenly, Michael Williams collapsed and died.

Devastated, Judy Williams, Michael's widow, gave up the game. She herself had served as Chelmsford's Lady Captain, but she felt she could never face the hole where Michael died.

Friends eventually persuaded her to come out, and the club picked her to represent Chelmsford in a team match. Understandably, she didn't look forward to playing the fourth hole.

From the ladies' tee, Chelmsford's fourth measured 104 yards. Reaching the tee, Judy drew her 6-iron and made solid contact, yet when she and her opponent stepped onto the green they saw no sign of her ball.

"We looked everywhere," she laughed, "and then I thought I might as well look in the hole."

There it was—the first hole-in-one of her career.

Reflecting on the irony of it all, Judy thought, "Michael was certainly looking out for me. And probably chuckling as well."

But not for long. She lost her match on the 20th hole.

Somewhere under the Rainbow

In a moment that bordered on mystic, Davis Love III birdied the 72nd hole at the Winged Foot Golf Club and won the 1997 PGA Championship under a rainbow.

Love shot 66-71-66-66—269, the lowest 72-hole score ever shot in a major championship at Winged Foot, and beat Justin Leonard by five strokes. A month earlier, Leonard had won the British Open.

There was a sense of mysticism about the finish, and of something missing. Davis Love, Jr., Davis's father, had died in the crash of a private plane near Jacksonville, Florida, in February of 1988 at the age of 53. Davis missed him; he had been his only teacher.

Heavy rain had fallen as Love and Leonard played the 15th hole with Love four strokes ahead, but it had stopped by the time they reached the 18th. As Love stepped onto the final green, a rainbow arched against the pale blue sky, the same sort of rainbow that had formed the day Love won the 1992 Players Championship.

Love had thought of his father all through that final day at Winged Foot, remembering he had been with him through three rounds of the 1987 Heritage Classic, at Hilton Head Island, South Carolina, the first tournament he'd won. Davis Jr. hadn't seen the finish; he'd left early because of teaching responsibilities at his club.

"It's hard to believe that's the only tournament he saw me have a chance to win," Davis reflected. "He was proud of me no matter what I did, but I sure would have liked to share one with him."

Casey Martin

Casey Martin could play a fine game of golf, good enough to succeed as a professional. It would be difficult, though. He had been stricken with Kippel-Trenauney-Weber Syndrome, a rare disease that withered his right leg to the size of a 3-iron and made walking difficult.

Denied the use of a motorized golf cart by the PGA Tour, Martin took the Tour to court and won his case.

Before his lawsuit had been settled, he entered the 1998 U.S. Open, but he would have to survive two rounds of qualifying to win a place in the starting field.

Breaking not only tradition but a precedent set in other rulings, the USGA granted Martin permission to use a motorized cart. When he qualified, the USGA had to determine how to accommodate him in the Open proper, scheduled for the Olympic Club, in San Francisco, a course with some serious hills.

Martin had been given a one-seater for the qualifying rounds, the kind seen commonly in supermarkets, but test rides around Olympic found them unstable. Instead, he rode a conventional cart.

Martin played reasonably well. He shot 291, just 11 strokes behind Lee Janzen, who won with 280, and 10 behind second place Payne Stewart, with 281.

Martin never qualified again.

Snatch and Grab and Run

Paul Simpson, an amateur golfer from Raleigh, North Carolina, had played Olympic's first nine in 33 strokes, two under par, during the first round of the Open, which wasn't at all bad. Another nine like that and he could lead the field.

Then fate and a sticky-fingered spectator ruined his day.

Simpson pulled his tee shot into a stand of eucalyptus trees alongside the the 10th, a 422-yard par 4, The ball hit a tree and dropped into the deep rough. Quickly, a swift spectator scooped it up and dashed off, clutching his souvenir.

Apparently no one at the scene saw Ali Baba steal the ball, and so Simpson and his caddie, along with a bunch of spectators, searched the ground but, naturally, didn't find it.

So far as Simpson knew, the ball was lost. Under the rules of the game, he had to return to the tee and play another ball, his third shot.

But the eagle eye of a television camera had caught the culprit in the act. Unfortunately, word didn't reach Simpson until it was too late. Had he known the circumstances, he could have put another ball into play at the spot where the original ball had landed, without penalty.

Evidently un-nerved by the incident, Simpson played the second nine in 43, shot 76-72, and missed the 36-hole cut by one stroke.

Duval's 59

On January 24, 1999, David Duval went into the last round of the 90-hole Bob Hope Desert Classic in 12th place, seven strokes behind Fred Funk, the leader, and assessed his prospects of winning as "slim to none."

At the end of the day he had picked up eight strokes, passed Funk and everyone else, and won the tournament.

When you shoot 59, it's easily done. Steve Pate shot a solid 66 but finished in second place, one stroke behind.

Duval's had been an astonishing round, one that left the rest of the field not only dazed but convinced he'd proved himself the best in the game. He shot his 59 with steady driving and exquisite iron play that consistently left him within reasonable holing distance. Yet he holed nothing outside 10 feet.

Beginning this memorable round by playing deadly irons, Duval birdied the first three holes, twice with approaches within three feet and the other within five feet. After missing from 15 feet on the fourth, he hit a 5-iron tee shot to five feet and birdied the fifth, then an 8-iron to the ninth and holed from eight feet, his fifth birdie of the first nine. Out in 31, he still lagged behind the leaders.

Now Duval turned on the heat. He birdied the 10th with a 3-wood and sand wedge to four feet, pitched to four feet again and birdied the 11th, a par 5, and followed with a 6-iron to two feet and another birdie, his fourth in succession and eighth of the round.

Duval's birdie string broke on the 13th when he missed from 12 feet and parred, but he tore a 5-iron from the rough to 10 feet and birdied the 14th, played an 8-iron inside a foot and birdied the 15th, a par 3, then nearly holed his approach to the 16th for an eagle. He birdied instead, his sixth on the second nine.

Duval's tee shot to the 17th, pulled up 20 feet from the hole, and he parred once again, then marched to the 18th tee needing an eagle 3 to break 60.

Through it all he had projected coolness rather than tension. After crushing his drive, he and Mitch Knox, his caddie, strolled toward the ball talking about a car apparently floating in a water hazard left of the fairway.

Setting himself, Duval rifled a 5-iron that carried across the hazard and braked about six feet behind the hole. David ran it in for the eagle he needed. Out in 31, he had come back in 28.

He had birdied six holes on the first nine, six more on the second, and closed with an eagle. His scorecard showed four 2s, six 3s, seven 4s, and just that one 5 on the sixth.

Asked about this remarkable round, David said only, "I don't know what to say. It's like pitching a perfect game. I stole this tournament from Steve Pate."

Duval's had been the third 59 played under tournament conditions at the game's highest levels in the United States, after Sam Snead in 1959 and Chip Beck in 1991. Eleven days before his forty-seventh birthday, Snead had shot 59 over the 6,317-yard Old White Course at The Greenbrier during the 1959 Sam Snead Festival, and Beck in the 1991 Las Vegas Invitational, over the Sunrise course in the Nevada desert. Snead could have shot 58, but after laying a 3-iron within five feet at the 17th, his putt lipped out.

The Tour credits Al Geiberger with a 59 in 1977, but he hadn't played under the rules of golf. Instead, because rain had left the ground wet, the tour had allowed players to lift their ball, wipe it clean, and place it somewhere other than where it had come to rest. Some call this process either lift, clean, and cheat, or lift, clean, and throw it.

Duval played under no such stigma.

Two years after Duval, Annika Sörenstam matched his score in an LPGA tournament at the Moon Valley Country Club, a 6,459-yard redesigned course in the Arizona desert outside Phoenix. While Duval shot his 59 in the last round, Annika shot hers in the second. It stood as the LPGA record.

Pia Nilsson, for 10 years the coach of Sweden's women's national team, once handed 19-year-old Annika a document that stated, "Every hole can be birdied." At 33, when she led off with eight consecutive birdies, Annika looked as if she might do it.

She began from Moon Valley's 10th tee and birdied 12 of her first 13 holes. On the 10th, a 535-yard par 5, she lobbed a sand wedge to nine feet and birdied, followed with a 9-iron to seven feet and birdied the 11th, a par 3, reached the green of the 12th, another par 5, with a driver and 7-wood and two-putted for another birdie, played a 4-wood and sand wedge to four feet and birdied the 14th, and birdied the 15th, a par 3, with a 7-iron to 11 feet.

Not through yet, Annika birdied both the 16th and 17th, holing putts of 10 and 18 feet, her longest so far. At that point she had birdied eight consecutive holes, but she would not birdie the 18th, a 404-yard par 4 where her 8-iron approach left her 30 feet from the hole. She took two putts and a par 4.

She had played the second nine in 28 strokes.

Word spread, and now other players began to watch. Amy Benz left the practice range for the course because, "I wanted to see history." Annika's father, Tom Sörenstam, called his wife in Sweden every two holes, at a rate of $12 each minute. In mid-round, others often stopped to see Annika putt, and Joanne Morely, another tour player, watched through binoculars from a guest house.

Starting her second nine holes from the first tee, she ran off four more birdies, the most satisfying on the second, a 169-yard par 3. Her 7-iron had left her 22 feet behind the hole. So far she had holed from 30 feet on the 12th and from 18 feet on the 13th and 17th.

"It was a tricky line," Annika said later, "up, then down."

She holed it, and when it was all over, she said, "When that went in I said, 'This is just my day.'" When Meg Mallon, who played in her group, patted her on the back, Annika warned, "Don't hit me too hard or I might wake up."

Then she added two more birdies, holing from 12 feet on the third and nearly chipped in for an eagle 3 on the fourth, a 511-yard par 5. Instead she birdied her 12th hole of her first 13. As far as anyone knew, her start had never been matched.

Annika's pace fell off then and she ran off three pars, but she made one final birdie. Moon Valley's eighth a shortish par 5, measures just 476 yards. She reached the green with a 7-wood second and two-putted from 25 feet. Now, with a 58 in sight, she missed from nine feet on the ninth and ended her extraordinary round with a par 4.

When the final putt dropped, she asked of anyone nearby, "I did shoot 59, didn't I?"

Others have shot 59 as well. Both Jack Nicklaus and Gary Player have done it, although Nicklaus shot his over the Breakers Hotel course in Palm Beach. Of course, if Jack had been playing his usual game, it would have been surprising if he hadn't. A very easy course, the Breakers measured just 6,008 yards with a number of par 4 holes he could drive. Fittingly, Jack had played in a group with Snead and the LPGA players Kathy Ahern and Kathy Whitworth.

Player had shot his 59 at the Gavea Golf Club, in Rio de Janeiro, during the second round of the 1974 Brazilian Open.

Both Zoran Zorkic and Sean Pappas shot 59s on a mini-tour, Zorkic at the 6,591-yard Terrapin Hills Country Club, in Fort Payne, Alabama, in 1990, and Pappas on the 6,336-yard Heartland Golf Course, in Bowling Green, Kentucky, in 1991.

And playing on the Japan Tour in September of 2003, 48-year-old Masahiro Kuramoto shot 59 at the Ishioka Golf Club, a man-sized course that measured 7,046 yards.

Then there was Homero Blancas with his 55 in the first round of the Premier Invitational, an amateur tournament in Longview, Texas. The course, however, measured just 5,002 yards.

The Guiness Book of Records credits a man named Monte Carlo Money with a score of 58 over the 6,601-yard Las Vegas Municipal Golf Course in 1981.

Four months after Duval's 59, the Japanese golfer Shigeki Maruyama shot 58 at the Woodmont Country Club, in suburban Washington, D.C. in a qualifying round for the U.S. Open.

When he finished, Billy Andrade, who had played in his group, fell to his knees and salaamed three times.

Old Acquaintance

Curtis Strange, who had won the U.S. Open in 1988 and 1989, renewed an acquaintance during the 1999 championship, at Pinehurst, the resort in North Carolina. Strange had attended college at Wake Forest, also in North Carolina, and had been handed a speeding ticket in nearby Lee county.

Chatting with some North Carolina Highway Patrolmen outside the Pinehurst locker room, Strange recalled the incident and said it might have happened about 15 years earlier.

"Yep," Garland Roth said, "It was 15 years ago, and I'm the one who gave it to you."

A trooper then, Roth had risen to sergeant, and during the Open had charge of the crew working the 18th green and scoring area.

Unanswerable Rules Question

When the U.S. Open came to Pinehurst, tales of the old resort's past spun in every corner of the clubhouse, especially those about Richard S. Tufts, whose family not only created the resort but the entire village of Pinehurst before the turn of the twentieth century.

A gentle man educated at Harvard, Richard Tufts had run Pinehurst for years, until other family members sold it, over Richard's protests.

Aside from his position at Pinehurst, Tufts had been a major figure in golf administration. An authority on the rules of the game, he had helped promote and organize the first conference on the rules with the Royal and Ancient Golf Club of St. Andrews, which in 1951 compromised on a single code for the game. He rose to president of the USGA in 1956.

A man with a sense of humor, Tufts like to tell of the time he had been called onto the no. 2 course to make a ruling during a tournament. When he arrived at the site he found two golfers who'd had more than a few drinks sitting on the edge of a bunker. As he moved closer he asked the nature of the problem.

One spoke up and said, "One of us is 1-up, but we don't know which of us it is."

Tufts, the man who wrote many of the rules golfers play by, had no answer.

The Death of Payne Stewart

Three men had each won two U.S. Opens during the 1990s, more than had won twice in any other decade. Ernie Els won in 1994 and 1997, Lee Janzen in 1993 and 1998, and Payne Stewart, the oldest of the three, in 1991 and 1999.

Stewart didn't live to defend his last Open championship. He died when an eight-passenger Lear Jet crashed in a South Dakota field on October 25, 1999. He had been part owner of the plane.

In June he had shot 279 over the no. 2 course at Pinehurst Country Club and edged out Phil Mickelson by one stroke.

Leaving Orlando on October 25, he headed for Dallas to discuss a golf course design project. Instead, the plane flew 1,600 miles up the center of the nation, making much of the trip on autopilot. Neither Michael King, the pilot, nor Stephanie Bellegarrigue, the co-pilot, transmitted a distress call. With Stewart on the plane were 46-year-old Robert Fraly, CEO of Leader Enterprises, Inc., a sports management group that handled Stewart's affairs; Van Ardan, the company president; and Bruce Borland, of the Jack Nicklaus design group.

The plane took off from Orlando at 9:19 A.M., heading northward, probably to avoid flying over the Gulf of Mexico, turned northwesterly and streaked across the sky, climbing ever higher, finally to an altitude of 48,000 feet, far above its assigned cruising altitude of 39,000 feet.

When the plane climbed on and failed to turn west, flight controllers in Jacksonville radioed the pilot, but neither he nor the co-pilot responded.

Working the sector control room of the Atlanta Air Traffic Control Center, Joe Hembrite tracked the plane past Gainesville, Florida, and across the Georgia-Alabama border. Hembrite alerted the Air Force, and planes scrambled to help from Tyndall Air Force Base, near Panama City, from Eglin AFB, in the Florida Panhandle, and then from bases in Tulsa, Oklahoma, and later from Fargo, South Dakota.

Flying an F-15, Captain Chris Hamilton saw the Lear first and flew close enough to radio a chilling report.

"There was ice on the cockpit windshield and the cabin windows were frosted over," he said. "I felt so helpless to know there was nothing I could do to help the people inside." He reported further that the plane seemed to be flying on autopilot.

His report indicated the cabin pressure had fallen, disabling both crew and passengers. All evidence indicated everyone on board had died long before Hamilton found the plane. There is no breathable oxygen at

48,000 feet, and the outside temperature registered from 50 to 60 degrees below zero.

At one stage the Air Force discussed shooting the Lear from the sky if it threatened a major population center but ruled against it because its course avoided congested regions.

Instead the plane flew on until it ran out of fuel at about 1:15 that afternoon. Its power gone, the plane stalled, spiraled downward through the clouds, and plunged almost vertically into the ground at an estimated speed of 600 miles an hour. At that speed, the plane buried itself so deeply that little more than the tail assembly showed above ground.

The most recognized golfer of his time, Stewart had run in a 15-foot putt on the Open's final hole, which is believed to have been the longest putt ever holed to win a U.S. Open. He died at the age of 42.

My Other Car's a Lamborghini

Taking a relaxing break before the 1998 World Cup, members of England's soccer team played a round of golf at the Mill Ride Golf Club, near Ascot, and invited Derek Lawrenson, the golf correspondent for London's *Daily Telegraph,* to join them. Since the automobile company that makes the Lamborghini owned the club, it offered one of its sleek roadsters to anyone who scored a hole-in-one on Mill Ride's 170-yard 15th.

When Lawrenson and his group reached the 15th tee, Paul Ince, a member of the England team, held the honor and asked Lawrenson which club he should play, a clear violation of the rules of golf. But, then, who cared? Lawrenson suggested a 4-iron. Ince misplayed the shot and left his ball short of the green.

Up next, Lawrenson surprised himself and hit a 3-iron flush. The ball carried to the green, ran left of the hole, then took the contours, swung right, and tumbled into the hole. Every member of the four-ball jumped up and down, pounded Lawrenson on the back, and shouted to everyone they saw that Derek had won the car.

Lawrenson drove it home, of course, and his wife bought a sticker for the back window of their little Metro that proclaimed "Our Other Car Is a Lamborghini."

Derek kept the Lamborghini for a while but eventually sold it. At a selling price of about $300,000, Lawrenson said it was worth more than his house.

He lost his amateur status, of course, but the Association of Golf Writers still accepted his entry into the organization's spring medal tournament over the East Course at Wentworth Golf Club.

Not only did Lawrenson not repeat his Mill Ride miracle, but on each of Wentworth's three par-3 holes, he lost a ball.

Jack's U.S. Open Farewell

Late on a Friday afternoon in June of 2000, Jack Nicklaus stepped onto the 18th tee at Pebble Beach and played his last hole in a U.S. Open. He turned it into a moment worth remembering.

Wearing a tight-lipped smile, he waved his cap to the gallery, then drove his ball into prime position in the fairway, past the two sentinel pine trees to the right and long enough to leave a shot of about 240 yards to the green.

Jack had played three previous Opens at Pebble Beach, but as he told his son Jackie, his caddie that day, he'd never once gone for the green with his second shot. Then he said, "Let's go for it."

Savoring the moment, Jack drew his 3-wood and ripped into the shot. He'd judged it perfectly. The ball streaked toward the distant green, hit dead center onto the narrow chute between two guarding bunkers, and skipped on.

The huge crowd that had massed to watch him finish roared. It had been an inspiring shot, as thrilling as Ted Williams's home run his last time at bat in Fenway Park

Grinning broadly now, Nicklaus waved to the crowd once again, and with his wife, Barbara, and his other sons Steve, Gary, and Michael, following along the ropes, Jack walked ahead to finish the last hole he would play in the Open. He had decided that at the age of 60, he'd had enough.

His eyes clouded by tears, either from the wind, as he insisted, or by emotion, as everyone hoped, Nicklaus three-putted. The man who had holed every putt of consequence he'd ever faced left his first two putts short, the first by ten feet, the second by just an inch or two.

Again the gallery roared.

Back down the fairway, waiting to play his approach, Tom Watson dropped his club and clapped his hands. David Gossett, the U.S. Amateur champion, and Don Pooley, who had played with Nicklaus, rushed to shake his hand, and Jackie threw his arms around his father. Then they walked off together to be with Barbara and the rest of the family.

With the par 5, Jack shot 82, missed the 36-hole cut, and the Open had seen the last of an enduring hero.

Nicklaus had begun his Open career as a 17-year-old prodigy in 1957 and had played in every Open since, a string of 44 years. No one had played in nearly so many.

He had won four of those Opens—in 1962, 1967, 1972, and 1980. No one had won more, and only Willie Anderson, Bob Jones, and Ben Hogan had won as many. He'd also placed second in four, third in one, fourth in two others, and from sixth through tenth in seven more.

No one had meant more to the Open.

The Tiger Slam

Over the 2000–2001 seasons, Tiger Woods established himself as a golfer for the ages.

From June of 2000 through April of 2001, he won all four of the modern major championships in succession—the 2000 U.S. Open, British Open, and PGA Championships, and the 2001 Masters Tournament. Others had set them as their goal, but only Ben Hogan, in 1953, had won as many as three.

His two Opens stand as monumental.

The U.S. Open

Woods shot rounds of 65-69-71-67—272, fully 12 strokes under par at Pebble Beach, and won by 15 strokes, not only a record-winning margin for the U.S. Open alone but the widest in the history of any of the four major competitions as well.

Woods birdied 21 holes and bogeyed five, but in the third round his approach entangled in high grass alongside the third green and he scored a 7 on a tempting par 4 of 390 yards.

He one-putted on 34 holes and did not three-putt a single green, a phenomenal achievement.

On the two driving holes statisticians recorded, he averaged 299 yards, the best in the field. In terms of accuracy, he hit 41 of the 56 fairways on driving holes, or 73 percent, and 51 greens, again, the best in the field. He needed 110 putts, six behind Nick Faldo, who led with 104.

His had been a record-setting performance.

His 65 ranked as the lowest opening round in the four Opens played at Pebble Beach.

His 134 for the first 36 holes matched the record held jointly by Jack Nicklaus (1980 at Baltusrol), T. C. Chen (1985 at Oakland Hills), and Lee Janzen (1993 at Baltusrol).

His six-stroke lead after 36 holes broke the record of five strokes set by Willie Anderson at Baltusrol in 1903 and matched by Mike Souchak in 1960 at Cherry Hills.

His 10-stroke lead at 54 holes broke the record seven-stroke lead set in 1921 by Jim Barnes at Columbia Country Club, in Chevy Chase, Maryland.

His finishing total of 272 matched the record set by Jack Nicklaus in 1980 and matched by Lee Janzen in 1993, both at Baltusrol.

His 15-stroke margin over Ernie Els and Miguel Angel Jimenez broke Willie Smith's record of 11 strokes set in 1899 at the Baltimore Country Club.

British Open

The month after destroying the competition at Pebble Beach, Woods portrayed The Terminator once again and won the British Open.

Playing nearly flawless golf, he shot 67-66-67-69—269 over the Old Course, the lowest score of all the 26 British Opens at St. Andrews, and won by eight strokes over Ernie Els once again and the Danish golfer Thomas Bjorn.

While it had been a record for St. Andrews, his 269 fell two strokes short of the 267 Greg Norman had shot at Sandwich in 1993.

At 24 years, six months, and 24 days, Woods was the third- youngest British Open champion since the Second World War. Gary Player (1959) had been 23 and Seve Ballesteros (1979) had been 22. Young Tom Morris had been just 17 years old when he won the first of his four British Opens at Prestwick in 1868.

Woods joined five others who'd held both the U.S. and British Open trophies in the same year. Bobby Jones had done it twice, in 1926 and 1930, Gene Sarazen won both in 1932, followed by Ben Hogan in 1953, Lee Trevino in 1971, and Tom Watson in 1982.

Woods's success at St. Andrews had been just as breathtaking as at Pebble Beach. In 144 holes over two punishing golf courses, he had bogeyed only nine holes, six at Pebble Beach and three at St. Andrews. Over those same 144 holes, he had three-putted only three greens, all at St. Andrews. Three three-putt greens over 144 holes of championship golf beggars belief.

Furthermore, while the Old Course may be pocked by 128 bunkers someone described as "just big enough for an angry man and his wedge," never did Woods set foot in a single one.

His scorecard bore out his uncanny precision. Except for a birdie 2 on the eighth hole of the third round, his scorecard showed only 3s, 4s, and 5s. He scored three of those 5s on the fifth, much the easier of the two par 5 holes. Every day he birdied the more difficult 14th, where the devilish Hell Bunker lies in ambush.

He birdied the 12th hole in every round as well. Els, on the other hand, played the 12th in even par and the 14th in one under, a difference of seven strokes.

Over the four rounds, Woods birdied 22 holes, not quite one in three.

The PGA Championship

After thoroughly whipping the fields in the U.S. and British Opens, Woods felt lucky to win the PGA Championship, played for the second time over Valhalla Golf Club's mediocre course, located in Louisville, Kentucky, a city known better for horse racing and baseball bats. The PGA of America owns Valhalla.

In an ironic twist, he was nearly beaten by Bob May, a fellow Californian who had set just about every state junior record Woods had broken. While Woods developed into the best player of his time, May never quite made it on the PGA Tour. Instead, he played steadily on the European Tour.

Given a special invitation into the PGA, May's career reached its peak in August of 2000 when he opened with 72, even par, then followed with three 66s, shot 270, tied Woods, but lost a three-hole playoff over Valhalla's 16th, 17th, and 18th holes.

Woods had been lucky to win. May had played him stroke for stroke throughout the fourth round, but Woods birdied the 16th, the first of the playoff holes, and moved one stroke ahead. Still ahead going into the 18th, a 540-yard par 5 with a sausage-shaped green, Woods pulled his drive so far left it tumbled down a hill toward dense bushes that might have left his ball unplayable. Suddenly, though, the ball reversed course, scampered backward along a paved pathway, gathering speed as it rolled ever farther from the green until, finally, it jumped off the path.

With a clear opening back to safe ground now, Woods reached the green with his third shot and made his par as May barely missed a birdie that would have sent the playoff into sudden death.

With that, Woods had won his third major championship of the 2000 season. He had matched Ben Hogan's record of 1953. Hogan had won the Masters and the two Opens. No one could have won more than three that year because dates for the British Open and the PGA conflicted.

The 2001 Masters

By dusk on Sunday, April 8, 2001, Tiger Woods could place the trophies for the U.S. Open, the British Open, the PGA Championship, and the Masters Tournament side by side on his coffee table. On that date he won the Masters and completed a string of four consecutive major championships.

He did it in grand style, fighting off strong challenges from both David Duval and Phil Mickelson by shooting 70-66-68-68—272, two strokes over his record-setting 270 of 1997. Where he had won by 12 strokes then, he won by two in 2001. Duval shot 274 and Mickelson shot 275.

Vijay Singh, who had won in 2000, shot 282 and tied for 18th place; Els, second to Woods in both the U.S. and British Opens in 2000, shot 277 and tied for sixth; Miguel Jimenez, tied with Els for second in the U.S. Open, shot 280 and tied for 10th; and Bob May, who had nearly denied Woods the 2002 PGA Championship, shot 293 and tied for forty-third place, 21 strokes behind.

Duval could have won. He led the field after birdieing the 15th, but he lost one stroke by overshooting the 16th green, then missed a 15-foot birdie putt on the 17th and another from six feet on the 18th. His challenge had ended, and Woods took the Masters.

Iron Man

Evidently, an unlucky few are accident-prone.

Tearing down the side of a mountain during a skiing trip to Switzerland before the 1999 golf season opened, Retief Goosen lost control and flipped over. His left arm hurt, but he'd been hurt before and shrugged it off.

Once the season opened and the arm didn't improve, he visited a doctor, who took an X-ray.

A good thing, too. Goosen had broken his arm.

Earlier in life, Goosen had been struck by lightning.

As he and some teenage friends played a round in his native South Africa, a storm gathered quickly, lightning flashed, and every one of the group fell to the ground. The bolt struck so close to Goosen it blew him out of his shoes and knocked him unconscious. He woke up in a hospital frightened but not seriously hurt.

Some years later, Retief Goosen won the 2001 U.S. Open at the Southern Hills Country Club, in Tulsa, Oklahoma.

Reprise

Back in 1977, police detectives mingled in the gallery following Hubert Green around Southern Hills during the U.S. Open. The local office of the FBI had been warned someone would try to shoot him during the final round.

Nothing happened, and Green won the championship.

Twenty-four years later, when the Open returned, the gendarmes descended on Southern Hills once again, scattering through the gallery, this time searching for an escaped convict. Sentenced to life in prison for a 1989 murder, Jerry Vernon had been hiding out since March after breaking out of the Central New Mexico Correctional Facility.

The Fugitive Warrants Team combed the gallery "Because of his (Vernon's) affection for golf," according to Officer Lucky Lamons.

But as it had in 1977, the search found nothing. Maybe Vernon watched on television.

Double Aces

Holes-in-one are rare enough, but how about two golfers holing-in-one on successive shots?

Playing the Glynhir Golf Club, in Ammanford, Wales, about 15 miles inland from the Bristol Channel port of Swansea, Richard Evans, a surveyor, assumed he had won the 192-yard third hole when his tee shot hit short of the hole, ran toward the flagstick, then dived in for a 1.

Smiling, he stood aside as Mark Evans, his opponent, drilled another smart-looking shot directly at the flagstick. His jaw dropped when that ball dropped into the hole as well.

"I just screamed my head off," Mark Evans said. (Evidently the two men are not related.)

So far as anyone knows, this is the first recorded incident in which successive players in the same match have holed-in-one.

The Hole-In-One Society of Great Britain had recorded nearly 9,000 holes-in-one since 1976, but never a double. It estimates the odds against it at 67 million to one.

A Testing Hole

Proud of their course, members of the Plainfield (New Jersey) Country Club, one of the best around, like to say a round there may ask a player to use every club in his bag. One of its members could swear to it.

Ange Paraskevas, a gynecologist, once used a different club to play each of his nine shots on the second hole alone. A 450-yard par 4, the second is the club's no. 1 handicap hole.

Paraskevas began by topping his drive under a pine tree just off the tee, then hacked it out with a 5-iron. Still in the rough, he played successive strokes with a 4-wood, a 7-wood, a 3-wood, and then flew a 9-iron into deep grass beyond the green. Digging it out with a sand wedge, Paraskevas ran the shot off the front of the green, then chipped to 20 feet with his 7-iron.

Mercifully, Ange holed the putt, a sliding downhiller on a dangerous green.

On the next hole, a par 3 over water, Paraskevas dumped his tee shot into the pond, dropped another ball behind the hazard, and holed his wedge. An easy 3.

Too Many Clubs

After three rounds of the 2001 British Open, Ian Woosnam, a pint-sized Welshman who barely reached 5-foot-4 but had the strength of an ox, shared first place with David Duval, Bernhard Langer, and Alex Cjeka. All three had played Royal Lytham and St. Annes in 207 strokes, six under par.

But Woosnam had a problem.

Experimenting on the practice ground to choose which of two drivers he would use, Woosnam rushed to the first tee to make his starting time, snatched a club from his bag, and teed up his ball.

Royal Lytham is unusual among major championship courses. It begins with a par 3 hole—a substantial par 3 of just over 200 yards, but a par 3 nonetheless. Woosnam ripped into a 6-iron and played a wonderful shot that hit just short of the green, rolled toward the flagstick, and for a heart-stopping moment looked as if it might fall for a hole-in-one. Instead it died within five or six inches of the hole. Woosnam holed it for a birdie 2, and fell to seven under par. Clearly in the lead now, he strutted to the second tee ready to take on the world.

When he got there he had a rude shock.

Woosnam routinely carried only one wood—his driver—so when he stepped onto the second tee, Myles Byrne, his caddie, spotted two head covers. He still carried that second driver.

His voice a little shaky, Byrne looked at Woosnam and told him, "You're going to go ballistic."

Suspicious and uneasy, Woosnam asked why.

"We have two drivers in the bag," Byrne answered. The rules of golf allow just 14 clubs. With the second driver, Woosnam carried 15. Instead of opening with a birdie 2, he added the two-stroke penalty, wrote down a bogey 4, and dropped out of first place.

While it is the player's responsibility, most tour players, Woosnam among them, leave it to their caddies to count their clubs. Furious, Woosie glared at Byrne, snapped, "I give you one fucking job to do, and you . . . ," then snatched the extra club from his bag and flung it away, shaft-first, with the form of an Olympian practicing his javelin throw.

Still seething and focusing more on his penalty than his game, Woosnam bogeyed two of the next three holes, yet still shot 71 and tied five others for third place at 278, just one stroke out of a tie for second. David Duval won, with 274.

Caddies Strike Back

After years of abuse, Jerry Osborne, a caddie on the Senior Tour explained:

"Players make mistakes. Caddies make blunders."

Woosie Wasn't Alone

Before Woosnam committed his sin, the European Tour disqualified Philip Parkin for carrying an extra club in the 1992 Italian Open.

He had learned he had made the field on Wednesday afternoon, rushed to the airport and arrived in Italy at 3:30 the morning of the first round. After playing the Montecastillo course in two over par, he took his clubs to the practice range. Picking through his bag he found his little son's golf club.

Parkin took it to David Garland, the tournament director, and asked if it qualified as a golf club. Taking out his tape measure, Garland told Parkin it would have to measure 18-1/2 inches to be considered a club. It measured 19 inches. Parkin had violated the 14-club rule all the way around and should have added four strokes to his score, the maximum under the rule.

Since he had already signed his scorecard and left for the practice ground, he was disqualified for turning in a score lower than he had actually shot.

Optional Vacation

For many years a sign beside the first tee at Bethpage State Park's Black Course warned golfers they were about to begin a very difficult course and should reconsider unless they play the game fairly well.

Impressed with its possibilities. the United States Golf Association arranged to stage the 2002 U.S. Open at Bethpage Black, the first ever played over a truly public municipal course.

Lots of work lay ahead.

First, in an era of 350-yard drives and 200-yard 5-irons, a course that tested the game's leading players had to stretch more than 7,000 yards.

Then, as with many municipal courses, maintenance wasn't the best. Dedicated and enthusiastic patrons of the course called it "ratty." Years earlier a nationally known golf course architect had been hired to assess all four Bethpage courses and recommend improvements. His report ran to 35 pages.

To polish this one-time gem to Open standards, the USGA pumped $3.5 million into the Black course and brought in Rees Jones to redesign holes, lengthen the course, and improve its overall quality.

When he looked it over, Jones saw the bunkers needed sand. In a spectacular six-month logistical *magnum opus,* truckers shuttled 9,000 tons of white sand from Maryland and another 1,000 tons from southern New Jersey right through New York City and 35-miles out to Bethpage, in central Long Island, scheduling their trips for dawn and dusk, times of lightest traffic.

During the renovation, Jones rebuilt some bunkers, uncovered others lost over time, killed off weeds and crabgrass, and reseeded every fairway with perennial ryegrass, sown at an annual rate of 600 to 800 pounds on each acre. Since the renovation took six years, crews spread roughly 3,500 to 4,800 pounds on each acre.

At the same time, Jones worked on adding length. He moved tees, sometimes to places Daniel Boone must have explored for him, and when he finished, he'd created a new Bethpage. Now it measured 7,214 yards, so long that that shorter hitters couldn't carry the ball far enough to reach the fairways. At the same time, he lowered par to 35-35—70.

The Open was won by Tiger Woods, the longest straight driver the game had ever known. Woods shot 277, three under par, and beat Phil Mickelson, another very long driver, by three strokes.

While the course was a nightmare to some, others among the most influential in the game loved it, and the USGA set about returning within a few years,

Paul Azinger, on the other hand, wasn't interested. Asked what he'd do if the USGA played the Open at Bethpage every five years, he said: "I would probably take every fifth year off."

Monty's Record Round

Colin Montgomerie had, without question, developed into the best English golfer since Nick Faldo. Unlike Faldo, he'd never played his best at the critical times—at the four most important competitions in the game, in chronological order as they were played: the Masters Tournament, the United States Open, the British Open, and the PGA Championship. Twice he'd come close to winning the U.S. Open, first in 1994, when he lost an 18-hole playoff to Ernie Els by playing the first nine in 42 strokes, and to Els once again in 1997. He'd had other moments as well.

Playing poorly in the first round of the 2002 British Open, Montgomerie shot 74, but he bounded back the next day and blistered tough old Muirfield by shooting 64, one of the lowest scores every recorded in any of the four major competitions. Back in the hunt now, he once again frustrated his fans. Montgomerie absolutely collapsed the following day.

Slogging through blustering winds, periodic rains driven sideways with such force each drop felt like a shot from a dart gun, and in temperatures so cold players could barely grip their clubs, Montgomerie and everyone else had a miserable day. Playing the fourth hole into the full force of the wind, Montgomerie hit a 3-wood to a 213-yard par 3, a hole he'd normally play with a 5-iron. He was short.

Montgomerie shot 84 that day, 20 strokes higher than his second round.

Never the British Open champion, he had added his name to the record book nonetheless. No one had ever played consecutive rounds of a 72-hole British Open in a difference of 20 strokes.

Top O' the World

Jay White, a 48-year-old engineer for the golf club and ball manufacturer Taylor Made, happened to be an accomplished mountain climber. He'd already scaled Mount Aconcagua, in Argentina, and Mount Elbrus in Russia, when he joined an expedition to Mount Everest.

White not only climbed those first two mountains, but when he reached their peaks, he teed up a golf ball—sort of—assembled his club, a specially made two-piece, screw-together shaft fitted with a hollow head, and let fly.

Calculating, as engineering whizzes do, White figured if he hit the ball 100 yards from the top of Everest, it would fall 5,000 vertical feet, a drive of approximately 1,666 yards, plus the horizontal flight of 100 more. Of course if the ball rolled all the way to sea level, he'd have driven about 9,766 yards.

Alas, White didn't make it. Climbing Everest's north face, he reached within 1,000 feet of the peak in May of 2003, but strong winds and snowstorms forced a retreat back to base camp, at 27,000 feet, special club and all.

Look Out Below

Dangerous holes usually mean those that can ruin an otherwise decent round, but one hole at a club in Florida turned out to be life-threatening.

Steering his golf cart cross-country to meet his 7:30 A.M. starting time at the Wildwood Country Club, Bill Self felt the ground give way. He, his cart, and his clubs plunged to the bottom of a 15-foot sinkhole.

"I was driving my cart down the 11th when the earth opened up," Self explained. "I thought I was going to die."

He climbed out with no more damage than torn cartilage in his chest. No word on his cart and clubs.

Now There's a Hole-in-One

Fergus Muir holed what the Guiness Book of Records considers, rightly or wrongly, the longest putt in the history of golf.

Playing the Eden Course with two friends on a windy day in St. Andrews, Muir stood by while both Peter Gillespie and George Fullerton overshot the fifth hole, which measured 125 yards from the yellow tees. With the wind at their backs, both Gillespie and Fullerton had played iron clubs.

A 66-year-old 13 handicapper who could use his imagination, Muir chose a different approach. To keep his ball out of the wind he played his tee shot with his putter. He nailed it.

The ball scampered along the ground, climbed onto the green, and dived into the hole. He'd holed a 375-foot putt, so says Guiness.

No matter what, though, he had indeed scored a real, although unorthodox, hole-in-one.

Les Girls

Back in 1945, the great woman athlete Babe Zaharias competed against men in the Los Angeles Open, the Phoenix Open, and the Tucson Open. While she made the cut in all three, she didn't win any money. Tournaments in those days routinely paid only 15 to 20 places, and she didn't finish that high.

After 57 years had elapsed, Annika Sörenstam tested her game against the men in the 2003 Colonial Invitational, at Colonial Country Club, in Fort Worth, Texas, a tournament Ben Hogan had won five times. She had been invited to play by the tournament's commercial sponsor.

As expected, her entry stirred controversy. Vijay Singh, among the three or four best men golfers of the time, said he hoped she'd miss the cut, and found himself widely condemned. Others praised her for testing herself against the best male golfers.

Newspapers requested credentials for 550 reporters, three times as many as ever before. They filed reams of copy that included stroke-by-stroke accounts of Annika's rounds. Photographers swarmed over the grounds shooting everything and everybody who moved or spoke.

In one week back home after the invitational in her native Sweden, Annika grew into the most popular celebrity of her generation, and was given more headlines than any athlete since the days of boxer Ingemar Johansson, the tennis ace Bjorn Borg, or skier Ingemar Stenmark.

At an LPGA tournament in Corning, New York, a huge television screen set up near a scoreboard showed Annika playing in the Colonial. Kris Tschetter, who shot 69, found herself peeking at Annika rather than concentrating on her own game, and noticed her caddie watching Annika rather than her.

Annika opened with 71, which doesn't sound bad, but she was traveling in fast company. When the round ended, she had tied for 76th place. Since only 70 and ties move on to the last two rounds, Annika had to play better the next day.

But she didn't. She shot 74, and with 145, missed the cut by four strokes.

Applauded for her effort, she left Colonial saying:

"This is way over my head. I'm going back to my tour where I belong."

Better than Forgotten

While Annika Sörenstam was given at least as much newspaper space and television time as the war in Iraq, Kenny Perry won the Colonial and revealed himself a philosophical man.

"I'll probably be remembered as the guy who won Annika's event," he said. "That's okay with me; at least I'll be remembered for something."

More Annika

Annika played the first two rounds of the invitational with Dean Wilson, a not-very-well-known player. Asked if he'd feel embarrassed if she beat him, Wilson responded, "Anybody can beat me if I play bad."

After shooting 71 in the first round and learning that a Las Vegas bookmaker had set her first-round over-and-under score at 77.5, she quipped:

"I should have bet on myself."

Months later, Annika stood in the media center at a tournament in Portland, Oregon, when someone asked asked her about the Colonial. She turned, gestured toward her arm, and said, "My hair still stands on end when I talk about it."

In response to, or perhaps retaliation for, Annika's playing in the Colonial, Brian Kontak, of the Canadian Tour, claimed he wanted to become the first male to play in the U.S. Women's Open.

Hearing this, John Daly offered advice.

"If you want to be a girl, go have a sex change and put on a skirt."

One young girl asked another what it must be like to be as well-known as Annika Sörenstam, to which the other replied:

"She can't be very famous. She's not on a bubble gum card."

More Unisex

Shortly after Annika Sörenstam's appearance in the Colonial, Suzy Whaley played in the Greater Hartford Open, in Connecticut, but instead of being waved in on a sponsor's exemption, Suzy earned her place in a qualifying round. She did, however, play from the ladies' tees.

Later, Laura Davies, the powerful British golfer who won the 1987 U.S. Women's Open, missed the cut in the 2003 Korean Open, and Jan Stevenson, a native Australian, tied for last place among those who survived the cut in the Turtle Bay Classic, part of the PGA Tour's senior tour.

Meantime, Rachel Teske, a member of the LPGA, declined an invitation to play with the men in Australia, her home country.

Les Young Girls

Morgan Pressel shot 70 on May 15, 2001, the lowest score among 120 hopefuls at the Bear Lakes Country Club, in West Palm Beach, Florida, and won a place in the U.S. Women's Open championship. The next day she showed up for her seventh grade classes at Omni Middle School, in Boca Raton.

At 13 years of age, she became the youngest player who ever qualified for the Women's Open.

In the meantime, at home, Kathy Pressel, Morgan's mother, fielded telephone calls from around the United States and from as far away as Sweden and Britain, all wanting a word from her daughter. She heard from Bryant Gumbel, the host of *The Early Show,* and Jay Leno, host of *The Tonight Show.*

A television crew interrupted her in class, and a classmate asked for ten autographs. A budding entrepreneur, he planned to sell them.

School over, she dashed to an orthodontist to have her braces tightened before meeting reporters at another club.

Morgan was a celebrity. But, then, it ran in the family.

Kathy Pressel was born Kathy Krickstein. She grew into a very good tennis player. Her brother, Aaron Krickstein, played better. A real tennis whiz, as a 16-year-old amateur he beat Vitas Gerulaitis, the 15th seed, in the 1983 U.S. Open tennis championship, turned professional shortly after, and won an ATP tournament, the youngest to have won a professional tournament up to that time. He also played golf with a 2 handicap. He caddied for Morgan at Bear Lakes.

The Women's Open field gathered at Pine Needles Golf Club, in Pinehurst, North Carolina, but Morgan didn't play as well as she had at Bear Lakes. With two rounds of 77, she missed the 36-hole cut by eight strokes.

Two years later two more 12-year-old Florida girls tried to qualify, but by then the USGA had added another 36-hole round to the process. Michelle Shin, of Cape Coral, lost four strokes to par over the last two holes, shot 79, and missed advancing to the sectionals by one stroke, and Alexandria Buelow, of Palm City, shot 81 by dropping eight strokes over the last three holes.

Michelle Wie, a 13-year-old phenom from Hawaii, had always hoped to attend Stanford University, in Palo Alto, California. Asked why, she had a quick answer.

"They have a really good shopping mall."

A big girl, standing six feet tall, Michelle qualified for the 2002 Women's Open and startled everyone with her 280-yard drives. Inexperienced in golf at this level, she annoyed others with perceived breaches of etiquette by apparently standing on an extended line of putt—on the other side of the hole from where the other player stood.

Given sponsors' invitations, Michelle tested her game against the men in both the Boise (Idaho) Open, on the minor-league Nationwide Tour, and the Bay Mills Open, in Brimley, Michigan, on the Canadian tour. She missed the cut in both.

The Masters under Siege

In June of 2002, Martha Burk wrote a four-paragraph letter to William W. (Hootie) Johnson, the chairman of Augusta National Golf Club, suggesting the time had come to accept women members.

Mrs. Burk wrote as chairman of the National Council of Women's Organizations, a group representing some prominent associations such as NOW (the National Organization for Women) and some others so unrenowned they've been described as "One women with a laptop and a cause."

Her letter said, in part, "Augusta National and the sponsors of The Masters do not want to be viewed as entities that tolerate discrimination." She wrote on, proposing that Augusta National accept women members "now, so that this is not an issue when the (Masters) tournament is staged next year."

Johnson read this passage as a dead threat and wrote back that he saw the letter as "offensive and coercive," and that "any further communication between us would not be productive."

Thus began the war.

The fight made the front pages of every major newspaper and televised news show, most supporting Mrs. Burk. Augusta National Members who were public figures felt pressure to quit the club. About to be named Secretary of the Treasury, John Snow gave up his membership, and others prominent in business left as well, either as a matter of principle or to protect their companies. Most didn't.

Recognizing the dilemma for CBS executives, Augusta National excused commercial sponsors from the Masters telecast, one of televisions treasured productions.

Going for the jugular, Mrs. Burk threatened massive demonstrations for Masters week, the first full week of April, promising to dress women demonstrators in green burkas to mock Augusta's traditional green blazers.

The New York Times jumped in with guns blazing, lining up against Augusta National as if its editors had just learned there are such things in this world as men's clubs. Its editorial page, edited by Gail Collins, practically demanded that Tiger Woods, the 2002 winner, withdraw from the 2003 tournament. When two columnists wrote opposing opinions, the editors quashed their columns. Word leaked out, and after several days of criticism, the paper relented and ran them.

Joining in, *USA Today* somehow found access to the club's membership list and over a two-page spread published every name.

The longer the controversy lasted the more bitter it grew. It blossomed into a major social and business issue that examined whether leaders of publicly traded companies practice sex discrimination by holding membership in all-male clubs. It questioned the boundary between the public and private lives of corporate executives.

Quoted by Cox News Service, Bradley Googins, director of the Center for Corporate Citizenship at Boston University, stated that a person's private life can't be divorced from his public life. Executives "increasingly are going to be held accountable for things like this," he said.

Then the city of Augusta had to be courted. The protesters envisioned parading along Washington Avenue, the road that runs past the club's entrance. Citing reasons of safety, the city flatly refused. Massive crowds would clog traffic on a busy thoroughfare and perhaps put the demonstrators in danger, officials said.

Where, then, could the demonstrators demonstrate? Not on private parking lots bordering the road; merchant owners rejected requests to lease them. Now the hunt for an assembly point got tougher for the protesters.

The city set aside a field a mile or so from the club entrance for Mrs. Burk to conduct her demonstration.

But attention spans often shrivel. As the months passed, the public lost interest, and when the great day came, more news reporters than demonstrators showed up. The Masters went on without a hitch. The telecast did, however, cost Augusta National lots of lost income, perhaps for years to come.

Did anyone win? Martha Burk believed she'd won. Interviewed for *The New Yorker,* she said:

"I've already won. Even if Augusta National never admits a woman, people will never look at it without thinking, 'Discrimination.' The club is already tainted, the tournament is tarnished, and that will remain. Once you've reached the highest levels of government . . . and it comes to the attention of the President of the United States, it's over, folks."

Perhaps, but the Masters has gone on, and at last accounting, Augusta National remained a men's club.

Katharine Hepburn's Ace

Katharine Hepburn died in 2003, the same year R. A. Scotti published her book *Sudden Sea,* the story of the devastating hurricane of 1938, a vicious storm that killed 682 people and left 1,754 injured throughout the northeast.

Miss Hepburn had spent the summer at her family's beach-front retreat in the Fenwick section of Old Saybrook, Connecticut, waiting to hear if she would play Scarlett O'Hara in the epic film *Gone With The Wind*. (The British actress Vivien Leigh got the part.)

On the morning of September 21, the day the swiftly moving storm struck, she went for a morning swim in Long Island Sound, then decided on a round of golf at the Fenwick Golf Club.

A strong wind had gathered increasing strength throughout the early holes. When she arrived on the tee of the ninth, a par 3, it blew stronger still. With the wind behind her now, she lashed into her tee shot and knocked her ball into the hole.

It was her first hole-in-one, and according to Miss Scotti's account, she played the nine holes in 31 strokes.

No Privacy

Fredrik Jacobson had a problem but no place to solve it. Feeling the urgent need of a Port-A-Potty during the 2003 British Open, Jacobson, a Swede, looked around Royal St. George's Golf Club and saw nothing but bleak landscape, no copses of trees, and precious few bushes high as a man's shoulders. No time to be choosy—this was an emergency—he ducked into a ditch beside some underbrush off the 14th fairway, hoping for seclusion.

Ah, but the press would not be denied. Loath to miss out on a scoop, a photographer followed him in, camera at the ready.

Jacobson explained, "I had missed my last chance and needed a bit of peace down by the ditch, but the cameraman wouldn't leave me alone. He kept zooming in even after I shouted at him."

The Prince and the Commoner

Langford Cook, an American golfer, had just played a drive and 4-iron into the 17th at St. Andrews, the infamous Road Hole, and stood waiting to putt. Glancing around he spotted a group of solemnly dressed men walking along the road that lends the hole its name, headed toward the Old Course Hotel, which flanks the 17th.

A friendly sort, Cook spoke to the youngest looking man in the group and said, "You're looking smartly dressed."

"We're just here for a meeting," the youngest member replied.

"Must be an important meeting."

"It's a meeting of the past captains of the R and A," the shortened version of the Royal and Ancient Golf Club.

Surprised a man so young could already have been a past captain, Cook asked, "When were you captain?"

"Actually," the youngest man said, "Not until tomorrow. Nice to see you," he said, and the Duke of York, Britain's Prince Andrew, went off to his meeting.

Because the R and A would celebrate its two-hundred-fiftieth anniversary during 2004 and the past captains wanted a member of Britain's royal family to preside over the year-long observance, they chose Prince Andrew to serve as captain. He would be the sixth member of Britain's royal family to fill this largely honorary office.

A carefully choreographed ballet leads up to a captain's canonization. At 8 A.M. on a Thursday morning in September, the outgoing captain leads his successor to the first tee of the Old Course, in front of the grayish sandstone building that houses the R and A, and after John Panton, the club's honorary professional, sets a ball on a tee, and as an ancient cannon booms, the new captain plays a shot toward a picket line of caddies lined at a distance they estimate the new man might manage.

When the Prince of Wales, who later became Edward VIII, stepped onto the tee, the caddies formed a line that an observer described as "disloyaly close to the tee."

The caddie who fields the ball then trots back to the new captain and exchanges the ball for a gold sovereign that cost the R and A £85.78 sterling (roughly $145).

A member of the Royal Liverpool Golf club (Hoylake), the Duke of York plays off a 6 handicap. With that in mind, his phalanx of caddies set themselves at a quite respectful distance.

Now there was a hitch; someone couldn't tell time. Five minutes before the big hand hit the 12, John Whitmore, the retiring captain, escorted Prince Andrew out to face his ordeal. While Prince Andrew noted that time waits for no man, the man, even though of the royal family, must wait for time.

After fidgeting for a while and swapping snappy comments with R and A members stacked three and four deep behind him, the tall grandfather clock inside the clubhouse tolled the hour in its deep, rumbling tone and Prince Andrew stepped up to his ball and killed it.

The ball soared over the first line of caddies, streaked along the ground through the rest, and ran on until a lucky caddie fell on it before it reached the Swilcan Burn, the narrow creek that winds across the combined fairways of the first and 18th holes. He ran back, exchanged the ball for his sovereign, and the Prince led an invited party of guests inside the clubhouse for the traditional champagne-fueled Captain's Breakfast.

Two hours after the driving-in ceremony, Prince Andrew stepped onto the first tee once again to play in the Autumn Medal, climaxing a three-day affair when members from around the world compete for the club championship.

Once again a sizable gallery stood by to see the prince match his sensational drive of the morning. Instead, he nearly knocked it out of bounds. It flew down the right side, and only a bend in the white-railed fence marking the course boundary saved him.

The new captain stumbled around in 94, a disappointing round for everyone, that included a bladed pitch toward the 18th green that scooted through another fence and out of bounds.

Besides John Whitmore, Prince Andrew also played with Bill Campbell, the only man ever to serve as President of the United States Golf Association and Captain of the R and A. Bill had the distinction that day of losing a ball on the 18th and still scoring a birdie 3. He lost it in the ball washer; it went in, but it wouldn't come out.

From 1951 through 1975, Campbell played for the United States in eight Walker Cup Matches, served as captain in 1955, and won 11 matches in both singles and foursomes. By winning seven singles matches against only one halved match, he has a winning percentage of .938, second only to Bob Jones's perfect record of five victories and no losses.

Campbell furthermore is one of the honored R and A members with lockers in the R and A's Big Room, in which no one sets foot without coat and tie.

Only Campbell's 1971 team lost to Great Britain and Ireland, although he played in 1965 when the teams halved the match at 11 points each. A week before Prince Andrew drove in, though, the British and Irish won their second consecutive match against the United

States. When the Walker Cup arrived back at the R and A clubhouse, someone set it atop Campbell's locker.

Surprise Entrant

Each year the Association of Golf Writers, a British organization, plays its championship at the Wentworth Golf Club, in Surrey, England, a London suburb. The championship is a convivial, rather private affair limited to its members and a few invited guests. In the fall of 2003, the writers learned an important guest would join them on the first tee. They'd know the guest's identity when he turned up.

Speculation circulated when two bodyguards cased the joint a week before the first ball was to be struck and declared Wentworth's East Course, the site of this annual gathering, free of dangerous elements.

On the day of the great affair, bodyguards showed up again, and one of them slipped into the woods, never to set foot on fairway until the round ended (he turned out to be useful in tracking down wayward golf shots, though).

Surprising everyone, a few weeks after assuming the captaincy of the Royal and Ancient Golf Club, Prince Andrew showed up to play, and he teed off with three golf writers.

He was somewhat shocked, though, to find he had broken new ground. After taking a few steps, the Prince asked, "Do all R and A Captains play in this event?"

"No, sir," he was told, "you are the first. The only function of ours the Captain attends is our annual dinner at the Open championship in July. You're a bit early for that."

Whereupon the Prince burst out laughing.

"Ane come to think of it," the AGW member went on, "why ARE you here?"

The Prince replied, "I was given a list of events by Peter Dawson (the R and A Secretary) and there was a tick against this one. I thought it was one I should attend."

When the Prince and his group stopped at a refreshment hut in mid-round, he declined any kind of drink and instead chose a very popular candy bar called a Yorkie, which seemed appropriate. Still, when he won the guests' prize—a bottle of champagne, he didn't turn it down.

Mike Blair won the members' championship, and opened his acceptance speech by saying, "Your royal highness, the Duke of York . . . Oh, if only my Mam could hear me say those words!"

Short Supply

By Saturday morning, the 2000 U.S. Open had turned into a test of patience, self-control, and endurance as well as skill. As Tiger Woods stood on the 18th tee at Pebble Beach, the clock had not yet struck 9

A.M., and he was about to finish his second round, not begin his third. Blanketed by thick sea mists, Pebble Beach had vanished, like the mythical Scottish village Brigadoon. The USGA suspended play in mid-afternoon on both Thursday and Friday and carried rounds into the next day.

Woods had opened with 65, the best first round ever shot at a Pebble Beach Open, and with a par 5 on the 18th, he would shoot 68, three under. He missed it by one stroke, but it turned out to be closer than he knew.

Ripping into his drive, Woods pulled the shot into Carmel Bay, a bight off the Pacific Ocean that follows all along Pebble's closing hole. When the ball soared toward Hawaii, Woods melted the television cables with a burst of language that would scorch the hide off a rhinoceros. Television's ubiquitous parabolic microphones picked up every syllable and beamed them from Azusa to Zanzibar.

Fuming, Woods called for a new ball from Steve Williams, his caddie. Williams took a ball from Tiger's golf bag, but before handing it over he suggested, "Why don't you play your two-iron?"

Woods declined, saying he would play the driver.

Now Williams picked up the bag and put it almost on Woods's feet, insisting once again that it would be more prudent to play the two-iron, not the driver.

Annoyed now, Woods took the ball, set it on the peg, and after Williams stepped away, pounded a drive that pierced the heart of the fairway. He made a 4 with the second ball, but with the two-stroke penalty he marked down 6 on his scorecard. Instead of 68, he shot 69.

All went well after that. Woods teed off again later, shot 71 in the third round, 67 the next day, 272 for the 72 holes, and won by 15 strokes over Miguel Angel Jimenez and Ernie Els, a record margin.

Tiger took a few weeks off after Pebble Beach, and when he played his next tournament, Williams approached and asked if he remembered their dispute at Pebble Beach.

"Yeah," Woods answered, "what was that all about?" Then, grinning, Williams told him.

"That ball I gave you? It was the last ball we had."

Index